Anthropocene Ecologies

Anthropocene Ecologies brings political ecology and tourism studies to bear on the Anthropocene.

Through a collective examination of political ecologies of the Anthropocene by leading scholars in anthropology, geography and tourism studies, the book addresses critical themes of gender, health, conservation, agriculture, climate change, disaster, coastal marine management and sustainability. Each chapter theoretically and empirically unravels entanglements of tourism, nature and imagination to expose the political-ecological drivers of the Anthropocene as a material and symbolic force and its deepening integration with tourism. Grounded in ethnographic and qualitative research, the volume is interdisciplinary in scope, yet linked in its shared focus on the political threat as well as the social potential of the Anthropocene and its imaginaries. This collection contributes to emerging scholarship on tourism, sustainability and global environmental change in the current geological epoch.

Anthropocene Ecologies will be of great interest to political ecology focused scholars of tourism, socio-environmental change and the Anthropocene.

The chapters were originally published as a special issue in the *Journal of Sustainable Tourism*.

Mary Mostafanezhad is an Associate Professor in the Department of Geography and Environment at the University of Hawai'i at Mānoa, Honolulu, USA. Her research examines the political ecology of tourism and socio-ecological change in the Asia-Pacific region.

Roger Norum is a University Lecturer in Cultural Anthropology at the University of Oulu, Finland. He studies the changing roles of mobility, media and the environment, with a particular emphasis on the everyday geopolitics of territory, time and labor, particularly among transient and precarious communities in the Arctic and South Asia.

Anthropocene Ecologies

Entanglements of Tourism,
Nature and Imagination

Edited by
Mary Mostafanezhad and Roger Norum

LONDON AND NEW YORK

First published 2020
by Routledge
2 Park Square, Milton Park, Abingdon, Oxon, OX14 4RN

and by Routledge
52 Vanderbilt Avenue, New York, NY 10017

Routledge is an imprint of the Taylor & Francis Group, an informa business

First issued in paperback 2021

British Library Cataloguing-in-Publication Data
A catalogue record for this book is available from the British Library

ISBN 13: 978-0-367-42908-9 (hbk)
ISBN 13: 978-1-03-208408-4 (pbk)

Typeset in Myriad
by codeMantra

Publisher's Note
The publisher accepts responsibility for any inconsistencies that may have arisen during the conversion of this book from journal articles to book chapters, namely the inclusion of journal terminology.

Disclaimer
Every effort has been made to contact copyright holders for their permission to reprint material in this book. The publishers would be grateful to hear from any copyright holder who is not here acknowledged and will undertake to rectify any errors or omissions in future editions of this book.

Contents

CONTENTS

Citation Information

The chapters in this book were originally published in the *Journal of Sustainable Tourism*, volume 27, issue 4 (2019). When citing this material, please use the original page numbering for each article, as follows:

Chapter 7
Friction in the forest: a confluence of structural and discursive political ecologies of tourism in the Ecuadorian Amazon
Annie A. Marcinek and Carter A. Hunt
Journal of Sustainable Tourism, volume 27, issue 4 (2019) pp. 536–553

Chapter 8
Tourism and community resilience in the Anthropocene: accentuating temporal overtourism
Joseph M. Cheer, Claudio Milano and Marina Novelli
Journal of Sustainable Tourism, volume 27, issue 4 (2019) pp. 554–572

Chapter 9
Tourists and researcher identities: critical considerations of collisions, collaborations and confluences in Svalbard
Samantha M Saville
Journal of Sustainable Tourism, volume 27, issue 4 (2019) pp. 573–589

Chapter 10
Entanglements in multispecies voluntourism: conservation and Utila's affect economy
Keri Vacanti Brondo
Journal of Sustainable Tourism, volume 27, issue 4 (2019) pp. 590–607

For any permission-related enquiries please visit:
http://www.tandfonline.com/page/help/permissions

Contributors

Keri Vacanti Brondo is a Professor and Chair of the Department of Anthropology at the University of Memphis, USA. Her research interests in conservation, development, local livelihoods, tourism, land rights, and nature-based volunteerism. Her publications include *Cultural Anthropology: Contemporary, Public, and Critical Readings* (Oxford University Press, 2016); *Land Grab: Green Neoliberalism, Gender, and Garifuna Resistance* (University of Arizona Press, 2013); and over 60 peer-reviewed articles, agency reports, reviews and commentaries. Her next book project, tentatively titled, 'Multispecies Entanglements in Honduras' Affect Economy', explores the relationship between conservation voluntourism, protected area management and local livelihoods.

Bram Büscher serves as a Professor and Chair at the Sociology of Development and Change group at Wageningen University. He also holds appointments as a Visiting Professor at the Department of Geography, Environmental Management and Energy Studies of the University of Johannesburg, South Africa, and a Research Associate at the Department of Sociology and Social Anthropology of Stellenbosch University, South Africa.

Joseph M. Cheer is a board member of International Geography Union (IGU) Commission on Tourism and Leisure and Global Change, and Steering Committee Member Critical Tourism Studies Asia Pacific. He was formally based at Monash University, Melbourne, Australia, and is currently at the Center for Tourism Research (CTR), Wakayama University, Japan. His research draws from transdisciplinary perspectives, especially human/economic geography, cultural anthropology and political economy. As a former practitioner and practicing consultant and analyst, he is focused on research to practice with an emphasis on resilience building, sustainability and social justice in tourism.

Matilde Córdoba Azcárate is an Assistant Professor at the Department of Communication at the University of California, San Diego, USA. Her research lies at the intersection between tourism development and political ecology in contemporary Southern Mexico. Webpage: https://quote.ucsd.edu/mcazcarate/.

Josh Fisher is an Associate Professor of Anthropology at Western Washington University in Bellingham, USA. He is the author of numerous publications on environmental and economic anthropology, focusing specifically on the politics of alternative development in Latin America. With Alex Nading (Brown University, Providence, USA) and Chantelle Falconer (University of Toronto, Canada), he is conducting a three-year study of Ciudad Sandino, Nicaragua, funded by the National Science Foundation (#1648667), about the intersection of different forms of value in urban political ecologies.

Robert Fletcher is an Associate Professor in the Sociology of Development and Change group at Wageningen University, the Netherlands. His research interests include conservation, development, tourism, climate change, globalization and resistance and social movements. He is the author of *Romancing the Wild: Cultural Dimensions of Ecotourism* (Duke University, 2014) and co-editor of *Nature^TM Inc.: Environmental Conservation in the Neoliberal Age* (University of Arizona, 2014).

Edward H. Huijbens is a Geographer, Scholar of tourism, Professor and Chair of cultural geography research group at Wageningen University, the Netherlands. Edward works on tourism theory, issues of regional development, landscape perceptions, the role of transport in tourism and polar tourism. He is the author of over 30 articles in several scholarly journals such as *Annals of Tourism Research*, *Journal of Sustainable Tourism* and *Tourism Geographies*, and has published three monographs in both Iceland and internationally and co-edited four books.

Carter A. Hunt is an Assistant Professor of Recreation, Park, and Tourism Management at Pennsylvania State University, USA. He received his PhD in Recreation, Park, and Tourism Sciences from Texas A&M University, College Station, USA. He is an Environmental Anthropologist who researches the intersections of tourism, biodiversity conservation, sustainable community development in Latin America and beyond.

Annie A. Marcinek received her BA in Anthropology in 2014 and received her MS in Recreation, Park, and Tourism Management in 2017 from Pennsylvania State University, USA. She is interested in the relationship between natural resource extraction and ecotourism, especially among rural communities in Latin America working to promote positive local social, economic and environmental outcomes.

Claudio Milano achieved a PhD in Social and Cultural Anthropology from Universitat Autonoma de Barcelona, Spain. He is a Lecturer and Researcher in Tourism at Ostelea School of Tourism & Hospitality at the University of Lleida (Barcelona, Spain). He has published in international peer-reviewed scientific journals, and he is a member of several international anthropological and tourism research networks. He recently contributed to a study on overtourism in European Union funded by the European Parliament as team member of a consortium. His research interests are focused on sociocultural impacts of tourism and the relation between tourism, social resistance and protest movements in rural and urban areas.

Amelia Moore has a BA in Environmental Biology from Columbia University, New York City, USA, and a PhD in Sociocultural Anthropology from the University of California Berkeley, USA. She has been conducting social research in The Bahamas since 2002.

Marina Novelli is a geographer with a background in economics. She is an internationally renowned tourism policy, planning and development expert, and a Professor of Tourism and International Development at the School of Sport and Service Management, which is an Affiliate Member of the UN World Tourism Organisation (UNWTO). She is Academic Lead for the University's Responsible Futures' Research and Enterprise Agenda, which is aimed at fostering internal and external interdisciplinary research and consultancy collaborations. She has advised on numerous international cooperation and research assignments funded by International Development Organisations (IDOs): The World Bank, the European Union, The UN (UNESCO, UNIDO, and UNWTO), the Commonwealth Secretariat, National Ministries and Tourism Boards, Regional Development Agencies and NGOs.

Jarkko Saarinen is a Professor of Geography at the University of Oulu, Finland, and a Distinguished Visiting Professor (Sustainability Management) at the University of Johannesburg, South Africa. His research interests include tourism and development, sustainability in tourism, Indigenous and cultural tourism, tourism–community relations, tourism and climate change adaptation and wilderness studies. Over the past 20 years, he has been working extensively in the rural and peripheral areas of northern (Arctic) Finland and southern Africa. His recent publications include co-edited books: *Tourism Planning and Development* (2018), *Political Ecology and Tourism* (2016) and *Cultural Tourism in Southern Africa* (2016).

Samantha M. Saville is currently a Research Assistant for Aberystwyth University, UK, working on the GLOBAL-RURAL project. Her doctoral thesis in Human Geography investigated the environmental politics and value contestations of conservation practices in Svalbard. She has previously worked as a Visiting Lecturer at Chester University, UK, and a Tutor and Researcher at the Centre for Alternative Technology.

Jundan Zhang is a Postdoc Researcher within the Department of Geography and Economic History at Umea University, Sweden. Her current research focuses on relationships between environmental discourses and environmental subjectivities in the context of tourism in peripheral areas, in particular how individuals translate and appropriate broad ideas on "nature" into their exchanges with other human and non-human actors through tourism activities, and how these translations and appropriations in turn contribute to the discourses of environment.

Foreword

Bram Büscher

The Anthropocene has an unbearable weight attached to it. Not just the increasing weight of the term's omnipresence in contemporary academic debates, but rather the imaginaries associated with it. Arguably the most prominent is that of the 'great acceleration' (Steffen et al., 2015; McNeill and Engelke, 2016): a set of 'hockey stick' graphs that show how many extractive, environmental and consumption indicators have rapidly increased over the last decades across the globe. What they show is the utter unsustainability of the current trajectory and that somehow we – all humans – are too blame. And despite the thorough debunking of the nonsensical idea that there is some general 'anthropos' responsible for the current crisis trajectory, many of us still feel this weight (Head, 2016).

Now compare this to the unbearable lightness of tourism imaginaries. On the one hand, the tourism industry sees itself as well positioned to profit from the weight of the Anthropocene. After all, it still prides itself on finding spaces and experiences to get 'away from it all' (West and Carrier, 2004). On the other hand, it has also been forced to adapt and change its – environmentally and socially rather weighty – practices, even if this is often only marginally so or by developing new business practices that do little to address the structural weight of the industry (Stroebel, 2014; Moore, this volume). Thus, even the way the tourism industry represents and deals with its own weight is rather light. Somehow, it seems to imagine that it is still outside the Anthropocene impacts that all of (the rest of) humanity is supposedly responsible for.

Clearly, the ways in which these unbearable phenomena come together are extremely diverse, as the fascinating contributions to the current volume vividly show. Indeed, bringing them together can be highly productive, as it opens up fresh ways of tracing, tracking and understanding newly emerging and stubbornly residual forms of power and politics, and how human and nonhuman actors relate to these. From Cancun, Mexico and Utila, Honduras to Svalbord, Norway, Niru, China and beyond, the rich studies in this volume delve deeply into "how human-environment relations are shifting in the current geological epoch" (Mostafanezhad and Norum, this volume) and how tourism – as an industry, set of actors, flow of resources and, especially, amalgam of imaginaries – drives and is driven by these (see also Gren and Huijbens, 2014; Moore, 2015).

Key across the contributions, according to the editors, is the political ecology of the encounter between tourism and the Anthropocene, especially in relation to broader dynamics of neoliberal capitalism. But as they also stress and the various chapters show, this statement always needs to be complemented with an emphasis on the 'astonishingly' diverse and creative ways in which capitalism renews itself in this encounter, as Fletcher (this volume) argues, and on the limits to this statement as neoliberal forces intersect with an equally astonishing plurality of dynamics, perspectives and scales, as Saarinen (this volumes) argues. Clearly, this leads to a broad set of concerns, places and foci in this one volume, many of which – following Marcinek and Hunt's (this volume) borrowing of the term from Anna Tsing – may harbor internal (theoretical, methodological and analytical) 'frictions'. Yet, I would argue, this is a good thing and represents the state and openness of political ecology more broadly. Friction, after all and if based on agonism rather than antagonism (Mouffe, 2005), can lead to productive (intellectual, analytical, practical and *political*) heat.

The question that still stands, however, is whether and how this heat can be translated into new forms of power and politics that may shift the path of the neoliberal Anthropocene and 'do tourism differently' (Mostafanezhad and Norum, this volume). After all, as McNeill and Engelke (2016: 209) conclude, the great acceleration "will not last long", and hence, great changes to life, power and politics must be expected. Several contributions in this volume study imaginaries that link into this, like Fisher's study of tourism and vivir bien in Latin America and Cheer et al.'s study of temporal overtourism in Australia. The first shows that forms of 'doing tourism differently' are as of yet difficult to discern, while the latter illustrates how the swift acceleration of tourism in some areas can undermine regional resilience.

In searching for an answer to this question, however, one element remains critical: the unbearable weight and lightness of the Anthropocene and tourism will always need to be brought down to earth through critical empirical research and historical-theoretical analysis. Only these can deconstruct the unbearable into more bearable elements that can be reconstructed into something altogether different and more promising. The present volume succeeds remarkably well in doing so and presents a solid grounding to continue the search. It attests to the promise and potential of political ecology in tackling the big questions of our time and hence deserves to be widely read, debated and build upon.

References

Gren, M. and E. Huijbens (2014). *Tourism and the Anthropocene*. London: Routledge.

Head, L. (2016). *Hope and Grief in the Anthropocene. Re-conceptualising Human-Nature Relations*. London: Routledge.

McNeill, J.R. and P. Engelke (2016). *The Great Acceleration: An Environmental History of the Anthropocene since 1945*. Cambridge, MA: Belknap Press.

Moore, A., ed. (2015). *Environment and Society: Advances in Research* 6, 1, special issue on "The Anthropocene: A Critical Exploration."

Mouffe, C. 2005. *On the Political*. London: Routledge.

Steffen, W., et al. (2015). The Trajectory of the Anthropocene: The Great Acceleration. *The Anthropocene Review* 2, 1: 81–98.

Stroebel, M. (2014). *Protecting Holidays Forever: Tourism, Climate Change and Global Governance*. PhD dissertation, University of Manchester.

West, P. and J. Carrier (2004). Ecotourism and Authenticity: Getting Away from It All? *Current Anthropology* 45, 4: 483–498.

Introduction

Mary Mostafanezhad and Roger Norum

At the risk of stating the obvious, the Anthropocene is having a moment. For easily a decade, the term Anthropocene has been popping up all over the place in the manifold outposts of academic production: journals, conferences, course syllabi, panels, podcasts and public punditry. Today, Anthropocene is the subject of hundreds of books and thousands of articles that run the gamut of fields in both the social and natural sciences, and the humanities. Academic disciplines have coalesced and mobilized around the term, broadening their research agendas and forging new cross-disciplinary fields (Castree, 2014a, 2014b; Häusler & Häusler, 2018). Since Crutzen's Anthropocene, initially framed in a conversation within the geophysical sciences, and today's Anthropocene, now engaged by writers and thinkers both inside and outside halls of scholarship, the topic of anthropogenic environmental change has consumed modern thought and public debate, giving way to unprecedented displays of planetary anxiety. Across its minute lifespan, Anthropocene has become a discussion about geological periodization, a conversation about environmental sustainability and a rethinking of the origins of ecological crisis (Moore, 2017). Indeed, it would appear that scholars from nearly every discipline have boarded the thematic bandwagon, an excitement that has thus far only marginally abated. As Tsing et al. (2017, p. 1) observe, "the word tells a big story". But what exactly is that story? Whose story is it? And how do the everyday practices of all of us figure into that story?

In this article, we argue that Anthropocene is fundamentally a social imaginary that is both shaped by and reshapes human life. Taking as a lens the industries, regimes and practices of tourism, we enroll the concept of the anthropocenic imaginary to describe how Anthropocene is symbolically, materially and discursively produced through human action. An entity unto itself, the anthropocenic imaginary perpetuates and challenges new and old forms of political and ecological inclusion and exclusion as it coauthors the script for humans in the Anthropocene story. Building on emerging research over the past several years on the relationship between Anthropocene and tourism, we question what role imaginations of human-environment engagement play in the tourism industry and encounters. How, for example, do people engage in and conceive of leisure in times of uncertain political, economic or environmental futures? In what ways do new tourism infrastructures structure our imaginative relationship with nature? How do people reconcile their growing recognition of climatic crisis with their desire to travel? A consideration of what the anthropocenic imaginary is and does inspire answers while raising new questions. It directs our attention to the ways in which nature has become a repository for social values, as the human imagination is triggered by fantasies of anthropogenic doom and destruction. It may be, too, that there is heuristic value here, since it is through emergent imaginaries that new and more complex societal solutions to environmental problems might be constituted.

To begin with the now-familiar origin story – the off-the-top-of-the-head coinage, mid-conference, of American biologist Eugene Stoermer and Dutch geochemist Paul Crutzen – Anthropocene is at its very root a proposed (and since validated) new geological epoch, one in which human activity and agency have become a planetary force acting at many scales across the planet. As an epochal change, Anthropocene denotes a specific marker in the geologic record – a geophysical shift in the planet's rock layer that is markedly distinct from the biogeological conditions of the Holocene (Boggs, 2016). While Anthropocene denotes a new era that is dominated by humans' transformation of the terrestrial and celestial aspects of the planet, it has also come to encompass a range of human activities and resultant environmental phenomenon (Hamilton et al., 2015). When the Stratigraphy Commission of the Geological Society of London met in 2008 to discuss whether a new epoch in planetary history had indeed begun, the empirically available signs of such a shift were aplenty. By then, climate change had already begun to notably (and noticeably) distemper seasons across the planet with extreme weather events that included droughts, cold freezes and hot spells – phenomena that have since been described as harbingers of the so-called sixth extinction (Barnosky et al., 2011).

Champions of the term Anthropocene seek to account for life on a planet that 500 years ago supported some 450 million humans and today struggles to support over 7.5 billion of us. They wield the term to explain that due to (capitalist) industrial progress and economic growth, each of us now has a vastly more consequential impact than we would have had a millennium ago. They demonstrate how societies and their relationships to the environment have indeed changed, perhaps irrevocably. Few would argue with the statement that we now live in a period of "rapid, unpredictable, and fundamental change that is literally planetary in scope" (Chester & Allenby, 2018). In many parts of the world, natural systems are effectively becoming another element of regional and global systems. Given society's penchant for coinages, acronyms and catch-all buzzwords, it makes sense that we would end up with a term whose lexical heft reflects both the gravity of the current situation and the momentousness of what it stands for, while also connoting its fair share of reference, paradox and irony for good measure.

But Anthropocene has morphed into more than just descriptive nomenclature. It has also become a mobilizing concept that has empowered scholars and scientists from across multiple disciplines to collectively brainstorm a pathway forward for humans and nonhumans alike (Haraway et al., 2016; Tsing et al., 2017). Now in place as an epistemological regime, Anthropocene has become a core debate about humanity and the planet while simultaneously being a critique of that very debate (Castree, 2014b), pointing out the problem of living on a planet damaged by the excesses of overaccumulated capital (Tsing et al., 2017). Perhaps more potently, it has become a political label intended to draw attention to superlative planetary change and the agentive role and responsibility of humans in such change in ways that suggest global uniformity and problematically elide difference and locality (Moore, 2016). As one of the world's biggest industries, tourism is a lead player in ongoing debates about the future of the planet. Thus, if Anthropocene is now a shorthand – if a mouthful of one – for the current climate of concern over socio-environmental issues, human excess and consumptive regimes, then tourism is part of both the problem and solution, and a fine place to begin thinking about what we can do to help see the planet successfully through to its next geological epoch.

Tourism and the anthropocenic imaginary

The anthropocenic imaginary reflects a discourse about the past that contains warnings about the future. It highlights how the world is pushing up against planetary boundaries into a zone of high turbulence and uncertainty. It also represents a break with ecological imaginations of a self-contained earth system. Andrew Reszitnyk can perhaps be attributed with first coining the term "anthropocenic imaginary", observing it as an instrumentalization and distortion of the language of earth sciences (Reszitnyk, 2015, p. 4). The discourse of Anthropocene, he argues, has perpetuated projects of depoliticization in ways that re-assign responsibility for the degradation of the earth and its atmosphere through discursive regimes of neoliberalism. As a political, economic and cultural ideology, neoliberal discourses wrapped in the cloak of Anthropocene naturalize environmental degradation as a collective problem that demands individual, privatized solutions. This has furnished opportunities for tourism development to facilitate a spatio-temporal fix for capitalism via the transformation of disaster into commodity form.

The narrative of Anthropocene, insofar as it can be pinned to any one narrative, evokes at once visions of a future environmental apocalypse while also speaking to past and present inequality and suffering. And yet the very use of the term itself invites critique that this hubristic belief that humans could (or should) wield power against the planet arguably employs the same logics that created the problem in the first place (Chakrabarty, 2016; Hamilton, 2016; Marcus, 1995). This discourse is particularly explicit in tourism, a regime in which imaginaries are traded on through narratives about the current state of the world, its people and its future. As a creative, flexible conceptualization, imaginaries accommodate the complex range of social and cultural practices that exist across multiple, embodied worlds, and are applicable across a range of individual and group experiences. The social imaginary, in the words of social theorist Charles Taylor, "incorporates a sense of the normal expectations that we have of one another, the kind of common understanding which enables us to carry out the collective practices that make up our social life" (Taylor, 2002, pp. 106–107). Social imaginaries do not just tell about how we see the world in our heads; they make that very world. Through social imaginaries, beliefs about reality are unconsciously shaped through representational practices (Althusser, 1971). As an industry driven in large part by practices of representation, tourism is compelled to circulate imaginaries that become vehicles for Anthropocene to take shape, which in turn reshapes the tourism industry itself. Food production, for instance, is one key site where the anthropocenic imaginary has taken hold. In the Caribbean, for instance, Amelia Moore (this volume) demonstrates how anthropocenic imaginaries push forward new commodity frontiers through the rebranding of tourism "products", such as a small island farm in the Bahamas. Moore's work demonstrates the ways in which anthropocenic imaginaries have reconfigured the role of the farm in both symbolic and material forms, providing fodder for the imagining and materialization of new leisure spaces (Moore, 2019).

Indeed, a gaggle of new leisure spaces have been developed globally, a feat confirming the anthropocenic imaginary's potency. Popular reenactments of Anthropocene are now thoroughly integrated into a range of leisure spaces. Beyond new forms of agricultural, dark and apocalyptic tourism experiences, museum exhibits, art shows and urban infrastructures speak directly to how Anthropocene is imagined, understood and (not) acted upon. In September 2018, the Carnegie Museum of Natural History appointed Nicole Heller as the world's first curator of the Anthropocene. According to the museum, Heller's role is to "tell the epoch's story, noting both the positive and negative ways that humans are changing the planet, how biotic organisms are responding, and what this may mean for the future" (Pickels, 2018). Anthropocene-themed exhibitions, brandishing the word everyone now loves to enunciate in dinner table conversation (e.g. The Smithsonian's "Anthropocene: Life in the Age of Humans"), are now on display in natural history and art museums around the world. These exhibits produce new imaginaries of humankind's fall from grace amidst a disruption the social order. The once-niche-now-popular genre of cli-fi (climate fiction) similarly reflects the broader assemblage of the anthropocenic imaginary.

Imagination is not opposed to reality; it makes reality a more creative human experience. As Tim Ingold argues, imagination is intimately entangled with perception and memory such that they cannot be considered independently (Ingold, 2013). Images and visual culture are integral to the ways we imagine, perceive and remember. The souvenir, the Facebook profile or the AirBnB reviews help us to trigger memories and sense the experience. There is a temporality to imagination in that it reflects the ways in which we always seem to be running ahead of ourselves and the materiality of our intentions. As Ingold notes, the experience of life is caught in the tension between the onward rush and the friction of materials, the pull of hopes and dreams and the drag of material restraint (Ingold, 1993).

Significantly, the meaning of "imagination" has inherited more than three centuries of European thought, since when it was associated with fantasy rather than fact (Ingold, 2013; Janowski & Ingold, 2016). The imagination has long been taken up by the humanities and social sciences; the topic has also been especially influential in transdisciplinary tourism studies (Salazar, 2012). Benedict Anderson, for instance, ascribed the significance of media in the development of collective imaginaries of the nation. His work demonstrated the ways in which the imagination shifts in accordance with the development of new media technologies; this is especially true in tourism where the development of social media has radically altered how tourists imagine, perceive and remember. Edward Said's analysis of Orientalism is particularly influential in tourism studies. His work demonstrates how, through early travel writing, the West's imaginary of the Arab world as backward, overly sexual and dangerous contributes to geographical imaginaries of the region. Through his work, tourism scholars have addressed how Orientalist stereotypes mediate tourism representations and encounters (Said, 1978; Hall & Tucker, 2004; Yoshihara, 2004; Bryce, 2007; Yan & Santos, 2009). Long before social media took hold, a number of anthropologists and media scholars described the imagining of homogenous cultures and the intertextual production of the state through new media technologies (see, e.g., Ginsburg et al., 2002).

The growing representation of indigenous imaginaries has reshaped the way we think through indigenous tourism experiences (Smith, 1999). Elsewhere, Erik Swyngedouw has been at the forefront of drawing attention to and theorizing political imaginaries of nature. For instance, he describes an apocalyptic imaginary which he argues, in the context of a neoliberal hegemony, forecloses asking serious political questions about possible socio-environmental trajectories (Swyngedouw, 2007). Such modes of apocalyptic thinking are now deeply embedded in the tourism industry and the way it is seen, understood, imagined and experienced (Tucker & Shelton, 2014). Together, these collections of manifold imaginaries become part of the discourses embedded in anthropocenic thought, now one of the most influential framings through which tourists experience the planet and its people and environment.

Political ecology: humanity and nature in the anthropocenic imaginary

Political ecology denotes an interdisciplinary framework characterized by a focus on place-based, historically situated and multi-scalar accounts of socio-environmental change. Political ecologists examine environmental matters from a broadly defined political economy perspective (Gössling, 2003; Susan et al., 2003; Peet & Watts, 2004; Bryant & Michael, 2006). Originating as a critique against an allegedly apolitical cultural ecology and ecological anthropology, political ecology illustrates the unavoidable entanglement of political economy with ecological concerns (Stonich, 1998, p. 28). Political ecological perspectives on the both the environment and the socio-environmental anxieties of Anthropocene illustrate, diachronically, how power and structural relations at different scales have held implications for local people's natural resource and land use practices. They also demonstrate how tourism is mediated by a range of political, economic and cultural relations of power. As a result of these relations, some ecological concerns become privileged while others are discarded of deemed unactionable. Marxist and post-structural perspectives have been particularly helpful frameworks for tourism scholars to make sense of the ways in which anthropocenic imaginaries perpetuate differential interests, benefitting some while marginalizing others. Thus, as Fletcher (this volume) notes, bringing Marxist and post-structural perspectives together affords an understanding of tourism development as the embodiment of a particular discursive perspective, an approach to human-environment relations and a political-economic process entailing the pursuit of capitalist accumulation.

Musings on Anthropocene in everyday tourism encounters become a well-positioned perch from which to observe the discursive separation of Humanity from Nature. Rooted in the Cartesian dualism that developed in early modern Europe, this separation is reified by the anthropocenic imaginary through an understanding of the social structures of capitalism as unhinged from nature. Humanity is subsequently framed as an undifferentiated whole which has become the bearer of responsibility for global environmental change rather than the socio-environmental relations of capitalist production (Haraway, 2015; Moore, 2015). This position depoliticizes and naturalizes all humans everywhere as equally responsible for global environmental destruction. Tourists and tourism practitioners perpetuate the Human/Nature dualism in both implicit and explicit ways. Anthropocenic imaginaries conceal the symbolic, material and often times bloody violence of this dualism which is typically backed by what Moore refers to as "imperial power and capitalist rationality" with the sole aim of capital accumulation (Moore, 2016, p. 78). In the wake of this realization, scholars such as Haraway, Moore and others have called for alternative concepts to Anthropocene so as to account for the ways in which planetary crisis is pushed forward by this blind drive for progress, growth and the accumulation of capital (Haraway, 2015; Moore, 2017). In the following section, we examine how anthropocenic imaginaries are appropriated in ways that build upon some of the core sociological categories on which tourism has often been predicated. We outline three conceptual imaginaries of anthropocenic thought that govern much of how Anthropocene is engaged with on discursive and experiential levels: posthuman, capitalist and ecological.

Post-human imaginaries: metaphor and materiality

The anthropocenic imaginary speaks to the renewed, transdisciplinary interest among scholars in human interaction with the material world (Miller, 2005; Anderson & Wylie, 2009; Haraway, 2015). This interest reflects new relations of humans with the environment, inviting in new materialities, socialities and positionalities. Environmental, development and social policy are being reconfigured through the anthropocenic imaginary, as contemporary experiences with forest, water, land, ecologies, technologies and viruses elucidate alternative ways of living – and perhaps more sustainable ways of being. Global ecological movements have been catalysts for the ontological turn in tourism studies. These are not merely social constructivist, and they are not restricted to humans. They give agency to nonhuman and more-than-human natures, what Haraway, Tsing and others refer to as a multi-species-ism, within which we are inextricably entwined (Haraway et al., 2016; Tsing et al., 2017).

Over the past decade, tourism scholarship has been repopulated with nonhuman beings in ways that account for their agency as social agents. It is largely through the popularity of actor-network theory that scholars of Science and Technology Studies (STS) have been integrally introduced into conversations with environmental scholars and philosophers, anthropologists and geographers (Callon, 1999; Latour, 2005). As van der Duim explains, the tourist researcher's task is to follow how meanings are co-produced in tourism practice. Drawing on Latour's actor-network theory, van der Duim describes "tourismscapes" as human and nonhuman entanglements dispersed across time and space (Van der Duim, 2007; van der Duim et al., 2013). Moving beyond social-constructivist approaches to the study of human-environment relations, tourism scholars now account for the agency of actors ranging from hair lice among volunteer tourists in Nepal (Benali & Ren, 2018) to trash in the lives of garbage pickers (Fisher, this volume). Such work that considers the multiple agentive roles played by the nonhuman has lent critical depth to tourism studies (Picken, 2010, p. 245). Indeed, this approach compels us to reconsider the very ontological positions of tourism research (Johannesson, 2005; Tribe, 2010; Cohen & Cohen, 2012; van der Duim et al., 2013). Thus, the gendered dimensions of tourism reinforce the Human/Nature binaries through representations of human and nonhuman relations. For instance, the rampant growth of whale-watching tourism in the Arctic reproduces a long-standing touristic cliché through its promotion of a pristine nature beyond the human realm (Kristoffersen et al., 2015). The widespread belief in the role of humans to "protect nature" through neoliberal practices of conservation, enclosure and regulation perpetuates the separation of humans and nature in ways that overshadow the structural drivers of environmental degradation and species extinction.

In tourism, the separation of nature from culture has long been perhaps most visible in the dualism of female and male. Anthropologist Sherry Ortner (1974) famously argued that women's universal subordination is a result of their discursive ties with nature (Ortner, 1974). Donna Haraway has latterly pointed out the ways in which Western regulatory fiction presumes that motherhood is natural, whereas fatherhood is cultural (Haraway, 2001). The representation of nature as feminine has a long-standing presence in the history of colonial and imperial ambitions. In tourism, this engagement often takes on gendered forms where "Nature" is seen as thoroughly female (Enloe, 1989; Besio et al., 2008). For instance, the overdetermined aspects linking nature with the female body have contributed to naturalizing the relationship between females and nature, and males and culture (Swain, 1995). In tourism, the binary opposition between male/female and nature/culture is reiterated in myriad ways where women's bodies become the metonymic signifier of "nature" in tourism destinations (Swain, 1995; Johnston, 2001).

Capitalist imaginaries: alienation and authenticity

Four decades ago, Dean MacCannell (1979) made the astute observation that tourists often seek out authentic experiences in places perceived to be beyond the realm of capitalist modernity. The ruse has long been to present a touristic experience that can be sold and consumed through its interpretation as unmediated by any capitalist exchange. This practice still holds true in many forms today, where desires for authentic, non-commodified, intimate experiences are widespread, and for which many tourists will still spend top dollar to realize (Conran, 2006). In this way, political ecologies of tourism are reimagined in the time of Anthropocene where alienation drives a desire for authentic cultures and environments that are seen to naturally exist in places perceived to have weaker links to neoliberal capital (Norum & Mostafanezhad, 2016). As Anna Tsing and Donna Haraway point out, "what thinking through capital means for knowing Anthropocene might be to consider the importance of long-distance investors in creating an abstract relationship between investment and property", which can be conceptualized as alienation (Haraway et al., 2016, p. 556). Marx originally conceived of alienation as a process via which the mechanistic nature of workers' labor leads to a separation of actor from self, such that she is unable to conceive of themselves as determinants of their own actions and desires. Transposed to the present-day encounters of the (capitalist) tourism industry, then, the tourist becomes ruled by his or her own desire (labor). People are in this way bound to becoming estranged from themselves and each other under the conditions of capitalist industrial process. Alienation is one of the clearest expressions of the Human/Nature dualism in tourism. This observation suggests new ways to think through the dynamics through which people, culture and identities become alienated resources. Alienation and authenticity play leading roles in the long-standing search by the tourist for redemption and self-fulfillment in and through other times, other places and other people. As a result, authenticity becomes an important discursive arena that shapes anthropocenic imaginaries in tourism.

The Human/Nature binary that drives the so-called search for authenticity has paved the way for the development of new tourism spaces. In Myanmar, for instance, spaces that have only recently been opened to tourism, such as post-conflict border regions, are widely perceived by tourists to stand in relative isolation from Western capitalism. They are frequently represented by popular and travel media, and timeless destinations that should be witnessed before it is "too late" (Norum & Mostafanezhad, 2016). Frontiers also evoke a range of spatial concepts such as borders, hinterlands and enclaves (Watts, 2017). States of insecurity characterize frontiers which are developed through "an influx and presence of non-native private actors in pursuit of the newly discovered resources" (Peluso & Lund, 2011, p. 1). Yet, land and resources are not the whole story, as Watts points out, because "The frontier is primarily a social space within which forms of rule and authority, and multiple sovereignties, are in question" (Watts, 2017, p. 480). Thus, capital accumulation can be described as a frontier process which is driven by "cheap Natures". Capitalist competition drives cheap Natures that materialize as new commodity frontiers (Watts, 2017, p. 479).

In tourism, enclosures are one core strategy through which new land is privatized and commodified (Buscher, 2013). Tourism frontiers are sites where national identities (e.g. settlers, indigenous, natives) are both homogenized and distinguished. The tourism encounter as a "contact zone" facilitates opportunities for native territory to be enclosed for tourism development. Sometimes through violence and sometimes through persuasion, the territorialization of new land for the pursuit of tourism profits is widespread globally. International peace parks (Buscher, 2013), safaris in Tanzania (Gardner, 2016) and conservation areas in Myanmar (Winton & Ocampo-Peñuela, 2018) become practices of territorialization through conservation which bear witness to long and often violent histories. Indeed, primitive accumulation, in which property-less laborers work for capitalists holding the means of production (Marx, 1976 [1867]), is integral to the tourism frontier making process. Through the strategic framing of capitalist anthropocenic imaginaries, new land, resource and commodity frontiers are developed in ways that further marginalize the landless while empowering the global bourgeois class.

Ecological imaginaries: commodification and nature

Tourism is typically one of the first industries to develop in frontier regions, in part because of the facility of appropriating "nature" and "culture" as commodity forms (Krakover & Gradus, 2002). The commodification of the intangible "products" of culture and nature produces new kinds of anthropocenic imaginaries that speak to our relationship to and understanding of planetary and human ecologies. Through tourism, material and social categories such as Nature (including human nature) are developed as commodities which are then attributed with exchange value. Such value is created through human and nonhuman social relations that exist outside capital, producing commodity forms out of non-capitalist forms of value (Tsing, 2013, p. 21). Mushrooms, the object of Tsing's own analysis, become valuable not through qualitative changes but through new social relations of exchange; in similar ways, culture and nature attain value through new practices of differentiation. Scholars have drawn upon Marx's conceptualization of "free gifts of nature" and the exploitation and commodification of unpaid work in order to demonstrate how capitalism depends on not-only capitalist modes of exchange (Gibson-Graham, 2006; Tsing, 2009, 2015; Moore, 2015, 2016). The commodification of the "free gifts of nature" includes human nature, the appropriation of which is widespread in the tourism industry. As Buscher and Fletcher (2017) describe, there are violent processes through which tourism products are made into capital. Their work lays bare the structural violence done to human and non-human nature through "the systematic production of inequalities, waste and 'spaces of exception'" (Buscher & Fletcher, 2017, p. 651). While scholars acknowledge that human nature, too, is exploited for capital accumulation, emerging scholarship demonstrates the ways in which these practices are shaped by and for tourism. Indeed, as Fletcher (this volume) shows, the emergence of an Anthropocene tourism is illustrative of the ways in which the tourism industry provides a spatio-temporal fix for capitalism through creation of new commodity frontiers. In this way, a particular form of disaster capitalism emerges, contributing to new ecological anthropocenic imaginaries (Klein, 2007). These imaginaries are perpetuated through commodity and land frontiers where the Human/Nature binary is the basis on which culture, identity and nature is commodified by tourism industry practitioners.

Sustainable tourism in the Anthropocene?

Anthropocene looms large in sustainable tourism. Yet, like the age of the Anthropocene itself, the concept of sustainability has a pervasive obscuring effect. Indeed, what is requisite here is the politicization of the environment predicated on the recognition of a multitude of potential socio-environmental futures and imaginaries. As Swyngedouw argues, a policy of sustainability is constructed around a single Nature, insofar as there are a multitude of natures and a multitude of existing or possible socio-natural relations, perpetuates a kind of post-political and post-democratic condition that forecloses the possibility of a real politics of the environment (Swyngedouw, 2007). While the call for a sustainable future is hardly contestable or controversial, it nevertheless presents a paradox (Krueger & Gibbs, 2007). It would be hard these days to lodge much argument with the economic, social, environmental and political ideas surrounding sustainability. Yet, the policies and practices that have developed around sustainability are frequently pursued within the broader context of neoliberal global capitalism – which, by definition, is throttled by a bottom line.

The ecomodernist Manifesto argues that since we are entering a new era of intervention, we have the responsibility, duty and capacity to change the environment using the tools available to us (Asafu-Adjayde et al., 2015). Such tools at our disposal include socio-technological capacities, such as nuclear power and carbon offsetting. Via a process of "selling nature to save it", green fixes can also perpetuate global injustices (McAfee, 1999). Though it hardly makes it into discussion about the UN's 17 Sustainable Development Goals, processes such as biodiversity protection, conservation and other forms of environmental protection through tourism frequently deprive local people of land and resources (West et al., 2006; Igoe & Brockington, 2007; Igoe et al., 2010). In order to have any marked effect on global environmental justice, the planet's current environmental crises must be framed to be as much geopolitical as geological (Grove & Chandler, 2017). In a similar way, so-called sustainable tourism is limited by the tourism industry's proverbial bottom line. As we write about the consequences of Anthropocene, when we discuss the "consequences" or "impacts" of tourism, we must remember that the kinds of questions we ask will shape the kinds of answers we create. Yet, as Jason Moore has asked, "But consequences of what? Of humanity as a whole? Of population? Of industrial civilization? Of the West? Of capitalism?" (Moore, 2016, p. 78).

A myopic focus on sustainable tourism can cause one to lose sight of the structural violence of accumulation processes on which the tourism industry depends. The relationship between the extent of environmental degradation in the so-called developing countries, poverty and the colonial experience is hardly happenstance. Still, as is the case with the recent climate change negotiations, many of the most critical scholars "will dislike their in-built assumptions that capitalism and nation-states are non-negotiable givens" (Castree, 2014b, p. 468). Part of teasing out these relationships is to pay attention to how the Human/Nature binary is appropriated and perpetuated in tourism practice where place-making via techniques of branding is a form of structural violence that is naturalized through the tourism industry (Buscher & Fletcher, 2017, p. 651). Tourism, as one of the world's largest industries – and as one of the planet's greatest dramaturg for encounters between humans and nature – plays a key, if underappreciated role in the historical development of the anthropocenic imaginary.

One must also reconsider what it means to speak about "tourism management" when the widening economic gap between the "hosts" and "guests" makes it all but impossible for many "hosts" to either remain strictly local or to even eventually become "guests" somewhere else. Thus, the widespread focus on tourism management and social impacts reveals the analytical bias towards the symptoms rather than the causes of the emergence of Anthropocene. The ways in which the anthropocenic imaginary informs our worldview drive our understanding and practice of tourism as both scholars and tourists. Like the abstraction of tourism, Anthropocene also finds value in its imprecise nature.

The concept of sustainable tourism has developed as an antidote to the Human/Nature binary and its ecological, social and economic implications. This binary links humans and nature as dead abstractions "that connect to each other as cascades of consequences rather than constitutive relations" (Moore, 2017, p. 5). The violence done by the Human/Nature binary is in the strategic removal of the relations of historical change through which contemporary tourism practice takes place. It thereby propagates extant systems of domination, exploitation and appropriation (Moore, 2017). This observation makes it clear how the very category of "sustainable" tourism is but duplicitous misnomer: its existence depends on a wider capitalist system of economic exchange that is driven by the tendency towards the overaccumulation of capital, necessitating expansion into new geographic spaces, diversified markets and further privatization and exploitation of both nature and humanity.

Efforts to attribute economic value to the environment, culture and/or identity of peoples and places through mechanisms such as cultural ecosystem services are one among many neoliberal strategies that threaten to turn the entire planet into a value to be traded, sold or stolen. Tourism plays a core role in such econometric processes. Such practices are echoed in conservation arenas where payments for ecosystem services, carbon trading and other such schemes are widely floated as viable options for ameliorating the ecological violence that poverty entails. Yet what these schemes all have in common is the shallow logic on which they are based. Telltale examples are not few in number. International efforts led by the Global North to enhance climate change resilience in places such as Vanuatu appear grossly disingenuous given that the U.S. and China alone account for a combined 45 percent of the global CO_2 emissions from fossil fuel. Sustainable tourism campaigns kindly ask that guests not to send their towels to the laundry in order to save water in Bali matter less when locals do not even have access to potable water for personal use in the first place (Cole & Browne, 2015). The idea of "sustainable" tourism often focuses on the symptoms rather than the causes of environmental, social and economic problems linked to over-, under- or just poor-managed tourism. The same might be said for the increasing calls for "peace through tourism", through which the frequently misguided efforts of tourism practitioners often merely fertilize the roots of inequality in egregious manner (Lisle, 2016).

Yet, all this is not to say that there is no alternative. Indeed, as others have pointed out, people have already begun developing non-carbon and stay-at-home tourism campaigns which challenge traditional associations of the industry (Huijbens & Gren, 2015). Anthropocenic imaginaries are deeply enmeshed in such alterative visions of tourism, and they facilitate new possibilities for thinking through human-environment relations. These ways of thinking about and representing humans' role in environmental change have also contributed to a number of forms of tourism in Anthropocene emerging from the loss of "natural" resources which take shape in forms of "self-conscious Anthropocene tourism" that also go by the names of, e.g., disaster, extinction, development and volunteer (Fletcher, this volume). If social scientists can agree on anything about Anthropocene, it might be that it has reshaped many of the theoretical and conceptual frameworks through which we have come to understand human-environment relationships. Tourism scholars have a unique role to play in this reassessment by asking tough questions about global sustainability and environmental ethics to enact "geo-ethically informed geo-hospitality" (Huijbens & Gren, 2015).

Thus, if as Amelia Moore argues, "our collective sense of 'environmental consequences has never been greater" (Moore, 2016, p. 78), then the role the anthropocenic imaginary plays in tourism forces us to address critical questions around how one ought to live in such a deeply vulnerable geological epoch. Simple, fundamental questions such as "How should we live?" (Castree, 2014b) must now be reconsidered within the broader context of neoliberal and other forms of globalization that increasingly drive economic, social, political and ecological practices. Sustainable tourism, often defined in opposition to an otherwise unsustainable tourism, is but one response to this call for how to position ourselves among human and nonhuman others. Yet, tourism scholarship's often narrow focus on the tourism industry itself can blind us to the broader system of neoliberal global capitalism within which the tourism industry exists. If, as the World Tourism Organization describes it, tourism "entails movement of people to countries or places outside of their usual environment for personal or business/professional purposes" (World Tourism Organization, 2015), then the very industry itself is deeply dependent on environmental degradation. For example, the tourism industry's contribution to greenhouse gas emissions is estimated to be as high as 8 percent with a forecasted global growth rate of 4 percent (Lenzen et al., 2018). Let us not forget that (a) most of this carbon footprint is exerted by high-income countries, and (b) the decarbonization of tourism-related technology is far from able to hold pace with the growing demand for tourism globally (Lenzen et al., 2018). It seems that what is needed is both global economic structural change and technocratic solutions, which require tourism industry practitioners and scholars to reach outside of the comfort of their disciplinary and epistemic homes.

Yet, what the anthropocenic imaginary does enable is the interdisciplinary conversations we are now having about the current and future state of the world. Anthropologists now work alongside ecologists to co-create new ways of knowing. Creative writers are in conversation with field biologists who tell stories of landscape and memory while measuring the particulate matter that swirls in the airstream around them. New fields in communication and journalism are developing a focus on science communication and public outreach. These are significant for how they integrate natural and social science with popular culture. Situated within the field's long history of interdisciplinary work, scholars who think about tourism are often well equipped with the analytic tools to think through relationships between people, capital and the environment that frame the anthropocenic imaginary. Still many of us are frequently confronted with our own lifestyles and its incongruencies. Seth Kugel describes the flight-shaming movement that has now spread throughout Europe and how "privileged but climate-conscious Americans are facing an August reckoning: Is your summer vacation destroying our planet?".[1] As we put the final edits this conclusion from opposite ends of the planet – Tallinn and Honolulu – we are both "packing our bags" to begin environmental-focused fieldwork several thousand kilometers away in different places, and then head to a large academic conference to engage with colleagues over what we as scholars (and as humans) can do about anthropogenic change. The issues involved are large and complex, but they are also simple and personal – and often paradoxical.

Conclusion

In the late 1990s, Donna Haraway asked: "What forms does the love of nature take in particular historical contexts?" (Haraway, 2013, p. 1). She was concerned with the genealogies and geographies through which nature was increasingly being constructed as and fetishistic object of intellectual desire. While Haraway's interest at the time was human-primate relations, her questions are ever more significant today. The ways in which tourists and tourism industry practitioners engage with nature are a driving force behind the ways in which we come to "love" nature which has both material and symbolic implications for humans and nonhumans everywhere. The means we have of understanding, discussing and representing contemporary human environmental change – our own anthropocenic imaginaries – exist as both irritants and salves for the growing number of societal and ecological problems we are faced with.

While the Anthropocene is best known to scholars as a geological demarcation of a temporal span, it is also fundamentally a social imaginary that governs and is shaped by a range of social and cultural phenomena. Anthropocenic imaginaries, we have argued in this introductory chapter, are appropriated in market-based solutions to environmental degradation that emanate from neoliberal contexts internal to the problem. Our aim, and that of the chapters which follow, has been to outline this notion of the anthropocenic imaginary and show how it relates to practices, encounters and regimes of tourism. In so doing, we call for more nuanced scholarship that leverages political ecology approaches to help identify exactly who Anthropos is when we can so swiftly place blame on Humanity at large the issues we face. Local communities are also making sense – perhaps their own, perhaps shared – of and within Anthropocene, and its concomitant post-human, capitalist and ecological imaginaries. We have highlighted how the concept of Anthropocene is shaping planetary social and cultural imaginations, while also changing practices across a range of human techno-social interventions. The Human/Nature binary, we have argued, is both appropriated and perpetuated by tourism industry stakeholders.

Tourism is a unique starting point from which to examine how human-environment relations are shifting in the current geological epoch. The chapters herein investigate distinct contexts and various understandings of habitation and habitability – the lifestyles people enact and the actions they take to ensure their continuance. They focus on the multiple and sometimes contradictory thinking, politics and practices that drive the anthropocenic imaginary implicit in both tourist practices and environmental subjectivities across a number of core themes that include gender, health, conservation, agriculture, climate change, disaster, coastal marine management and sustainability. They build on emerging work that brings a political ecology lens to tourism (Mostafanezhad et al., 2016; Nepal & Saarinen, 2016), pushing this agenda forward by integrating the roles played by conceptual regimes that both embody and produce anthropogenic planetary socio-environmental change. The chapters also contribute new theoretical and empirical insights to nascent scholarship that links assemblages of political ecological action with globalizing processes of tourism. Taken individually, each of the papers theoretically and empirically integrate linkages between tourism practice and political ecological responses to tourism through the lens of Anthropocene. But as an integrated collection of engaged research that speaks to one another, they identify and problematize core themes, concepts and issues for critical scholarship, asking how thinking about and doing tourism differently might spawn individual and communal planetary action. By attending to the everyday lives of tourism actors across a range of spaces that include nature parks, ecotourism attractions, agritourism initiatives and green urban spaces, among other tourism geographies, we can consider how tourism helps us to theorize key questions about development, environment and modernity. And in bringing tourism studies and political ecology to bear on all that is now anthropocenic, we hope that this collection will start new conversations and interventions into sustainable tourism and practice along the Anthropocene tour.

Note

1 Kugel, Seth. August 24, 2019. How Guilty Should You Feel About Your Vacation? New York Times. https://www.nytimes.com/2019/08/24/opinion/sunday/tourism-climate-flight-shaming.html?searchResultPosition=1.

References

Althusser, L. (1971). Ideology and ideological state apparatuses. In L. Althusser (Ed.), *Lenin, philosophy and other essays* (pp. 123–192). New York, NY: Monthly Review Press.

Anderson, B., & Wylie, J. (2009). On geography and materiality. *Environment and Planning A*, 41(2), 318–335.

Barnosky, A. D., Matzke, N., Tomiya, S., Wogan, G. O. U., Swartz, B., Quental, T. B., … Ferrer, E. A. (2011). Has the Earth's sixth mass extinction already arrived? *Nature*, 471(7336), 51.

Benali, A., & Ren, C. B. (2018). Lice work: Non-human trajectories in volunteer tourism. *Tourist Studies*.

Besio, K., Johnston, L., & Longhurst, R. (2008). Sexy beasts and devoted mums: Narrating nature through dolphin tourism. *Environment and Planning A*, 40(5), 1219–1234.

Boggs, C. (2016). Human niche construction and the anthropocene. *RCC Perspectives*, 2, 27–32.

Bryant, R. A. G., & Michael, K. (2006). A pioneering reputation: Assessing Piers Blaikie's contributions to political ecology. *Geoforum*, 39(2), 708–715.

Bryant, R. L., & Bailey, S. (1997). *Third World Political Ecology*. New York, NY: Routledge.

Bryce, D. (2007). Repackaging orientalism: Discourses on Egypt and Turkey in British outbound tourism. *Tourist Studies*, 7(2), 165–191. doi:10.1177/1468797607083502

Buscher, B. (2013). *Transforming the Frontier: Peace Parks and the Politics of Neoliberal Conservation in Southern Africa*. Durham, NC: Duke University Press.

Buscher, B., & Fletcher, R. (2017). Destructive creation: Capital accumulation and the structural violence of tourism. *Journal of Sustainable Tourism*, 25(5), 651–667.

Callon, M. (1999). Actor-network theory—the market test. *The Sociological Review*, 47(1_suppl), 181–195.

Castree, N. (2014a). The Anthropocene and geography III: Future directions. *Geography Compass*, 8(7), 464–476.

Castree, N. (2014b). The Anthropocene and the environmental humanities: Extending the conversation. *Environmental Humanities*, 5(1), 233–260.

Chakrabarty, D. (2016). Whose Anthropocene? A response. *RCC Perspectives*, 2(2), 101–114.

Chester, M. V., & Allenby, B. (2018). Toward adaptive infrastructure: Flexibility and agility in a non-stationarity age. *Sustainable and Resilient Infrastructure* (Vol. 1), 1–19.

Cohen, E., & Cohen, S. A. (2012). Current sociological theories and issues in tourism. *Annals of Tourism Research*, 39(4), 2177–2202.

Cole, S., & Browne, M. (2015). Tourism and water inequity in Bali: A social-ecological systems analysis. *Human Ecology*, 43(3), 439–450.

Conran, M. (2006). Beyond authenticity: Exploring intimacy in the touristic encounter in Thailand. *Tourism Geographies*, 8(3), 274–285.

Enloe, C. (1989). *Bananas Beaches and Bases: Making Feminist Sense of International Politics*. London: Pandora.

Gardner, B. (2016). *Selling the Serengeti: The Cultural Politics of Safari Tourism*. Atlanta: University of Georgia Press.

Gibson-Graham, J. K. (2006). *A Postcapitalist Politics*. Minneapolis: University of Minnesota Press.

Ginsburg, F., Abu-Lughod, L., & Brian, L. (Ed.). (2002). *Media Worlds: Anthropology on New Terrain*. Berkeley: University of California Press.

Grove, K., & Chandler, D. (2017). Introduction: Resilience and the Anthropocene: The stakes of 'renaturalising' politics. *Resilience*, 5(2), 79–91.

Hall, C. M., & Tucker, H. (2004). *Tourism and Postcolonialism: Contested Discourses, Identities and Representations*. New York, NY: Routledge.

Hamilton, C. (2016). The Anthropocene as rupture. *The Anthropocene Review*, 3(2), 93–106.

Hamilton, C., Gemenne, F., & Bonneuil, C. (2015). *The Anthropocene and the Global Environmental Crisis: Rethinking Modernity in a New Epoch*. New York, NY: Routledge.

Haraway, D. (2015). Anthropocene, capitalocene, plantationocene, chthulucene: Making kin. *Environmental Humanities*, 6(1), 159–165.

Haraway, D., Ishikawa, N., Gilbert, S. F., Olwig, K., Tsing, A. L., & Bubandt, N. (2016). Anthropologists are talking – about the Anthropocene. *Ethnos*, 81(3), 535–564.

Haraway, D. J. (2001). "Gender" for a Marxist dictionary: The sexual politics of a word. In Women, Gender (Ed.), *Religion: A reader* (pp. 49–75). New York: Palgrave Macmillan.

Haraway, D. J. (2013). *Primate Visions: Gender, Race, and Nature in the World of Modern Science*. New York: Routledge.

Häausler, H. (2018). Did Anthropogeology anticipate the idea of the Anthropocene? *The Anthropocene Review*, 5(1), 69–86.

Huijbens, E. H., & Gren, M. (Eds.). (2015). *Tourism and the Anthropocene*. New York, NY: Routledge.

Igoe, J., & Brockington, D. (2007). Neoliberal conservation: A brief introduction. *Conservation and Society*, 5(4), 432–449.

Igoe, J., Neves, K., & Brockington, D. (2010). A spectacular eco-tour around the historic bloc: Theorising the convergence of biodiversity conservation and capitalist expansion. *Antipode*, 42(3), 486–512.

Ingold, T. (1993). The temporality of the landscape. *World Archaeology*, 25(2), 152–174.

Ingold, T. (2013). Dreaming of dragons: On the imagination of real life. *Journal of the Royal Anthropological Institute*, 19(4), 734–752.

Janowski, M., & Ingold, T. (2016). *Imagining the Forces of Life and the Cosmos in the Kelabit Highlands, Sarawak Imagining Landscapes* (pp. 157–178). New York, NY: Routledge.

Johannesson, G. T. (2005). Tourism translations: Actor-network theory and tourism research. *Tourist Studies*, 5(2), 133–150.

Johnston, L. (2001). (Other) bodies and tourism studies. *Annals of Tourism Research*, 28(1), 180–201.

Klein, N. (2007). *The Shock Doctrine: The Rise of Disaster Capitalism*. New York, NY: Macmillan.

Krakover, S., & Gradus, Y. (2002). *Tourism in Frontier Areas*. Lexington: Lexington Books.

Krueger, R., & Gibbs, D. (2007). *The Sustainable Development Paradox: Urban Political Economy in the United States and Europe*. London: Guilford Press.

Latour, B. (2005). *Reassembling the Social: An Introduction to Actor-network-theory*. Oxford: Oxford University Press.

Lenzen, M., Sun, Y.-Y., Faturay, F., Ting, Y.-P., Geschke, A., & Malik, A. (2018). The carbon footprint of global tourism. *Nature Climate Change*, 8(6), 522–528. doi:10.1038/s41558-018-0141-x

Lisle, D. (2016). *Holidays in the Danger Zone: Entanglements of War and Tourism*. Minneapolis: University of Minnesota Press.

MacCannell, D. (1973). Staged authenticity: Arrangements of social space in tourist settings. *American Journal of Sociology*, 79(3), 589–603.

Marcus, G. E. (1995). *Technoscientific Imaginaries: Conversations, Profiles, and Memoirs* (Vol. 2). Chicago, IL: University of Chicago Press.

McAfee, K. (1999). Selling nature to save it? Biodiversity and green developmentalism. *Environment and Planning D: Society and Space*, 17(2), 133–154.

Miller, D. (2005). *Materiality*. Duke: Duke University Press.

Moore, A. (2015). Tourism in the Anthropocene park? New analytic possibilities. *International Journal of Tourism Anthropology*, 4(2), 186–200.

Moore, A. (2019). *Destination Anthropocene: Science and Tourism in the Bahamas* (Vol. 7). Critical Environments: Nature. Berkeley: University of California Press.

Moore, J. W. (2015). *Capitalism in the Web of Life: Ecology and the Accumulation of Capital*. New York, NY: Verso Books.

Moore, J. W. (2016). *The Rise of Cheap Nature*. Binghamton: The Open Repository, Binghamton University.

Moore, J. W. (2017). The Capitalocene, Part I: On the nature and origins of our ecological crisis. *The Journal of Peasant Studies*, 44(3), 594–630.

Mostafanezhad, M., Norum, R., Shelton, E. J., & Thompson-Carr, A. (2016). *Political Ecology of Tourism: Community, Power and the Environment*. New York, NY: Routledge.

Nepal, S. K., & Saarinen, J. (2016). *Political Ecology and Tourism*. New York, NY: Routledge.

Norum, R., Kramvig, B., & Kristoffersen, B. (2015). Arctic whale watching and Anthropocene ethics. In Gren, M., (Ed.) *Tourism and the anthropocene. Contemporary geographies of leisure, tourism and mobility* (pp. 94–110). Oxford & New York: Routledge.

Norum, R., & Mostafanezhad, M. (2016). A chronopolitics of tourism. *Geoforum*, 77, 157–160.

Ortner, S. B. (1974). Is female to male, as nature is to culture? In Rosaldo and Lamphere (Eds.), *Women, Culture and Society* (pp. 68–87), Stanford, CA: Stanford University Press.

Peet, R., & Watts, M. (Eds.). (2004). *Liberation Ecologies: Environment, Development, Social Movements* (2nd ed.). New York, NY: Routledge.

Peluso, N. L., & Lund, C. (2011). New frontiers of land control: Introduction. *Journal of Peasant Studies*, 38(4), 667–681.

Pickels, M. (2018). Carnegie Museum's Anthropocene curator featured in fall events. *Trib Live*. Retrieved from: https://triblive.com/aande/museums/14072454-74/carnegie-museums-anthropocene-curator-featured-in-fall-events

Picken, F. (2010). Tourism, design and controversy: Calling on non-humans to explain ourselves. *Tourist Studies*, 10(3), 245–263.

Reszitnyk, A. (2015). *Uncovering the Anthropocenic Imaginary: The Metabolization of Disaster in Contemporary American Culture*. Dissertation. Department of English and Cultural Studies at McMaster Univeristy, Canada.

Said, E. (1978). *Orientalism*. New York, NY: Vintage Books.

Salazar, N. (2012). Tourism imaginaries: A conceptual approach. *Annals of Tourism Research*, 39(2), 863–882.

Smith, L. T. (1999). *Decolonizing Methodologies: Research and Indigenous Peoples*. New York, NY: Zed Books.

Stonich, S. C. (1998). Political ecology of tourism. *Annals of Tourism Research*, 25(1), 25–54.

Susan, P., Lisa, L. G., & Michael, W. (2003). Locating the political in political ecology: An introduction. *Human Organization*, 62(3), 205–217.

Swain, M. B. (1995). Gender in tourism. *Annals of Tourism Research*, 22(2), 247–266.

Swyngedouw, E. (2007). Impossible 'sustainability' and the postpolitical condition. In R. Krueger & D. Gibbs (Eds.), *The sustainable development paradox: Urban political economy in the United States and Europe* (pp. 13–40). New York: Guilford Press.

Taylor, C. (2002). Modern social imaginaries. *Public Culture*, 14(1), 91–124.

Tribe, J. (2010). Tribes, Territories and networks in the tourism academy. *Annals of Tourism Research*, 37(1), 7–33. doi:10.1016/j.annals.2009.05.001

Tsing, A. (2009). Supply chains and the human condition. *Rethinking Marxism*, 21(2), 148–176.

Tsing, A. (2013). Sorting out commodities: How capitalist value is made through gifts. *HAU: Journal of Ethnographic Theory*, 3(1), 21–43.

Tsing, A. L. (2015). *The Mushroom at the End of the World: On the Possibility of Life in Capitalist Ruins*. Princeton: Princeton University Press.

Tsing, A. L., Bubandt, N., Gan, E., & Swanson, H. A. (2017). *Arts of living on a damaged planet: ghosts and monsters of the anthropocene*. Minneapolis: University of Minnesota Press.

Tucker, H., & Shelton, E. (2014). Traveling through the end times: The tourist as apocalyptic subject. *Tourism Analysis*, 19(5), 645–654.

Van der Duim, R. (2007). Tourismscapes an actor-network perspective. *Annals of Tourism Research*, 34(4), 961–976.

Van der Duim, R., Ren, C., & Thor, J. G. (2013). Ordering, materiality, and multiplicity: Enacting actor – network theory in tourism. *Tourist Studies*, 13(1), 3–20.

Watts, M. (2017). Frontiers: Authority, precarity, and insurgency at the edge of the state. *World Development*, 101, 477–488.

West, P., Igoe, J., & Brockington, D. (2006). Parks and peoples: The social impact of protected areas. *Annual Review of Anthropology*, 35(1), 251–277.

Winton, R. S., & Ocampo-Peñuela, N. (2018). How to realize social and conservation benefits from ecotourism in post-conflict contexts. *Biotropica*, 50(5), 719–722.

World Tourism Organization (2015). *Glossary*. Retrieved from: http://cf.cdn.unwto.org/sites/all/files/docpdf/ glossaryen-rev.pdf

Yan, G., & Santos, C. A. (2009). "China, forever": Tourism discourse and self-orientalism. *Annals of Tourism Research*, 36(2), 295–315.

Yoshihara, M. (2004). The flight of the Japanese butterfly: Orientalism, nationalism, and performances of Japanese womanhood. *American Quarterly*, 56(4), 975–999.

Selling Anthropocene space: situated adventures in sustainable tourism

Amelia Moore

ABSTRACT

The Anthropocene is a proposed technical term for a new geological timeframe, but it is also a conceptual tool with the potential to redefine the stakes of contemporary environmental politics. One facet that is often overlooked is that the Anthropocene is a concept with commercial potential, even if the term itself has not been widely adopted. This article presents an investigation of the commercial potential of the Anthropocene idea through the lens of self-described sustainable tourism ventures in The Bahamas. These examples demonstrate some of the ways in which Anthropocene imaginaries participate in the recreation, redesign, and rebranding of specific spaces as emergent "tourism products", specifically the small island farm and the anthropogenic coral reef. The goal is twofold: (1) to explore the symbolic and material creativity of the Anthropocene idea as its themes are used to extend capitalist innovation, and (2) to examine the Anthropocene idea as a strategy that builds upon existing histories of inequality to enable transnational accumulation in particular locales. As a situated adventure, this article articulates a reflexive mode of political ecological research for the Anthropocene that is equipped to critically articulate emergent practices at the intersection of postcolonial tourism, environmental conservation, and sustainable development.

Introduction: from Paradise island to Anthropocene island

It's an unusually overcast day on the island of New Providence, the most populated island of the Bahamian archipelago. I find myself far from the cruise ships and congestion of Nassau, and far from the beaches and hotels that skirt the coast. I am "way out west" on, of all places, a small farm, my feet planted firmly in the dirt. The air smells faintly of arugula, and it is still quite warm despite the clouds. I find myself far from the beaten path out on the farm because I am exploring emergent forms of self-described "sustainable tourism" in The Bahamas to understand the early effects of recent geographical imaginaries on the country's physical space and cultural politics.

To know how far this farm is from the standard form of Bahamian "tourism product", one needs to know about The Bahamas and Bahamian tourism.[1] The Bahamas is a former British colony and Caribbean archipelago of 700 islands and cays, with a population approaching 380,000. In 2010, ninety percent of the population identified as black, descended from enslaved peoples.

Over 250,000 people live on New Providence, the seat of the nation's urban capital, Nassau.[2] New Providence is also the site of the majority of tourist arrivals to the country, numbering in the millions annually, and the site of the country's largest hotels and resorts.[3] In The Bahamas, tourism comprises 48% of the GDP, making it the ninth most tourism dependent country in the world relative to its size (WTTC, 2017). This is the culmination of decades of strong marketing campaigns made by the Bahamian tourism industry, institutionalized within the Bahamian government in the form of the Ministry of Tourism (Cleare, 2007).

Howie and Lewis explain that, "the idea of geographical 'imaginaries' is an attempt to capture not only that there are multiple geographical imaginations at large in the world, but that they do work in framing understandings of the world and in turn making our different worlds, and that particular imaginaries are willfully put to work with political affect and effect" (2014, p. 132). Marketing campaigns in the United States, Canada, and Europe have historically sold The Bahamas as a specific kind of geographic imaginary: the paradise island. This imaginary has always been imbued with colonial tropes of smiling black servitude, the segregation of whiter privileged populations within resort enclaves, and tropical Edenic nature that exists outside of the civilized world. The paradise island imaginary has long been the basis of the Bahamian tourism brand in the global travel market, a brand that Bahamian scholar Ian Strachan calls "paradise and plantation" (2002), but that is referred to as "sun, sand, and sea" within the tourism industry itself (Cameron & Gatewood, 2008).

Since at least the 1950s, this imaginary-as-brand has animated the spread of exploitative capital in the archipelago via tourism, leading to the development of large hotels, the dredging of waterfront for large cruise ships and yachts, and the expansion of the nation's international airport to accommodate more and larger planes. In addition to the growth of the industry, the Bahamian scholar Angelique Nixon argues that this travel imaginary has supported a white, upper class, heteronormative, and Christian traveler as the most desired traveler for the Bahamian tourist market (2015). Thus, the paradise island imaginary has been an effective tool for continuing the colonial segregation of space via white supremacy in New Providence, even after independence in 1973. Wealthier, whiter residents and visitors dominate stretches of coastal territory in resorts and gated communities, while the majority of the black and less-white population live in land-locked subdivisions, aspiring to "good hotel jobs" serving that coastal flux of visitors (Johnson, 1997).

As a result of the success of the paradise island imaginary, most tourists do not yet think of farming when they think of The Bahamas (Cleare, 2007). Both Ian Strachan and Angelique Nixon argue that the standard Bahamian (and Caribbean) tourism product is evacuated of history, obscuring the environmental impacts of mass visitation while enabling neocolonial relations of servitude and mastery between island "hosts" and visiting "guests". And yet, as members of the travel industry observe, the Bahamian paradise island brand may be weakening in the face of competing beach destinations entering the global and regional market for travelers (Moore, 2010). After drops following the World Trade Center attacks of 11 September 2001, and the recent Great Recession, visitor numbers are stable, but they are not substantially growing (Trading Economics, 2016).[4]

Recent events show that the paradise island brand has further cracks. The Bahamas, like many small islands, is already experiencing the stresses of global environmental change. These are the "Anthropocene challenges" that are increasing the country's vulnerability and decreasing its resilience: sea level rise, shifting weather patterns, increased storm intensity and frequency, overfishing, coral degradation, dependence on petroleum-based energy, over-development, loss of fresh water, loss of species, increased presence of regional migrants and refugees, and population increase (Moore, 2016). For example, sea level rise and coastal erosion leads to shrinking coastlines, causing the government to spend on sand replenishment at popular beaches (Campbell, 2012). Further, as local seafood becomes scarcer due to the consumptive habits of tourists and locals, prices for seafood products rise, rapidly outpacing that of imported seafood in grocery stores and disappearing from hotel menus (personal observation and communication from Nassau residents and members of the hotel industry). These are just some ways that the paradise island imaginary-as-brand is threatened by the realities of global environmental insecurity.

But instead of being overtaken by these realities, fears of global anthropogenic change are rearticulated within some tourism ventures that are strategically utilizing such realities as opportunities for more tourism-based enterprise. And in addition to the all-inclusive resort with its extreme consumption of resources and energy, its extreme output of waste, and its importation of industrial scale labor, materials, and food, we now have examples of "sustainable" hotels and visitor experiences that are designed to impart a more place-based authenticity. This is not necessarily ecotourism [which is explicitly based on environmental and cultural preservation and education (Honey, 1999; Weaver, 2001)], but it is intentionally greener tourism than mass tourism, branded under the sign of sustainability and implicitly framed by the Anthropocene idea. What is now known as "sustainable tourism" in this context emerges from a central irony: the expansion of tourism into new spaces exacerbates global environmental change, and at the same time the tourism industry creates products and imaginaries that stem from ideas about global environmental change to accumulate more space for more tourism.

In light of such events, scholars of tourism point out that international tourism is evolving in creative ways (Mostafanezhad, Norum, Shelton, & Thompson-Carr, 2016). Thus, the need to understand the significance of rebranded tourist imaginaries for emergent Anthropocene inspired travel markets – like the small island farm – is what brought me to the center of New Providence, my resort wear exchanged for sturdy shoes and shorts. I have been studying events at the intersection of ecology and tourism in The Bahamas as an anthropologist for over a decade. My observations here stem from accumulated research visits since 2007 and on specific ethnographic experiences with a dive voluntourist program concerning coral restoration in 2014 and on an organic agritourist farm in 2016.

This piece is a conceptual exploration beyond the standard research article, although it does provide ethnographic evidence. The work presented here is the result of a "situated adventure" in emergent practices, an attempt to disarticulate adventure and exploration from their colonial referents (including anthropology) and reclaim them as tools for decolonizing tourism studies. Adventures are journeys of inherent risk and uncertainty in which the outcome is not known at the outset. For tourism studies, adventures also imply the commodification and domestication of risk and exoticism within experiential business ventures that attract tourist dollars. Situated adventures (adapting Haraway, 1997) are, therefore, a mode of engagement with destinations that allow the tourism scholar to experience given tourism products and ventures (the line between scholar and tourist has always been quite blurry after all) while necessarily observing how such products align with asymmetrical neocolonial realities to transform local space and place into something new in ways that risk reinforcing those asymmetries. Situated adventures force readers to rethink and relearn their vacations. Political ecologists of contemporary tourism should experiment with this reflexive mode of engagement.

While inspired by a number of literatures, this article most immediately builds off of the work of Ian Strachan and Angelique Nixon who, among others, have been instrumental in decolonizing the paradise island imaginary in the Caribbean, demanding that scholars recognize the colonial legacies embedded in that tourism product while pushing to create alternative realities and imaginaries. The discussion builds off of this work of cultural analysis, combining a critical reading of tourism with a political ecological interest in the inequities stemming from particular intersections of nature and capital. Coinciding with the goals of this special issue, this article examines local imaginaries, materialities, and opportunities recreated in a tourism dependent economy in an era of global environmental change. The goal is to demonstrate a reflexive mode of political ecological research for the Anthropocene equipped to tackle emergent practices and ironies at the intersection of postcolonial tourism, environmental conservation, and sustainable development.

The remainder of this article explores Anthropocene space as an emergent travel product stemming from the Anthropocene island imaginary-as-brand. These concepts are grounded in two emergent "farm" examples from the island of New Providence, exploring them both above and below the surface of the sea. The article concludes with a discussion of the implications of these adventures in sustainable tourism for The Bahamas, the Caribbean, and tourism in general while returning to Strachan and Nixon's concern about alternatives to neocolonial tourism in the region.

Reimagining and rebranding space

The Anthropocene is a technical term generated by Earth scientists to label the ubiquitous impacts of human activities on the planet's biogeochemical systems (Crutzen & Schwägerl, 2011; Crutzen & Stoermer, 2000). The idea demands the recognition that humans are now the primary force behind most planetary change across scales. The scientific relevance of the term is something geologists have been debating for seventeen years, and they are approaching a conclusive vote as to whether the Anthropocene will replace the Holocene as the designation for the planet's geological present (Carrington, 2016). But beyond the immediate significance of the vote to scientifically validate the term, the idea itself is proliferating, allowing for multiple framings of the stakes and multiple possible responses. In other words, there are multiple Anthropocenes at work in the world today (Moore, 2015c). What matters here is the creative work of the *idea* (in all its guises from climate change to biodiversity loss to the global fresh water crisis) to raise awareness about anthropogenesis as a major component of our current reality. To put it bluntly, "wild" is dead. "Pristine" is passé. "Untouched" is unreal.[5] And crucially, the Anthropocene idea has helped propel the widespread shift in understandings of the relationship between nature and culture currently underway [albeit a highly uneven shift (see Chakrabarty, 2013; Haraway, 2016; Latour, 2013)].

As this special issue shows, there are multiple approaches to the intersection of tourism and the Anthropocene. One under-appreciated event that links the Anthropocene idea to tourism is the emergent phenomena of Anthropocene travel imaginaries as the basis for place-based travel brands. Generally, ecotourism is a popular mode of tourist travel tied to a political environmentalism that has produced familiar travel imaginaries based on viewing wilderness and wildlife (West & Carrier, 2004). The end goal is to market spectacular imaginaries to "save" pristine wilderness (and pristine cultures) from destruction via their entry into tourist markets [though results have often missed the mark (also see West, 2006)]. Similarly, emergent modes of sustainable travel, development design, place-based travel, labor practice, and spatial re-imagination are now implicitly tied to the Anthropocene idea via new kinds of travel imaginaries. But the purpose of these travel products is not salvage. Instead, "sustainable tourism" is purported to uphold development and economic growth itself in this increasingly precarious world [see the United Nations, which declared 2017 the Year of Sustainable Tourism (UNWTO, 2016)]. It is now possible in this context to sell time-sensitive adventures with new "change adapted" practices in locales of heightened anthropogenic significance. In other words, the travel industry can now use revised spatial imaginaries to brand Anthropocene space in markets for sustainable travel.

Along with the paradise island, the *small* island and its vulnerable systems have recently become alternative tourism imaginaries for the Anthropocene. Islands have played an important role in the geographic materializations of science and capital, especially in the colonial context where islands served as "natural" laboratory spaces for experiments with social and ecological systems, trade, and various forms of production and consumption (Baldacchino, 2006; Grove, 1996; Mintz, 1985). The geographic imaginaries that traditionally articulate tropical islands for

Euro-American publics stem from the age of scientific exploration and subsequent colonial accumulative practices, and these imaginaries have branded islands with a sense of heroism, adventure, exoticism, utopianism, and Edenic mastery. The paradise island is one form of island imaginary that proved to be particularly seductive for the Post WWII international tourism industry, especially as a brand platform for tropical island regions (Sheller, 2003). However, the recognition of the Anthropocene idea is inspiring an update to the paradise island tourism product. Once valued as laboratories for evolutionary processes or the exploitation of resources and imagined as an exotic locale for travel, small islands are now also valued as vulnerable geological formations, supporting vulnerable forms of human and nonhuman life in the face of anthropogenic change (Moore, 2010).

Revised imaginaries can, therefore, recreate space in the Anthropocene, but this is not the wholesale transformation of the imaginaries that animate island travel. Islands are still envisioned as Romantic laboratory spaces in the postcolonial context of anthropogenic global change. However, the Earth sciences are the prime movers of this Anthropocene awareness, acting across scales, though most recognizable in bodies like the Intergovernmental Panel on Climate Change (IPCC). Their work is politically institutionalized in international climate negotiations through small island voting blocks, such as the Alliance Of Small Islands States (AOSIS, 2015). Small islands are now characterized by their vulnerability, social and ecological fragility, and lack of security (Lazrus, 2012; Moore, 2010) and re-imagined in policy circles as material formations of socioecological risk. They have become a biogeographic imaginary for the Anthropocene – an Anthropocene space.

There are other examples of the reformulation of spatial knowledge, geographic imaginaries, and material practices at the intersection of the Earth sciences and transnational policy, including the Arctic and Antarctic, large forested areas such as the South American Rainforest, Andean and Himalayan glaciers, the deep oceans, and many more. These biogeographic formations now signify the force of anthropogenesis, risk, and uncertainty just as they are remade to signify resilience, adaptation, and sustainability in the face of anthropogenic change. As an emergent phenomenon, small island spaces are now designed for multiple forms of entrepreneurial Anthropocene enterprise.

These generalities are a scaffold for understanding current events, but they require more specification at the local level. The Bahamian cases described below don't involve a monolithic "tourism industry" or the direct involvement of the Ministry of Tourism or tourism developers. Instead, the examples stem from a small-scale form of entrepreneurial tourism wherein individuals and organizations attempt to capitalize on and reinvent emergent travel trends. These actors are on the vanguard of the larger travel industry, reshaping preferences that may eventually become aspects of mainstream travel brands, even as they reinforce some of the most problematic conditions of Caribbean tourism. Further, the term "Anthropocene" is not one that comes up often in The Bahamas and it is not yet a term used within the tourism industry, even at the narrow cutting edge of new travel markets. Instead, terms like "sustainability", "adaptation", and "innovation" are far more common. This language is not new, but when applied to emergent tourism products, such as the farms described below, these terms help to discursively reframe the character of a destination and physically reshape its visitable space within an Anthropocene imaginary.

In sum, the specific Bahamian examples that follow are two early attempts to sell Anthropocene space as a sustainable tourism venture. Once we start to look, we will find many destinations all over the world utilizing scientifically-informed Anthropocene imaginaries, even if this term itself is not yet widely used. These reimagined spaces are gradually becoming more mainstream brand components for tourism.

Farm fantasies: agritourism and coral nurseries

To sell Anthropocene space, potential travel entrepreneurs must first tap into the global conversation about anthropogenic change and align their product through design or marketing with a recognizable Anthropocene imaginary, such as the small island space and its attendant vulnerabilities. One visible area of international policy interest within the context of small islands and global change is food security (Ganpat & Isaac, 2015; Samoa, 2014). Islands have long been associated with crises of food security, but since anthropogenic global change reached the top of the international policy agenda in the late twentieth century, many novel solutions to the problems of food security, self-sufficiency, vulnerability, and sustainability have been proposed for islands in the face of climate change, overfishing, population sprawl, and other Anthropocene issues. Interventions in island food security range from accumulative "green grabbing" to place-based social movements for equitable relationships within food systems (Aragon, 2011; Torgerson, 2010). In The Bahamas, some tourism entrepreneurs have settled on farming as their entrée into the nascent market for Anthropocene-animated "sustainable" travel products.

Farms – with imaginaries of localization, cultural heritage, sustainability, healthy processes and products, and community – are primed to be the latest thing in niche-to-mainstream travel products. The term "farming" is used loosely here to mean both terrestrial ventures in agriculture as well as marine ventures in coral restoration. These practices have many differences, but in both instances farming is a practice that requires the material cultivation of living produce via the maintenance of physical and social space that has been organized for that purpose. Farm spatial products rely on the design and maintenance of a complex set of interconnected relationships with nonhuman organisms, organic and inorganic processes, and social dynamics including postcolonial formations of race and class in the Bahamian context. Terrestrial agriculture is an emergent island tourism product and a good place to begin.

Agriculture is not usually an explicit part of most paradise island imagery, but it is of course a part of the familiar paradise *imaginary*.[6] As Ian Strachan points out, the tourist expects (implicitly) to experience the lifestyle of the plantation from the perspective of the white plantation owner, with the accompaniments of smiling, local hotel staff who cater to their needs and prepare their meals (2002). As mentioned, international tourist arrivals to The Bahamas number in the millions, annually, and visitors expect Bahamian dishes to be available in hotel restaurants and restaurant buffets that seem to feature local fruits, vegetables, and animal products, even when everything may be imported. Beyond the edges of tourist perception, the island resort has come to stand in for the island plantation.

Bahamian agriculture has also been implicated in tourism in the form of heightened food insecurity. Due to regional trade imbalances, the expense of farm land, the diminished social value of local products compared to imported products, and the high input costs of farming, agriculture has been in a decades-long backslide in The Bahamas with low entry into the sector as youth pursue service-based careers (Boyce, 2014; Hedden, 2011). The country cannot feed its population on local farm produce, let alone the influx of tourist arrivals. This means that hotel restaurants primarily feed tourists imported "tropical" foods: pineapples from Hawaii, mangoes from Mexico, limes from South America (personal observation). Food imports are now approaching $1 billion USD annually (this number comes from several interviews with Bahamian agriculture officials), resulting in a related carbon output into the atmosphere stemming from all the shipments of consumable goods that arrive to the country via cargo ship and jet plane. This also means increased waste as imported foods are packaged in plastics that become pollution in the ocean and wetlands if they manage to avoid the brimming (and occasionally burning) New Providence landfill (Dorsett, 2017). Inadequate local production only exacerbates processes of anthropogenic global environmental change.

The linkages between agriculture and tourism in The Bahamas are obvious, but there are not many positive linkages. Tourism consumes large amounts of food and water, and it requires the importation of large amounts of foods that fit tourist expectations for quality, variety, and familiarity. Cultural critics point out that the country's dependence on tourism, coupled with postcolonial feelings of national inadequacy, have privileged imported products over local products, resulting in a Bahamian preference for imported foods and tastes (Bethel, 2008). This has lead to the privileging of island space for tourism development over farm production, especially on New Providence, and to a disinterest in local products.[7] Local produce (onions, thyme, tomatoes, etc.) is sold in some grocery chains, but it is not often labeled, and its origins are often times unidentifiable.

Some farmers promote local farms and produce for Bahamian consumption, hoping to break the dependency on imports. These farms participate in farmers markets and healthy food campaigns. One farm in particular has moved in this direction while creating a very direct linkage with the tourism industry, offering farm space and produce as a destination itself. The farmer who owns this farm believes that his island farm is primed for sustainable agritourism.

Back to my cloudy day "out west" on the farm. On that day I learn that this small acreage is owned by a Bahamian entrepreneur from a prominent white family who made his fortune in finance in the United States years ago. In an interview, he tells me that he now lives "back home" in The Bahamas as a gentleman farmer on New Providence, utilizing land purchased from a wealthy white colonial landowner. I observe that the farmland is close to two infamous gated communities, one a historical repository of white wealth and holdover from colonial days, the other a more recent high-end second (or third) home destination designed to be a modern version of the first enclave. Residents and tourists from these locales, along with yachting tourists passing through the country, are the target clientele. They learn about the farm via promotional stories in glossy regional lifestyle magazines, and especially from travel websites, social media, and word of mouth.

This farmer explicitly doesn't grow the staple crops of the traditional Bahamian farm (pigeon peas, goat pepper, thyme, plum tomato, banana, pumpkin, citrus, etc.). Instead, he grows luxury produce for wealthy expatriates and tourists seeking lunch and something a little different from the typical restaurant. When I first meet the farmer we stand in a field of organic arugula in view of his compost bins, hydroponic micro-greens, expanding aquaponic green houses, and farm-to-table restaurant. During our interview he proudly explains that his micro-greens sell for $60 USD per pound. I discover for myself that a lunch of fresh juice, salad, grains, and fish or chicken protein costs nearly $30, a high price even for a tourist destination, compared to the places many Bahamians frequent, matched only by places on the island that cater to affluent crowds. "This is the future of agriculture in these islands", the farmer told me adamantly as we sat at an al-fresco table, "growing high-value crops for an upscale market and welcoming visitors directly to the farm as agricultural tourists". He was incredibly serious, looking me in the eye while stating, "there is no other way to sustain local food production here".

This well-informed farm owner markets his island farm as an Anthropocene space (without using the term himself) by acknowledging the fraught conditions of small island food security in an era of planetary uncertainty and promoting his innovative solutions to the problem of island underproduction. Remotely, via the farm website, and in person, via the farm restaurant, visitors tour the aquaponic green houses to marvel at the closed system engineering in which tanks of red tilapia fertilize water pumped to vertically grown produce with minimal waste or chemical input and the capacity to produce far more product per square foot than conventional farming. One article about the farm in a regional online tourism promotional website states, "They're living your dream: the beach at their door, lots of sun, and their very own organic farm. And they've invited you along for the day – and for dinner" (source not provided to protect anonymity).

In our earnest conversation, the farmer stressed that the growing island population could only hold out for days or weeks without receiving shipments of imported foodstuffs because no one is farming what land there is, and there is not much undeveloped land left on New Providence due to the intensifying population density. "What do you think will happen if the boats stop coming?" he asked. The farm is self-sustaining, but "they will come here with guns and take everything we have". His attitude is that if the government would only take his success seriously – and encourage other farmers to adopt his methods while building linkages to the luxury tourism industry – then the country might yet be saved. His farm tourism product is a precarious oasis of sustainable innovation in a dystopian, anthropogenic, postcolonial world. Welcome, affluent visitors. Get the micro-greens while you can.

However, this cynicism is not evidenced from other members of the island's agricultural community who mentioned this farm to me during interviews when asked for examples of agritourism and sustainable farming. Members of the government tour the farm looking for inspiration for agricultural planning, paying attention to see if wealthy visitors will vote with their feet and leave gated enclaves for local sustainable produce. Within the NGO community on New Providence, this farm is the most commonly-cited example of innovation and the future of island-adapted agriculture combined with tourism as bedfellows.

If tourists do not yet think directly of farming when they think of The Bahamas, they are not yet likely to think about work. "Sun, sand, and sea" is synonymous with the "Isles of June" mythology of indolent occupants of the tropics (Bell, 1934). Yet when I started to ask around about coral reef restoration as a new tourist activity, I received several requests to work: specifically, requests from local NGOs and dive operations to volunteer my labor, or even to pay to labor, in coral nurseries. And I do work.

One blazing and cloudless Sunday, I find myself on the New Providence coast, even farther out west. I sit on the dock of a large local dive shop along with a willing student, assisting a representative of a local NGO in the material preparation for a coral restoration project. Further down the dock we see dive boats coming in and out of the marina, loading and unloading small groups of tourists in swimming gear. The boats are bedecked with tanks and wetsuits. Dive instructors corral clients onto the boats, coordinating gear while keeping spirits and enthusiasms up. Tourists of all ages arrive in groups and leave the dive shop via large vans emblazoned with the shop's name and logo. These vans are familiar sights around the large hotels of the north shore.

As volunteers, we have been tasked with building coral nursery "trees" out of PVC pipe, fishing filament, and blue plastic glue. The NGO employee demonstrates how to produce a tree in an assembly line process: glue the pre-cut PVC arms to the pre-cut PVC trunks, attach the pre-cut fishing line to the arms, thread a rope through the center of the trunk, and attach it to a small buoy, making sure all knots are secured with glue. The finished trees, also known as "coral propagation units", will be anchored to the sea floor by more volunteers in a defined site with amenable conditions, and hard-coral fragments will be hung on loops of fishing line. Forty coral fragments can grow on a coral tree at one time, and the NGO plans to anchor dozens of trees offshore in a predesignated nursery area marked by buoys on the surface of the water. The main nursery for this project is sited on the southwestern end of New Providence, in an area near the industrial pier where the nation's fossil fuel supply is regularly delivered.

Like the terrestrial farm, this nursery is not located in a traditional tourist area. The nursery is sited in a working coastal marine-scape shaped by cargo, near-shore fishing, and historical littoral relations. In the Pre-Colombian past, Arawak, Lucayan, and Tiano populations settled in the same area, relying on the fruits of the ocean. During the colonial slave period, African slaves from nearby plantations utilized the near-shore reefs and coasts for subsistence fishing (Clifton Heritage National Park, 2016). In the early 2000s, the coastal area was saved from real estate development and converted to a national historical site to preserve its historical plantation ruins, eventually opening for visitation in 2009, while the offshore area was preserved within a marine management area in 2015. The nursery is submerged within that management area.

As we work we sweat and talk about the relationship between the NGO and the dive shop that has developed around coral restoration and "voluntourism". "The hard-corals here are critically endangered", the NGO rep explained. "If you have any capacity to help at all, you should be helping". His fervency stems from the decline in hard-coral species that hold Caribbean reef aggregations together, like Staghorn (Acropora cervicornis) and Elkhorn (Acropora palmata) in recent decades. He is Bahamian, also from a prominent white family, and he too lived abroad for years, returning home to stay close to the marine environment he loves. He believes coral decline is anthropogenic, echoing the rhetoric of international coral scientists and the nascent coral restoration industry (Coral Restoration Foundation, 2017; Hughes et al., 2017).

Like declines in food security tied to the effects of planetary anthropogenesis in increasingly insecure locales, the decline of coral reefs has been tied to human activities, including tourism. Rising ocean temperatures and subsequent ocean acidification are exacerbated by an increase in marine debris and siltification from dredging and coastal building as well as offshore runoff, waste, and pollution. As a result, there is an internationally recognized coral reef crisis, and corals are believed to be decreasing, bleaching, and generally declining in health and resilience (Madin & Madin, 2015). Hard-corals are said to be in retreat around New Providence, and the large reef stands and coral gardens of childhood memory are gone. "We have killed off a lot of our corals with coastal development", the NGO employee observed as he glued pipes, "and now it's time for tourism to help restore the reefs. And this is a great idea", he went on, "because its value added for the industry". In other words, as a result of the international conversation about anthropogenic coral decline, along with local recognition that this decline affects the coral-based marine tourism industry, there is an emergent market for coral restoration-based marine tourism products. Tourists can now pay to labor to maintain coral nursery spaces to restore reefs that future dive tourists will one day pay to explore.

The dive shop sells restoration certification dive packages for about $200 USD per day. Tourists book certification days on the shop website, featuring images of divers recording data about coral fragments and posing in masked selfies with coral trees. Dive tourists come in small groups from the coastal hotels for prearranged sessions with a certified trainer. Travelers who might otherwise choose to dive on reefs with variable degrees of health (whose presence may exacerbate declining reef health), or on the submerged wreckage of ships or airplanes, can now choose to become "coral care specialists". My student volunteered to take the course (I paid), and she explained that it consisted of a morning classroom session about coral biology and con-servation followed by an afternoon of cleaning algae off of coral trees with sturdy brushes in the underwater nurseries managed by the NGO. She confirmed that the dive tourists that sign up for these certification days are relatively affluent, generally EuroAmerican, and much whiter than the Bahamian population as a whole.

Potential tourist products like the agritourist farm and the coral restoration dive package tap into international conversations about anthropogenesis and into prevailing geographic imagina-ries-as-brands by advertising designed experiences, like touring aquaponic greenhouses and div-ing in forests of coral propagation units, primarily online and via client social media.[8] The farm offers locally produced, organically farmed meals *in situ* at the site of production while the dive shop offers a place-based, hands on, educational dive experience. They both sell a "sustainable tourism" product in an alternative island space, far from the beaches and gated communities of the coast, in such a way that their products now continually stabilize and reshape the space for tourist visitation. The dive operation does this by advertising their hard-coral nursery (via web-site) as the only place in the Caribbean (a vulnerable small island region) to offer such a volun-tourist experience. In this marketing strategy, The Bahamas (a vulnerable small island nation) becomes a significant example of "eco-friendly conservation" and innovation within anthropo-genically modified socioecological systems. The dive shop becomes the only means for tourists to access that example. These Bahamian farms are now spatial products that have a calculable value (thousands of farm lunches sold and dozens of coral restoration certifications awarded annually) within an imaginary-as-brand shaped by emergent understandings of anthropogenesis and a travel market for meaning.[9]

These two brief cases of emergent Anthropocene tourism products in New Providence, Bahamas, exemplify how space can be reproduced and sold in reimagined island markets. These island "farms" (terrestrial and marine) are important local sites of material and symbolic connec-tion between larger practices and processes that span scales; they are active experiments with neoliberal capital in the form of tourism, local sovereignty, complex island socioecologies, and social justice and ethics on multiple levels. Both Strachan and Nixon might observe that these spaces, as promoted and used by the farm and dive shop, ignore and, therefore, erase colonial and slave history in The Bahamas, even as they benefit from the tourism cachet of The Bahamas that is one legacy of that history. And so, while both sites attempt to sell Anthropocene-worthy alternatives to the mass tourist imaginary of paradise, even these emergent versions of sustainable tourism do not address issues of race, class, and colonialism in their sites in such as way as to make Anthropocene Island spaces viable ethical alternatives to the typical Caribbean tourist product.

Conclusion: adventures in and out of sustainable tourism

The conveners of this edited volume are right to recognize the emergent linkages between the Anthropocene and tourism. This is an understudied area of thought and action that has real consequences for reshaping tourist modes of production and consumption, social relations between populations of human and nonhuman beings, and spatial politics. Further, a political ecology of tourism for the Anthropocene must do what political ecology does so well: follow forms of power as they manifest across scales, assessing the changing relationship between politics and the more than human world (Biersack, 2006).

This article and this special issue should stand as examples of a political ecological approach to the study of tourism in and for the Anthropocene capacious enough to recognize how emergent destinations interact with existing dynamics in specific locales. This article has characterized this approach as a situated adventure in sustainable tourism. The brief examples included here signal that we are encountering an emergent phase of tourist travel in which circulating imaginaries tied to the Anthropocene idea have real material effects in specific locations. Again, this is not strictly ecotourism, or at least not the mainstream brand of ecotourism of the 1990s. Instead, the farmer and the NGO employee see their farms as innovative examples for the sustainable redesign of the tourism product of The Bahamas. They are remediated sustainable tourism schemes for the Anthropocene. Situated adventuring (again, always a fraught business) through tourism spaces means exploring traveling ideas stemming from authoritative realms like global change science (the Anthropocene) as they combine on the ground with geographic imaginaries (the paradise island) to become revised imaginaries-as-brands (the small island), enabling material recreations of space, including the people and life forms that constitute that space (terrestrial and marine island "farms").

A political ecology of the Anthropocene must also pay close attention to the nexus of science and capital that drives so many interventions in the lives of others (human and non). Other examples of this nexus include biotech, geoengineering, and any number of commercially viable systems modeling projects. The nexus that links science and tourism has long been studied as it pertains to nature conservation via park enclosure and earlier forms of ecotourism, but the sustainable tourism products of the Anthropocene provide an opportunity for closer scrutiny. The examples described here are best understood as collaborative synergies producing forms of value within Anthropocene tourism products.

This article repeatedly refers to "re" stabilization or "re" imagination because the Anthropocene idea reworks prior spatial imaginaries and stabilizations of place. As implied above, islands are now prized by Earth scientists as the "canaries in the coal mine" of global change, revising earlier understandings of islands as microcosms of global processes and imbuing these understandings with a sense of peril.[10] The paradise and plantation model of sun, sand, and sea tourism already relied on colonial island exoticism and the sense that islands are exceptional spaces of ease and encounter, but marketing strategies and brand campaigns (not to mention funding calls for scientific research) must be continually revised lest they become stagnant, and the Bahamian Ministry of Tourism continually explores the revamping of its national tourism product.

Beyond the early Anthropocene ventures described above, sustainable tourism is now a strategy that looks likely to bear fruit for the Ministry. The Director General for Tourism said recently that "as a small island developing state, The Bahamas must consider the increasing competition in the global tourism market, and act to ensure that the tourism industry remains strong and competitive. We expect that the … focus on sustainability, along with tourism management, strategy and marketing, will enhance the future of our tourism industry" (Deveaux, 2015, sentence five). Sustainability here has multiple meanings and purposes.

The overt meaning of sustainability as it is used by the Ministry aligns with the definition of sustainable development popularized in the Brundtland Report: "Sustainable development is development that meets the needs of the present without compromising the ability of future generations to meet their own needs" (International Institute for Sustainable Development, 2016). But a political ecologist must repeat the oft asked question: sustainability for whom? In the trend setting cases I explore here, what is made sustainable are island travel markets and divisions of space, life, and labor for the tourism industry. In other words, the Anthropocene idea can be an accumulation strategy for the tourism industry that perpetuates powerful forms of entrenched inequity under the guise of global sustainability.

The dive shop is selling a place-based product: the small forest of "trees" that is their offshore coral nursery. That product is reliant on the spatial imaginaries of the global change sciences that situate The Bahamas and the Caribbean within a framework of small island vulnerability, endangerment, and resilience. The NGO that hired the representative to collaborate with the dive operation in the creation of the nursery is part of an international network of scientists and researchers whose mission is to evaluate the anthropogenic origins of global coral degradation and to devise novel means to ameliorate that degradation. This is not in and of itself the accumulation of value, which comes in many forms. However, these networks can collaboratively enable accumulation strategies when they promote global narratives over local relationships and advocate for market-based solutions without any deep understanding of the social and historical context of markets.

Political ecologists studying coral restoration as a form of sustainable tourism anywhere should mark how historical coastal activities like reef fishing and other complex local ties to coral beings and reef-based processes are erased within conservation rhetorics that amplify the "degradation of critically endangered hard-corals". They should notice how working marine places become marked as de facto dive space where independent fishing livelihoods and post emancipation subsistence practices are slowly replaced with consumptive tourist "labor". They should explain how selling coral restoration dive packages to tourists becomes a paradoxical solution to the ecological damage caused by tourist populations and infrastructure. In sum, what is enabled by the imaginaries of global change animating nursery spaces for coral restoration is the potential accumulation of coral knowledge, marine territory, local spatial imaginaries, the meaning of work, and prior local relationships to coral reefs. This is just a taste of what can be accumulated. Erasure or repurposing of these prior forms of value helps to generate capital from the coral restoration dive packages for the dive operation and for the owners of regional tourism products. Money in the form of profit is just one materialization of these integrated forms of accumulation.

On regional agritourist farms the story is similar. Political ecologists studying this form of tourism in the Caribbean should follow how farmers tap into circulating narratives about small island vulnerability and food security in the face of global environmental change. They should examine how visitors experience red tilapia linked to kale beds in closed aquaponic systems and consume fresh salads of micro-greens as forms of innovation. While tourists are visiting highly designed places, touted as sustainable farms, they are likely not informed that this farm tourism product utilizes migrant labor (Haitian, Jamaican, Central American) and expensive land purchased by the white colonial elite long ago as an extension of their segregated enclaves. For example, Bahamians cannot labor on farms for a living wage, and they certainly do not have access to land like this. Most Bahamian farmers do not own their land at all, instead leasing it from the government, losing their leases after death. This is a pattern repeated across the region. What is accumulated on the agritourist farm is of course, then, land, but also less-tangible histories of subsistence farming practices within the plantation system, generational relationships to well-loved but low-value crops, and the capacity for farmers who are not wealthy landholders to pass down farms to descendants. The small island farm tourism product has value in part because it symbolizes these histories and romantic notions of self-sufficiency, but a political ecologist knows that this is an enabling illusion that aligns with the prevailing branding of Anthropocene space being sold in markets for sustainable products.

The Anthropocene, as an acknowledgement of multiple drivers of anthropogenic change, is a challenge to the expansion of capital. But it is simultaneously an *outcome* of the historical expansion of capital and an *invitation* for further capital expansion. This is one of the great ironies of the Anthropocene. And the ironies proliferate. Yes, the recognition of the Anthropocene is an ethical challenge for the expansion of tourism (Gren & Huijbens, 2014), but it also invites the creation of new tourist products and markets that can suffocate the radical potential of the idea. This irony is true of all tourism in the context of global environmental change. Local agriculture and local food are prime sites for the expansion of agritourism just as coral restoration and coral nurseries are prime sites for the expansion of voluntourism. "Farms" like this work within prevailing imaginaries as tourism products, regardless of any actual contribution to functional island food systems or coral reef ecologies. These proliferating ironies are evidence that directly linking agriculture or reef restoration and tourism will not automatically lead to improved island security. These ironies must not be ignored. The cases presented here show that if the "island farm" product continues to expand in the Caribbean it could indeed successfully stabilize food systems, produce, and terrestrial and marine space *for the tourism sector*, while potentially leaving islanders across the region in an even more precarious and dependent position than before as more and more space is stabilized for the enjoyment of visitors over the use of local residents.

At this point, it should be clear that neither farmers nor dive operators nor scientists who conduct Anthropocene research are *intentionally* creating tourism products that redesign and repurpose land, knowledge, and relationships towards accumulative ends. The next phase of this research will explore the specific mechanisms by which members of the tourist industry from small scale "farms" to large resort hotels to government tourism offices translate circulating Anthropocene imaginaries into brands and products. While this article has focused primarily on the symbolic and material manifestation of the Anthropocene idea as a reformulation of space for tourism in The Bahamas, the examples presented here also show that these mechanisms are driven in part through processes of virtualism and prosumption tied to the pervasive use of the internet and social media in tourist source populations (Buscher & Igoe, 2013; West & Carrier, 2004). The point here is that political ecologists of tourism seek scholarly adventures that focus on inequities and ironies to counter widespread neocolonial and neoliberal narratives, and political ecologists know that hegemonic structures of accumulation rarely operate on the level of explicit intention. However, Strachan and Nixon remind us that we must demand more from tourism, especially in postcolonial contexts like the Caribbean and The Bahamas.

Strachan's (2002) analysis of the conjoined tropes of paradise and plantation that undergird Bahamian mass tourism within the paradise island imaginary (described above) highlights the inexcusable way that historical forms of inequity are perpetually reconstituted within neocolonial and neoliberal capitalist ventures. Nixon (2015) extends this critique by exploring alternatives to the paradise island brand in the Caribbean, arguing that, paradoxically, tourism can actually combat the denigrating effects of tourism, but *only* if alternative ventures are explicit in their resistance. Her examples of ethical tourism include artists working within a mega-resort complex who produce work that directly represents the fraught relationships Bahamians have with their national industry, a small-scale cultural tourism venture that pairs tourists with citizens to learn about current social manifestations of past forms of slave resistance, and an annually occurring educational workshop that recruits limited numbers of selected tourists to participate in an exchange of knowledge about colonial history and the African diaspora. Unlike the examples presented above, all of these alternative ventures are led by Caribbean intellectuals who self-identify as black.[11] Following Deborah McLaren, Nixon argues that "any rethinking of tourism must challenge the travel industry at every level" (p. 143).

These two examples of the reproduction of space for sustainable tourism ventures in The Bahamas are indeed a kind of alternative to mass tourism products sold via the paradise island imaginary-as-brand.[12] But as Strachen and Nixon show, they are not yet alternative enough to offer real resistance to the accumulative, exploitative, and supremacist capacities of postcolonial tourism. In order to move beyond this contextual vortex, Caribbean and Bahamian sustainable tourism ventures would have to directly acknowledge the *conjoined ironies* at the heart of tourism itself, educating visitors about their contribution to anthropogenic environmental change as consumers in a given Anthropocene space *as well as* educating them about their contribution as tourists to the perpetuation of histories of racial segregation and class inequality in the region.

The Anthropocene idea is both an implicit brand platform for emergent sustainable tourism products and simultaneously a powerful conceptual enabler of the familiar processes of accumulation, white supremacy, and relationships of visible/invisible labor tied to the global expansion of neoliberal capital in which tourism and science are major players. That fact must not be overlooked in the heady rush to analyze the Anthropocene. These brief examples of future-oriented tourism (small laboratory experiments for tourism we might say, following Magubane (2003) who in turn follows Cooper & Stoler, 1997) have shown that the circulation of new spatial imaginaries generates and stabilizes emergent Anthropocene spaces with real material consequences for the way people do business and the way business effects social relations. And business, as a means of living with and relating to human and nonhuman others by redesigning land and marine space, is always a highly *recreative* process.

Notes

1. For a definition of tourism products, see Jefferson and Lickorish "1991).
2. Bahamian demographic information comes from the CIA Factbook, last updated in 2015 https://www.cia.gov/library/publications/the-world-factbook/geos/bf.html
3. In 2014, tourist arrivals to New Providence were recorded at 3.5 million (Tourism Today, 2014).
4. Current projections do point to modest growth in tourism arrivals (WTTC, "2017), although these projections are based on the development of the Baha Mar Resort and Casino project that has had a very fraught history (Vora, "2017) and which is not yet contributing significantly to growth at the time of this writing.
5. The existence of "wilderness" has been debated for some time, but it takes new shape in the debates around the utility of the Anthropocene idea ((Graef, 2016; Kareiva, Marvier, & Lalasz, 2012; Purdy,((2017)). Anthropocene fever. Aeon. Retrieved from https://aeon.co/essays/should-we-be-suspicious-of-the-anthropocene-idea"2017).
6. This paradise imaginary in The Bahamas has also been documented by Bahamian scholar, Krista Thompson, as "a domesticated version of the tropical environment and society" also known as the "Caribbean Picturesque" created through visual processes of "tropicalization" (Thompson, (2007). An eye for the tropics: Tourism, photography, and framing the Caribbean Picturesque. Durham, North Carolina: Duke University Press. "2007, quoted in Nixon, (2015). Resisting paradise: Tourism, diaspora, and sexuality in Caribbean culture. Jackson, Mississippi: University Press of Mississippi. "2015, p. 126)
7. The Bahamian island of Andros has been designated by the government as a site for industrial scale agricultural production and agricultural education, thus far with relatively limited results in terms of offsetting reliance on imports.
8. I do not show or quote actual advertising materials for either product because this would further risk revealing the identities of these operations.
9. There is now an underwater sculpture garden adjacent to the coral nursery that also can be accessed outside of the purview of the dive shop. The nursery itself was damaged in Hurricane Matthew in late 2016 and has yet to be fully restored.

10. There are of course other emblems of anthropogenic global change, for example mammals like the polar bear and whales, and these certainly have a tradition of spurring environmental protest and action (e.g. see Kristoffersen, Norum, & Kramvig, 2016 on the "new whale").
11. The examples Nixon (2015) cites are The Current Art Gallery run by John Cox at the Baha Mar Resort and Casino in The Bahamas (circa 2014), the Bahamian Educulture organization founded by Arlene Nash-Ferguson, and the Blackspace program designed by Erna Brodber in Jamaica.
12. Other examples of Anthropocene spatial products are still few and far between in The Bahamas, but they include ventures like "the Development" in Abaco (Moore, 2015a, see also Moore, 2015b) and the infrastructural design of large resorts that have made attempts to innovate in sustainable energy use and waste recycling.

Declaration statement

No potential conflict of interest was reported by the author.

Funding

Fulbright Scholar Program; Wenner-Gren Foundation.

References

Alliance of Small Island States. (2015). *Alliance of small island states: Twenty-five years of leadership at the United Nations*. Retrieved from http://aosis.org/wp-content/uploads/2015/12/AOSIS-BOOKLET-FINAL-11-19-151.pdf

Aragon, L.V. (2011). Distant processes: The global economy and outer island development in Indonesia. In B. R. Johnston (Ed.), *Life and death matters: Human rights, environment, and social justice* (pp. 29–54). United Kingdom Routledge.

Baldacchino, G. (2006). Islands, island studies, island studies journal. *Island Studies Journal, 1*(1), 3–18.

Bell, J. (1934). *Bahamas: Isles of June*. London: Williams & Norgate.

Bethel, N. (2008). *Essays on life* (Vol. 1). Morissville, North Carolina: Lulu.

Biersack, A. (2006). Reimagining political ecology: Culture/power/history/nature. In *Reimagining political ecology* (pp. 3–40). Durham, North Carolina: Duke University Press.

Boyce, R. (2014). *Cultivating a culture of agriculture: Are Caribbean youth up to the challenge?* Caribbean Development Trends. Retrieved from https://blogs.iadb.org/caribbean-dev-trends/private-sector-and-entrepeneur-ship/2014/11/02/cultivating-culture-agriculture-caribbean-youth-challenge/

Buscher, B., & Igoe, J. (2013). Prosuming conservation? *Journal of Consumer Culture, 13*(3), 283–305.

Cameron, C. M., & Gatewood, J. B. (2008). Beyond sun, sand, and sea: The emergent tourism programme in the Turks and Caicos Islands. *Journal of Heritage Tourism, 3*(1), 55–73.

Campbell, K. (2012, February 28). *Contracts signed for second phase of improvements to saunders beach*. The Government of The Bahamas, Press Releases.

Carrington, D. (2016). *The Anthropocene epoch: Scientists declare dawn of human-influenced age*. The Guardian. Retrieved from https://www.theguardian.com/environment/2016/aug/29/declare-anthropocene-epoch-experts-urge-geological-congress-human-impact-earth

Chakrabarty, D. (2013). *History on an expanded canvas: The Anthropocene's invitation*. The Anthropocene Project Keynote Presentation: HKW. Retrieved from https://www.youtube.com/watch?v=svgqLPFpaOg

Cleare, A. B. (2007). The *history of tourism in the* Bahamas: A *global perspective*. Bloomington, Indiana: Xlibris.

Clifton Heritage National Park. (2016). *History*. Retrieved from http://www.cliftonheritage.org/history.html

Cooper, F., & Anne S. (Eds.). (1997). *Tensions of empire: Colonial cultures in a Bourgeois world*. Oakland, CA: University of California Press.

Coral Restoration Foundation. (2017). *Threats to reefs*. Retrieved from https://coralrestoration.org/threats-to-reefs/

Crutzen, P. J., & Schwägerl, C. (2011, January 24). *Living in the Anthropocene: Towards a new global ethos*. Environment, 360.

Crutzen, P. J., & Stoermer, E. F. (2000). The 'Anthropocene'. *Global Change Newsletter, 41*, 17–18.

Deveaux, N. (2015, February 5). *Bahamas launches phase one of major sustainability program. The Bahamas Weekly.* Retrieved from http://www.thebahamasweekly.com/publish/ministry_of_tourism_updates/Bahamas_launches_phase_one_of_major_sustainability_program_printer.shtml

Dorsett, S. (2017, March 6). *Fleeing the fire.* Tribune 242. Retrieved from http://www.tribune242.com/news/2017/mar/06/jubilee-gardens-residents-flee-fire/

Ganpat, W. G., & Isaac, W.-A. P. (2015). *Impacts of climate change on food security in small island developing states.* Hershey, Pennsylvania: IGI Global.

Graef, D. J. (2016, September 30). *Wildness. Theorizing the contemporary.* Cultural Anthropology. Retrieved from https://culanth.org/fieldsights/965-wildness

Gren, M., & Huijbens, E. H. (2014). Tourism and the Anthropocene. *Scandinavian Journal of Hospitality and Tourism, 14*(1), 6–22.

Grove, R. H. (1996). *Green imperialism: Colonial expansion, tropical island Edens and the origins of environmentalism, 1600–1860.* Cambridge, England: Cambridge University Press.

Haraway, D. J. (1997). *Modest_witness @second millennium. femaleman _meets _oncomouse: Feminism and technoscience.* United Kingdom: Routledge.

Haraway, D. J. (2016). *Staying with the trouble: Making kin in the chthulucene.* Durham, North Carolina: Duke University Press.

Hedden, J. (2011). *Bahamian agriculture, an overview.* Nassau, Bahamas: The Nassau Institute. Retrieved from https://www.nassauinstitute.org/files/Hedden-%20BahamianAgriculture.pdf

Honey, M. (1999). *Ecotourism and sustainable development.* Washington, D.C: Island Press.

Howie, B., & Lewis, N. (2014). Geographical imaginaries: Articulating the values of geography. *New Zealand Geographer, 70*, 131–139.

Hughes, T. P., Barnes, M. L., Bellwood, D. R., Cinner, J. E., Cumming, G. S., Jackson, J. B. C., … Scheffer, M. (2017). Coral reefs in the Anthropocene. *Nature 546*(7656), 82–90.

International Institute for Sustainable Development. (2016). *Sustainable development.* Retrieved from http://www.iisd.org/topic/sustainable-development

Jefferson, A., & Lickorish, L. (1991). *Marketing tourism: A practical guide.* Harlow: Longman.

Johnson, H. (1997). *The Bahamas from slavery to servitude, 1783-1933.* Gainesville, Florida: University Press of Florida.

Kareiva, P., Marvier, M., & Lalasz, R. (2012). Conservation in the Anthropocene: Beyond solitude and fragility. *Conservation Biology.* Retrieved from https://thebreakthrough.org/index.php/journal/past-issues/issue-2/conservation-in-the-anthropocene

Kristoffersen, B., Norum, R., & Kramvig, B. (2016). Arctic whale watching and anthropocene ethics. In M. Gren & E. H. Huijbens (Eds.), *Tourism and the Anthropocene* (pp. 94–110). United Kingdom: Routledge.

Latour, B. (2013, July 3). *Facing Gaia: Six lectures on the political theology of nature.* Figure/Ground Communication. Retrieved from http://figureground.ca/2013/07/03/bruno-latours-gifford-lectures-facing-gaia-a-new-inquiry-into-natural-religion/

Lazrus, H. (2012). Sea change: Island communities and climate change. *Annual Review of Anthropology, 41*, 285–301.

Madin, J.S., & Madin, E. M. P. (2015). The full extent of the global coral reef crisis. *Conservation Biology, 29*(6), 1724–1726.

Magubane, Z. (2003). Simians, savages, skulls, and sex: Science and colonial militarism in nineteenth-century South Africa. In D. S. Moore, J. Kosek, & A. Pandian (Eds.), *Race, nature, and the politics of difference.* Durhman, North Carolina: Duke University Press.

Mintz, S. (1985). *Introduction. Sweetness and power: The place of sugar in modern history.* London, England: Penguin Books.

Moore, A. (2010). Climate changing small islands: Considering social science and the production of islands vulnerability and opportunity. *Environment and Society: Advances in Research, 1*, 116–131.

Moore, A. (2015a). Islands of difference: Design, urbanism, and sustainable tourism in the Anthropocene Caribbean. *Journal of Latin American and Caribbean Anthropology, 20*(3), 513–532.

Moore, A. (2015b). Tourism in the Anthropocene park? New analytic possibilities. *International Journal of Tourism Anthropology, 4*(2), 186–200.

Moore, A. (2015c). The Anthropocene: A critical exploration. *Environment and Society: Advances in Research, 6*, 1–3.

Moore, A. (2016). Anthropocene anthropology: Reconceptualizing contemporary global change. *Journal of the Royal Anthropological Institute, 22*(1), 27–46.

Mostafanezhad, M., Norum, R., Shelton, E. J., & Thompson-Carr, A. (2016). *Political ecology of tourism: Community, power, and the environment.* Jackson, Mississippi: Routledge.

Nixon, A. (2015). *Resisting paradise: Tourism, diaspora, and sexuality in Caribbean culture.* Jackson, Mississippi: University Press of Mississippi.

Purdy, J. (2017). *Anthropocene fever.* Aeon. Retrieved from https://aeon.co/essays/should-we-be-suspicious-of-the-anthropocene-idea

Samoa P. (2014). *SIDS accelerated modalities of action pathway outcome document*. Retrieved from http://www.sids2014.org/index.php?menu =1537

Sheller, M. (2003). *Consuming the Caribbean: From Arawaks to Zombies*. United Kingdom: Routledge.

Strachan, I. (2002). *Paradise and plantation: Tourism and culture in the Anglophone Caribbean*. New World Studies. Charlottesville, Virginia: University of Virginia Press.

Thompson, K. (2007). *An eye for the tropics: Tourism, photography, and framing the Caribbean Picturesque*. Duke University Press.

Torgerson, A. (2010). Fair trade banana production in the Windward Islands: Local survival and global resistance. *Agriculture and Human Values, 27*(4), 475–487.

Tourism Today. (2014). *Foreign arrivals to the Bahamas, 1998-2014*. Retrieved from http://www.tradingeconomics.com/bahamas/tourist-arrivals

Trading Economics. (2016). *Bahamas visitor arrivals, 2004-2016*. Retrieved from http://www.tradingeconomics.com/bahamas/tourist-arrivals

UNWTO. (2016). *United Nations declares 2017 as the International Year of Sustainable Tourism for Development*. Retrieved from http://media.unwto.org/press-release/2015-12-07/united-nations-declares-2017-international-year-sustainable-tourism-develop

Vora, S. (2017, April 17). In the Bahamas, a long awaited opening for Baha Mar Resort. *The New York Times*. Retrieved from https://www.nytimes.com/2017/04/17/travel/bahamas-baha-mar-resort-nassau-hotel-casino-opening.html

Weaver, D. B. (2001). *The encyclopedia of ecotourism*. CAB International. New York: CABI Publishing.

West, P. (2006). *Conservation is our government now: The politics of ecology in Papua New Guinea*. Durham, North Carolina: Duke University Press.

West, P., & Carrier J. G. (2004). Ecotourism and authenticity: Getting away from it all? *Current Anthropology 45*(4), 483–498.

WTTC. (2017). *Travel and tourism economic impact 2017 Bahamas*. London, England: World Travel & Tourism Council.

Nicaragua's Buen Vivir: a strategy for tourism development?

Josh Fisher

ABSTRACT

Although often framed as an emerging anthropocenic socio-ecological imaginary, the Latin American paradigm of Buen Vivir has provided a broad base of support for tourism development in the region. This article focuses on Nicaragua's Buen Vivir, a national development campaign entitled "Live Clean, Live Healthy, Live Beautiful, Live Well" (called Vivir Bonito, Vivir Bien) The campaign was launched in 2013 as a multi-pronged approach to integrated development in distinct areas including employment, public health, waste management, education, urban aesthetics and national pride. However, it has also had the effect of opening up opportunities for tourism development not only in the capital city of Managua but also alongside other mega-projects such as the planned interoceanic canal. This article draws upon the example of Vivir Bonito, Vivir Bien to illustrate the variety of tourism development strategies currently emerging at the intersection of a left-turn toward Socialism of the 21st Century and Buen Vivir in Latin America, on the one hand, and a post-neoliberal context in which political economic projects of the past continue to leave their mark.

Introduction

Latin American development has taken an environmental turn. At first glance, the trend reflects a familiar imperative: To address significant social, economic and ecological challenges of unchecked market growth with sustainable solutions. Upon closer inspection, however, the situation is more complex. As Uruguayan ecologist Eduardo Gudynas (2016) argues, a more robust set of alternatives to neoliberal development are emerging across the region – "varieties of development," as he calls it, that include the neo-developmentalism of centralized progressive governments, new forms of South–South cooperation and even Chinese state capitalism. The obvious political and philosophical differences expressed by these platforms have led some to situate global development itself as a heterogeneous assemblage rather than a single project (Escobar, 2012; Li, 2007; Mosse, 2005). And yet, for many Latin American governments, tourism development remains a common platform for national development.[1] The simultaneous mobilization and preservation of natural and cultural resource frontiers for tourism-related services and experiences has been celebrated as a self-evident alternative to exploitative and extractivist strategies of the past (Bebbington & Bebbington, 2012; Fletcher, 2011).

Still uncertain is the place that tourism development will have within a more heterodox approach that is gaining ground across Latin America: *Buen Vivir* (good living, living well, or collective wellbeing).[2] Inspired partly by the socio-ecological philosophies or "cosmovisions" of Indigenous and Afro-descendent movements, and partly by the critical work of networks of activists and intellectuals across the region, Buen Vivir purports to announce a *post-neoliberal* era because it displaces the market as the primary mechanism for provisioning well-being (Calisto Friant & Landmore, 2015; Radcliffe, 2012). It has also been called *post-developmental* because it favors a more expansive notion of progress as socio-ecological harmony (Escobar, 2010; Quijano, 2011; Walsh, 2009). By contrast, tourism development – once identified with the collapse of viable industry and the rise of neoliberalism – might seem wholly incompatible, yet this is far from the case. As a plural and sometimes contradictory space, not a single movement or set of policies (see Gudynas, 2011; Thomson, 2011; Villalba, 2013), Buen Vivir has also found a place for tourism.

In this article, I explore the contradictions of tourism development in contemporary Nicaragua, where the broad discursive frameworks of sustainability and Buen Vivir have taken root in the country's social policy and development agenda. My findings come from ethnographic research in Nicaragua between 2008 and 2017 and include the observation of public events, analysis of official statements, journalistic accounts and social media discussions, as well as interviews with public officials, tourism industry experts and other stakeholders. My principle focus is a permanent national campaign, launched in 2013 by the Ortega government and the FSLN (*Frente Sandinista de Liberación Nacional*, or Sandinista National Liberation Front), entitled *Vivir Limpio, Vivir Sano, Vivir Bonito, Vivir Bien* (Live Clean, Live Healthy, Live Beautiful, Live Well). As political strategy, the campaign signals a twenty-first century "greening" of the Sandinistas' platform, one in which the iconic red and black imagery of the 1980s has been replaced by softer motifs of turquoise, pink, lime green and canary yellow (Figure 1). Spearheaded by First Lady and Vice President Rosario Murillo, the move was widely regarded as an expression of the FSLN's revised Christian socialist philosophy – sometimes called Sandinismo 2.0 – which sought to refigure the revolutionary party's base and to address, if only nominally, questions of social and environmental citizenship. From another angle, however, Vivir Bonito, Vivir Bien also reflects long-standing practical concerns about the environment in Nicaragua's interlinked crises of capital investment, infrastructure, public health, employment and economic development. Hence, in the capital city of Managua, the campaign has spearheaded a comprehensive project to revitalize urban space by formalizing waste economies, promoting environmental education and encouraging local tourism. Meanwhile, sustainable tourism continues to play a central, if still contradictory, role in national development. Despite potentially devastating environmental impacts, for example, state and capital interests have billed the mega-project of constructing of an interoceanic canal through the jungle interiors of the Río San Juan region as an opportunity for ecotourism.

Nicaragua's Buen Vivir thus provides the opportunity to reflect on a second order of questions: The diverse political and socio-ecological imaginaries that drive tourism development, and the complex ways in which those imaginaries shape relationships between hosts and guests as well as communities, cities, states, and local and global environments. Geo-scientific indices such as the Anthropocene, which recognize human agency writ large as an epochal force, tend to ignore the specific social, cultural and economic relationships that cause planetary ecological change. By contrast, Latin American debates about Buen Vivir signal an emerging alternative socio-ecological imaginary, inspired by – and perhaps still aspiring to – the cosmovisions of its authors. These instead foreground the region's own cultural–ecological patrimony as well as the differential historical impacts of the global North and South. And yet, across Latin America, both have also been powerfully shaped and reshaped by neoliberal processes underway for decades (Escobar, 2010, p. 8). State-sponsored tourism development in Nicaragua thus sheds light on a complex remapping of political terrain. In Nicaragua, that is, sustainability is not merely giving way to Buen Vivir. Rather, seen through the lens of Nicaraguan tourism, these distinct socio-ecological imaginaries operate simultaneously on different scales and in complex and contradictory ways: The former is outwardly engaged with global discourses of sustainable development via ecotourism and the like, while the latter is inwardly oriented toward cultivating new kinds of political actors and relationships between families, communities, cities, guests, and the natural and built environs. Likewise, tourism development does not merely indicate the persistence of neoliberal imaginaries in the face of Buen Vivir, but instead operate an interface between them.

Figure 1. Pamphlet distributed at the launching of the Live Clean, Live Healthy campaign.

To signal that complexity, as well as the tensions that arise from trafficking in multiple socio-ecological imaginaries, I build upon Gudynas' (2016) thesis and argue for consideration of the "varieties of tourism development." The framework calls attention not only to the historical particularity of tourism economies, formed as they are through the confluence of multitudinous social, political and ecological processes, but also to tourism's emerging role in ushering in, and capitalizing upon, new socio-ecological imaginaries.

Latin American tourism at crossroads

Debates about Vivir Bonito, Vivir Bien in Nicaragua are made possible by several distinct currents in Latin American development over the last few decades that are worth outlining.

The first involves the spread of foundational philosophies of Buen Vivir, which are inspired in part by Indigenous modes of thinking and being. Modernist ontologies separate nature and culture and enshrine development as the transcendence of the former by the latter (Escobar, 2011; Latour, 2013; Quijano, 2000). By sharp contrast, Andean and Amazonian "cosmovisions" (*cosmovisiones*) – culturally and historically distinct socio-ecological philosophies including *Sumak Kawsay* in Kichwa, *Sumaq Qamaña* in Aymara, *Ametsa Asaiki* among Peruvian Amazonian peoples and *Ñandereko* in Guarani – make no such distinction.[3] In these "relational ontologies" (Escobar, 2010), life is instead posited as an uninterrupted and all-connected web of relations. "Living well," understood here as a harmonious community extending beyond humans to include other-than-humans, thus entails a common fund of perceptions, values, and socio-ecological and ethical practices, but not necessarily an aspiration to government (de la Cadena, 2015, p. 285; Oviedo, 2014, p. 276; Walsh, 2010).

The emergence of Buen Vivir as a political framework is instead the outcome of a second current. This Buen Vivir is energized by decades of critical, intellectual debates among indigenist and Afrodescendent movements, peasants, women, students, theologians and other activists (Acosta, 2010; Becker, 2011; Hidalgo-Capitán et al., 2014, pp. 35–36). Such groups have challenged "the civilizational model" (*el modelo civilizatorio*) as well as modernity/coloniality as the ultimate source of the looming global economic, environmental and climatic crisis (Escobar, 2010; Quijano, 2011). By generating a "conceptual rupture" (Acosta, 2009, p. 309) in six decades of development thought, they have also proposed that a complete transition to a new model of society is necessary, inevitable and in some cases already underway. Of course, glossing diverse Indigenous concepts with the Spanish term *Buen Vivir* is problematic, in no small part because, as political practice, the movement mostly falls short of a radically alternative ontology. However, as an alternative social *imaginary* – energized by currents in Western environmentalism, critical development studies, Marxist political economy and decolonial thinking, to name a few – Buen Vivir has succeeded in establishing new terrains for political debate and new possibilities for cultural–environmental citizenship (Acosta, 2010; Artaraz & Calestani, 2015; Walsh, 2009).

To the extent these processes can be separated out from one another, Latin America's Buen Vivir – and Nicaragua's Vivir Bonito, Vivir Bien in particular – are also the product of a third current: Latin America's so-called "pink tide" (*la marea rosada*). In the final decades of the twentieth century, Latin America provided original testing grounds for neoliberal ideologies. The recent resurgence of a progressive state, by contrast, has marked a significant political shift toward a philosophy of "Socialism of the 21st Century" (*Socialismo del Siglo XXI*). In place of market solutions, this new generation of new political platforms has emphasized integrated social development (*desarollo integral*), popular mobilization through direct democracy, renewed state administration of natural resources and the encouragement of regional political and economic integration (Escobar, 2010; Vanhulst, 2015).[4] Although neoliberalism continues to leave its imprint on that political terrain, post-neoliberal politics have nonetheless carved out new spaces for political struggle, engagement and opportunity (Burdick, Oxhorn, & Roberts, 2009; Silva, 2009; Vanden, 2007). They have also generated a contradiction in which progressive social policies – land redistributions, cash transfers, alternative energy programs and civic engagement campaigns – have forced progressive governments to reinvest in past strategies, including ecologically intensive forms of resource extraction and labor-intensive service industries like tourism (Bebbington & Bebbington, 2011; Luna & Figueira, 2009; Radcliffe, 2012).

Latin America is thus at a "crossroads," as Escobar (2010) writes, between Buen Vivir and a plurality of other socio-ecological imaginaries.[5] Within the framework of the Bolivarian Revolution, for example, Venezuelan President Hugo Chávez sought to establish a "new geometry of power" through community councils, technical committees and social missions at the same time that his administration trafficked in a neo-developmentalist "oil imaginary" (French, 2009). In Bolivia under President Evo Morales, the Movement for Socialism (MAS) sought systemic change to socio-ecological relations – called *Pachakuti*, or "world reversal" – that drew upon Indigenous notions of Suma Qamaña as Vivir Bien and sought to recognize the environment as a collective subject of public interest.[6] In practice, however, despite aspirations to establish a pluralistic power structure, MAS has enlisted the state in a kind of neo-developmentalist project to modernize industries such as mining, energy, agriculture and tourism (Postero, 2007). In Ecuador under President Rafael Correa and the *Movimiento Alianza PAIS*, the Citizen Revolution (*Revolución Ciudadana*) aimed to bring about a new intercultural and plurinational model of society, in part through a 2008 constitution that codified the country's distinctly biocentric approach to "Buen Vivir" and extended rights to nature or *Pachamama*, or "world mother" (Gudynas, 2009). Yet, working against those aspirations is Correa's somewhat conventional developmentalist strategy, which privileges expert knowledge, information technology and extractive industry (SENPLADES, 2007, 2009). In the bio-culturally diverse Yasuní National Park, for example, Correa sought to finance national development by calling in the ecological debt owed by the global North and by requesting that the international community sponsor a moratorium on oil extraction. When the gambit failed, his administration began to aggressively market the preserve as a destination for community-based ecotourism (Calligaris and Bellini, 2015; Caria & Dominguez, 2015; Smith, 2013). Likewise, in El Salvador, President Sánchez Cerén of the FMLN has also proposed a national development plan organized around a philosophy of Buen Vivir, interpreted in its national context as a profound cultural change in relations between society, the state, the economy and the environment that would create the conditions for personal and collective dignity, inclusivity, security and harmony with the environment (SETEPLAN, 2015).[7] In strategy so far, however, the project has pursued a rather mainstream approach that emphasizes sustainable economic growth, development of human capital through education in science and technology as well as investments in sustainable domestic and international tourism (ibid, p. 110).

Eduardo Gudynas (2016) proposes the framework of "varieties of development" for situating Buen Vivir in relation to the wide range of strategies and imaginaries currently in play across Latin America. The terminology is helpful in two respects. On the one hand, it illustrates the diversity and hybridity of each of these strategies, including the coexistence of sustainable and human development, progressivist state-led neo-development, South–South cooperation, as well as Chinese state capitalism (an important trading partner and emerging political force in the region). On the other, it points to the boundaries of development as such and thus to "alternatives to development," cultural–political projects that break with development's core precepts (Escobar, 2011). Gudynas considers Buen Vivir a "truly Southern alternative" (2016, p. 772) and places it in the latter category. Indeed, a compelling case may be made on the level of pure discourse, given that the projects explicitly reject the market as the dominant institution and capital as its official logic. However, by imagining a "core" development project and then placing Buen Vivir outside of it, Gudynas downplays how different expressions of Buen Vivir overlap with other varieties of development, how these paradigms operate simultaneously on different scales as well as how inherent contradictions, including the persistent gap between political discourse and material practice, may actually serve the purposes of power (Ettlinger & Hartmann, 2015). In many of the cases outlined above, in fact, purportedly post-neoliberal or post-developmental imaginaries promoted by Buen Vivir have also become driving forces behind neoliberal and developmentalist projects. Some endeavor to conserve and market a country's cultural and ecological patrimony for tourism-related services and experiences.

To illustrate how contemporary tourism development might be animated by neoliberal and post-neoliberal imaginaries alike, including Buen Vivir, I suggest the framework of "varieties of tourism development." A great deal of scholarship has chronicled the historical particularity of tourism economies in Latin America, as well as the political economic contexts from which they emerge (Babb, 2010; Barkin, 2002; Berger, 2006; Galinier & Molinié, 2006). Another growing corpus is examining the complex relationship between tourism development and local/global political ecologies – the complex feedbacks between political economic systems and human and non-human ecological processes (Mostafanezhad, Norum, & Sheldon, 2016; Nepal & Saarinen, 2016; Stonich, 1998). Yet, tourism, more than most industries, relies on the construction of particular imaginaries, or implicit schemas for interpreting the events or experiences (Salazar, 2012; Salazar & Graburn, 2014; Strauss, 2006, p. 329; Wilk, 1995). By varieties of tourism development, I seek to open up another line of questioning about the connection between these historically generated political ecologies of tourism economies and the diverse political and socio-ecological imaginaries that they both usher in and capitalize upon. Latin American tourism development thus serves as a window on the many political projects currently shaping relationships between communities and their environments, citizens and states, as well as hosts and guests. Drawing on anthropocenic imaginaries of planetary change, discourses of sustainability bring hosts and guests together in shared responsibility for the environment. Meanwhile, projects like Buen Vivir re-politicize socio-ecological relations, valorize Indigenous knowledge and call attention to structural inequalities such as the North–South carbon divide. The tensions between the two are not always destructive. In many cases, in fact, they have created new opportunities for tourism development by offering more ecologically responsible alternatives to mainstream development as well as by pushing governments in Ecuador, Bolivia, Nicaragua, and elsewhere to invest in an ecologically conscious "national brand" (*la marca país*, see Villalba, Lastra Vélez, & Yánez Velásquez, 2014).

In what follows, I map the horizons of Nicaraguan tourism development at the intersection of two major projects: The revitalization of Managua via Vivir Bonito, Vivir Bien and sustainable ecotourism development via the proposed Interoceanic Grand Canal.

Rebranding Managua through Vivir Bonito, Vivir Bien

In August 2016, fellow anthropologist Alex Nading and I sat down with Raymundo, a high-level bureaucrat in the mayor's office of Ciudad Sandino, Nicaragua. We hoped to secure the official blessing of local government to conduct an experimental study of the political ecology of that city, where we have each done ethnographic fieldwork for over a decade. We also wanted to invite government officials to participate in the project, given that Vivir Bonito, Vivir Bien resonated so closely with our project's focus on the links between economic, infrastructural, public health and aesthetic dimensions of urban development.

Raymundo spoke at a caffeine-driven clip and was prone to interrupting. After clarifying that we were not, in fact, investors wanting to sink capital into the city but rather mere academic researchers, he launched into a well-rehearsed speech about Nicaragua's Buen Vivir. Despite its poor reputation within the country – and complete lack of reputation beyond that frame – Ciudad Sandino is nevertheless blessed with immense resources: "In El Salvador, they have vistas that tourists will come and pay 100 dollars just to see," he insisted, "We have vistas that would easily rival those. It's an untapped resource." One of the policy positions of Nicaragua's Buen Vivir, he continued, is that while such resources should not be exploited to the point of destruction, neither should they go un-utilized for reasons of national pride. "That's why every office in this building has to answer to our campaign, including tourism" (see Figure 2). In the moment, it was surprising to us that tourism development, of all things, was on Raymundo's radar given the more immediate issues raised by Vivir Bonito, Vivir Bien. Admittedly, it was not immediately on ours. But it is not so surprising when one considers the changing face of its closest neighbor, the capital city of Managua, in whose orbit Ciudad Sandino has long been trapped.

In the 1950s and 1960s, Managua was perhaps the most modern city in Central America, replete with high-rises, leafy promenades and a thriving business district. Old Managua was a particularly vibrant well-known part of that city because of its colonial architecture, nightlife and its very own world-famous archaeological site, Acahualinca, a UNESCO World Heritage site where one can view some 2000-year-old human footprints preserved in volcanic ash and mud. It was also here, along the

Figure 2. Department of Tourism in the Ciudad Sandino Mayor's Office.

shoreline of Xolotlán (Lake Managua) that de facto ruler and President-for-life Anastasio Somoza Debayle dreamt of building a boardwalk resembling Miami Beach (Whisnant, 1995, p. 447).

In 1972, however, disaster struck. A powerful earthquake, centered directly under the city, leveled nearly 600 city blocks, killing over 10,000 and leaving another 500,000 homeless. The rubble was carted off to a site along the shoreline, immediately adjacent to Acahualinca, where it would for the next 40 years serve as Managua's unofficial, unregulated, open-air dump, La Chureca (a Nahuatl word meaning "old rag"). In the wake of the earthquake, popular opposition to the Somoza government mobilized under the banner of the FSLN, but not before Somoza himself could redirect significant international aid, earmarked for reconstruction, to his personal ends (Walker & Wade, 2016, p. 40). What did not end up his family's bank account was channeled into paving semi-private roads to one of his many estates outside the city, including Montelimar, now a popular beach destination and the location of one of Nicaragua's most luxurious all-inclusive hotels (Babb, 2010, p. 47).[8] Somoza also used the disaster as an excuse for the aggressive eviction of squatters living in neighborhoods surrounding the *malecón* – the future site of the Miami Beach-style boardwalk – who were resettled to cotton fields to the west, generously purchased from close political allies, the Blandón family (Barreto, 2001). "Permanent National Emergency Operation 3" (OPEN-3), so named because it represented the third such wave of resettlement from Managua, would become Ciudad Sandino.

In the decade of the 1980s, the FSLN-led government of Daniel Ortega cultivated a very different relationship with tourism. Although the government was primarily concerned with pressing issues of national development and sovereignty – land redistribution, cooperative development, literacy campaigns and fending off Contra invasions funded by the Reagan administration – the Sandinistas nevertheless attracted international attention through their mixed-socialist experiment which captured the imaginations of activists, intellectuals and journalists. These *internacionalistas* (solidarity travelers) established hostels throughout Managua, their hub and even collaborated with the FSLN's Ministry of Tourism to write some of the country's first travel guides. It was also during this decade, for instance, that a small New England food co-op called Equal Exchange cut its teeth on a "fair trade" coffee called Café Nica, known as "the forbidden coffee" because it was imported in protest of the economic embargo (Equal Exchange, 2015).

Although the first Sandinista experiment would not continue into the decade of the 1990s, the influences of solidarity tourism nevertheless persisted as networks of internacionalistas continued to pour into the country to offer support. Meanwhile, a new government with a distinctly neoliberal agenda, led by Violeta Chamorro and later Arnoldo Alemán, launched an urban revitalization project titled *Nueva Managua* (New Managua) that sought to erase the public symbols of the country's revolutionary past and to make the city more palatable to private investors (Babb, 1999). They painted over murals, patched over the bullet-ridden facades of buildings, and installed a modern highway system that would allow tourists and business elite to proceed directly to their destination via a "fortified network" of private spaces, and thus bypass peddlers, squatters and beggars (Rodgers, 2004, p. 120). Global franchises such as McDonald's, Dominos, Pizza Hut and Subway set up shops in defensible spaces, including posh new malls like the one at Metrocentro and later Santo Domingo, in order to cater to that demographic (Rodgers, 2008). As Babb (2010) chronicles, that project took a particularly interesting turn in the early 2000s when the Nicaraguan Institute for Tourism (INTUR) began to appropriate images from the country's revolutionary past, albeit in the politically sanitized form of a national brand that served to distinguish Nicaragua from competitors like Costa Rica.

The environmental turn in tourism development in the capital city of Managua, Nicaragua was thus a long time in the making. But a series of events that occurred a mere year and a half after President Daniel Ortega started his third term in office, following a 16 years of absence, was an important catalyst for its emergence. In March 2008, trash pickers blockaded the dumps in both Managua and Ciudad Sandino (see Fisher, 2016; Nading & Fisher, 2018). When municipal garbage trucks arrived to dump trash along their daily routes, trash pickers pelted them with rocks. When police tried to drag them to jail, they set fire to makeshift barriers fashioned out of old car ties. In so doing, these marginalized groups effectively halted the waste-management infrastructure for several days, prompting a city-wide economic and public health crisis that would be called *el churecazo* (the "fiasco" in La Chureca). Trash pickers were tired of watching salaried government employees cut into their bottom line by taking valuable recyclables out of the waste stream and then dumping the rest. They were also upset about their increasingly stigmatized work. Virtually every major city in the region, if not the world, has populations that sift through waste in search of items with residual value. In 2007, however, a Spanish magazine singled out Managua's La Chureca as one of the "Twenty Horrors of the Modern World" – number three, to be precise, just ahead of the burning of the Amazon (Cabrera, 2007). "Like flies to dung" (*como moscas a mierda*), in the words of one recycler I interviewed in 2014, development organizations, Christian groups and other poverty tourists were suddenly drawn to the site, evidently hoping to bear witness to the unmediated aesthetics of Nicaraguan poverty and to bring home with them photographic testimonials of injustice (Zapata Campos, 2010).

Meanwhile, Nicaraguan politicians began to see the fallout of the churecazo as a public relations fiasco. They sought to delegitimize trash-pickers' claims by comparing them to *zopilotes* (vultures) feeding off the detritus of society. They also resolved to address the numerous infrastructural and environmental issues that the crisis laid bare. In late 2008, the Ministry of Health (MINSA) embarked on a municipal-wide campaign called the Plan Chatarra, with the intention of pushing unwieldy garbage and recycling economies – along with bugs, rodents and other pathogen carriers – out of population centers (Nading, 2014, p. 68). Months later, the Vice President of Spain, María Teresa Fernández de la Vega, visited Managua and pledged 60 million dollars on behalf of the Spanish Agency for International Cooperation and Development (AECID) to close the dump and to convert it into a modern waste-processing, recycling and methane extraction facility. As the project director of the Acahualinca Development Programme (the official name of the La Chureca project) told me during a 2012 interview, there were numerous interested parties pushing the plan behind the scenes – mostly against the wishes of recyclers – including the Ministry of Health (MINSA), the mayor's office, the INTUR, corporate interests such as the Pelas Group (makers of Flor de Caña rum, among other things), and of course President Daniel Ortega and First Lady Rosario Murillo.

The conversion of La Chureca was completed in February 2013, though the crisis had already brought into sharp relief a much larger set of issues regarding Managua's degraded urban environment. In anticipation of that important event, a month earlier in January 2013, Rosario Murillo announced the comprehensive, permanent and high-priority national development campaign: *Vivir Limpio, Sano, Bonito, Bien*. The resemblance to discourses of Buen Vivir in the southern cone, commentators argue, is not accidental, nor are the resonances between the FSLN's re-visioned Sandinismo – "the second stage of the revolution" that has evidently been in the making since Ortega's reelection – and twenty-first century socialisms (Bautista Lara, 2017; Equipo Envío, 2013; Jarquin, 2015). Indeed, like its counterparts, Vivir Bonito, Vivir Bien has manifested simultaneously as an ideological proposal, a political strategy and a legal text, each of which has had significant implications for the country's approach to tourism development.

The campaign's ideological proposal is most clearly expressed in the half-million copies of a 20-page treatise, encompassing 14 action points, called the "Basic Guide for Living Clean, Living Healthy, Living Beautiful, Living Well...!" (*Guia Básica para Vivir Limpio, Vivir Sano, Vivir Bonito, Vivir Bien...!*) that have been distributed since the campaign's announcement (Murillo, 2013). Some of those action points are concrete suggestions for improving the socio-ecological lives of generations of Nicaraguans, like planting trees, flowers and medicinal herb gardens in urban, suburban and rural areas and strengthening the country's culinary culture (Point 4); creating and maintaining public spaces to encourage healthy family and community dynamics (Point 3) and observing good governance with regard to environmental, public health and labor regulations (Point 5). Others are more broadly philosophical and encourage habits and practices of "Cleanliness, Hygiene, Order, Aesthetics, Loving Care, and Permanent Solidarity" (Point 1); cultivating "the basic norms of respect for life" (Point 2); living a simple life "without waste or ostentation" (Point 11) and "[joining] forces in order to prevent and deal with this plague of so-called 'Modernity'" (Point 7).

As an ideological statement for "transforming our culture of everyday life" (Murillo, 2013, p. 3), Vivir Bonito, Vivir Bien thus starts with questions of *conciencia* (consciousness or conscience), *autoéstima* (care for the self), *éstima* (care for others), and *la estética* (aesthetics), but it ultimately emerges as a comprehensive proposal for a broad-based, holistic social development (*desarrollo integral*). In this proposal, encounters with nature – characterized in the text as "Gifts of God, Temples for the replacement of Energies, the renewal of Power and Physical and Spiritual Strength, and Harmony and Human Comprehension" (p. 17) – carry significant weight. In the text, they appear matters not only of profound social and cultural import, but also economic consequence. Indeed, cultivating the habits and attitudes of respect for nature and community are also opportunities to "present our best face to the world" (Point 6) through tourism encounters, which are strongly implied but never explicitly discussed.

Political and legal changes with profound effects for tourism followed the publication of the guide. The first such move occurred in February 2013 when then Director of Communications (and First Lady) Rosario Murillo moved to grant legal status to a new cabinet position, the Offices of Family, Community and Life.[9] Shortly thereafter, Murillo publicly addressed waste management in Managua and outlined a new set of policies that would target infrastructural shortcomings as well as problematic cultural habits and attitudes:

> When we convert these areas into trash heaps, when we make our immediate surroundings and public areas ugly, that is aggression — it's a lack of love and we are called to mind the beauty, for our sense of self-esteem and respect, to respect ourselves, to love ourselves, and to care for ourselves, which is also about caring for harmony and for the aesthetics of daily life beyond poverty; we can be poor but honorable, with community and nature. (Quoted in Equipo Envío, 2013)

I witnessed some of the immediate material and political effects of the call to action beginning in July 2013, when various government offices and FSLN organizations, working on diverse issues such as the environment, education, economy, infrastructure and citizenship, sprang into coordinated action. There were immediate implications for Managua's tourism economy. During that summer, municipal workers cleared the city's causeways, clogged by trash and debris after years of heavy downpours. They began construction on new waste and recycling infrastructures – trash bins on street corners, collection substations and expanded collection services to peri-urban spaces, where most urban dwellers live (Parés Barberena, 2006; Zapata Campos & Zapata, 2013). Around the same time, the FSLN political apparatus also started to build a new educational platform with the goal of revamping cultural–environmental citizenship. The Sandinista Youth (*Juventud Sandinista*, or JS19) organized "ecological brigades" (*brigadas ecológicas*) and sought to act, in keeping with historical precedent, as "multipliers of knowledge" (*multiplicadores de conocimiento*, MINSA 2012, p. 15) by distributing literature and demonstrating around the city in order to raise consciousness about environmental issues. The Ministry of Education (MINED) also began organizing its own student "clean-up days" (*jornadas limpiezas*) as well as teacher training workshops, based on an extensive new curriculum that was codified in a 58-page text entitled "Grand National Campaign of Affection in the Daily Coexistence of Nicaragua's Families" (*Gran Campaña Nacional de Cariño en la Convivencia Cotidiana de las Familias Nicaragüenses*). The mandatory curriculum, the dictates of which are still in place as of this writing, extends from kindergarten to university and emphasizes pedagogy for highlighting the connection between cleanliness, aesthetics, respect, self-esteem and citizenship. A similar tack appears in a new generation of urban social policies, which have been enforced in public spaces with a heavy hand. In late 2013, for example, Managua's city council passed an unprecedented rule that would issue fines to those who damage the environment, either through illegal dumping or through various other forms of visual or noise pollution. They also installed green telephones (*líneas verdes*) around the city to encourage public reporting of such crimes.[10]

As a political platform, then, Nicaragua's Vivir Bonito, Vivir Bien – much like Sumak Kawsay in Ecuador or Suma Qamaña in Bolivia – can be situated at the confluence of several political tides. Some spring from within Nicaragua and concern with the political jockeying that has transpired since the FSLN first relinquished power (see Walker & Wade, 2016). Others originate in broader Latin American political processes that gave rise to twenty-first century socialist and Buen Vivir. Still others are global reverberations of neoliberalism, which have powerfully shaped Nicaragua in the decades since the first Ortega administration was forced to sign onto structural adjustment. In the summer of 2013, I interviewed Dora María Téllez – a well-known Nicaraguan historian as well as historical figure, known as *Comandante Dos* ("Second Commander") during the 1979 popular revolution – about this confluences of forces that have come together to make Vivir Bonito, Vivir Bien. She painted a very cynical portrait. On the one hand, the project resonates – albeit in a novel cultural–environmental key – with classically Orteguista ideological projects, including efforts to construct a new socialist citizen-subject called the "New Man" (*Nuevo Hombre*) as well as early attempts to re-codify structures of gender and family through the 1981 Family Code (see Borges, 1985; Lancaster, 1992; Montoya, 2007). On the other hand, the campaign has also served the more instrumental purpose of providing political cover for the current Ortega government, whose agenda took a significant hit when Hugo Chávez died and Nicaragua could no longer count on Venezuela's financial support for its social and economic programs. The Ortega administration has since been forced to take a different approach: to invest in "the spectacle of development" (*el espectáculo del desarrollo*) by patching over their most obvious shortcomings while also devolving responsibility for community well-being and development to the citizenry itself. Or as Téllez later wrote in a Facebook post, the campaign's slogan might as well be: *No me pidás que yo te resuelva los problemas, cambiá vos de actitud* ("Don't ask me to solve your problems, change your attitude").

In his research in the Acahualinca barrio of Managua, Hartmann (2016) makes a similar point about how Nicaragua's particular brand of post-neoliberalism has given rise to the "responsibilization" of public health, a decidedly odd phenomenon that simultaneously individualizes responsibility and reasserts the central role of the state. The tensions he illustrates between post-neoliberal imaginaries and neoliberal realities are equally evident in tourism development in Managua, in which the state has played an important role in cultivating urban public spaces that are attractive to citizens, investors and tourists alike. Parks around Managua, including a number of new constructions, have been upgraded with new landscaping, free public wireless internet and playground equipment pained in the campaign's signature color scheme of turquoise, lime green, pink and canary yellow. Transportation routes have been updated, such as the intersection formerly known as Rotunda Colón, which has received a makeover of the overtly political variety and now serves as a memorial to Hugo Chávez sitting atop the same Mesoamerican symbol used elsewhere in the Vivir Bonito, Vivir Bien campaign (Figure 3). Alongside his likeness, and along virtually every other thoroughfare in old Managua, are some of the most obvious changes to the landscape: dozens of illuminated metal *Árboles de Vida* ("Trees of Life"), the image and title of which is borrowed from work of late nineteenth century Austrian modern artist Gustav Klimt (see Prado Reyes, 2014). Traveling through the city, these colorful installations radiate out from Puerto Salvador Allende, the *malecón* that has also undergone renovations that might have even pleased Somoza. And indeed, the poetics of that transformation are profound (Morales & Cruz, 2016; Revels, 2014). Upon the debris of the 1972 earthquake – arguably the death-knell of the Somoza regime – the Ortega government has created a vibrant space for dance clubs, restaurants, ice cream parlors, playgrounds and a go-cart racing track, among many other activities. To the east of the central plaza, the Paseo Xolotlán continues with more restaurants, a water park, a decommissioned Boeing 737-200 airplane and miniaturized installations of Nicaragua's many cathedrals, a replica of the pre-earthquake Managua city-center and replicas of late nineteenth and early twentieth century homes that also serve as a museum for the famous poet Rubén Dario (Figure 4).

I toured the site in July 2017 with Yessenia, a FSLN supporter and representative of INTUR. We discussed the political and aesthetic aspects of its design, including the influence of the Vivir Bonito, Vivir Bien national campaign. Yessenia points out that tourism development is, of course, not the

Figure 3. Rotunda Hugo Chávez, formerly Rotunda Colón, in Managua, Nicaragua.

Figure 4. Replica of downtown Managua prior to the 1972 earthquake, in the Paseo Xolotlán.

only purpose of the campaign, but it is nevertheless a key influence. The boardwalk, she says, is a space for local, regional and international tourists alike. For the former, the place provides a safe family space that is also accessible to those with less than ample means. Daily entrance to the waterpark, for example, is only 30 Córdobas (about one dollar) and patrons are welcome to bring their own picnic food. Likewise, getting a glimpse of the inside of a Boeing airplane may not attract jet-setters, but it draws interest those for whom air travel is prohibitively expensive. Meanwhile, for international tourists, the comparatively expensive clubs, restaurants, boat tours, museums and other installations are compelling precisely because they are tried and true tourist attractions – i.e. a mix of culture, history, dancing and rum.

Running through the whole project are explicitly socialist and environmentalist themes. Regular trash and recycling bins, a rarity elsewhere in the city, are common here. Concrete benches sport Vivir Bonito, Vivir Bien campaign slogans like *Yo Vivo Bonito* ("I Live Beautiful," Figure 5). Wooden signs also implore visitors and residents to "Conserve Our Trees," "Protect Our Fauna," "Don't Contaminate the Water," "Respect Mother Earth," "Love Nature…Love Your Country" and – elaborating an classic American environmentalist slogan – "Reduce, Reuse, Recycle, Revalue, Restructure, Redistribute." Environmentalist art made by schoolchildren was also on display near the entrance of the park. One illustration, replete with stick figures of a mother, father, siblings and dog, all surrounded by puffy green trees, read in children's handwriting: "Take care of our planet, my family and I live there" (*Cuidar nuestra planeta, mi familia y yo vivimos allí*). Yessenia helped me to contextualize these images within the political and socio-ecological imaginary currently under construction in Nicaragua. The point, she says, is to present an image of Nicaragua – to citizens and visitors alike – that is clean, healthy and aesthetically pleasing not only because doing so is good for the economy, but also because it helps the country to fulfill its own social and environmental responsibilities. Tourism thus

Figure 5. A concrete bench sporting the Vivir Bonito, Vivir Bien slogan in Puerto Salvador Allende.

serves a dual role in the sense that it carves out spaces for families and communities to engage with semi-natural spaces (albeit in a highly structured way) *and* generates the capital to fund such initiatives.

One consequence is that Puerto Salvador Allende has received TripAdvisor.com's Certificate of Excellence and is listed as the "#1 Thing to Do in Managua," unseating "Leave Managua" via one of the city's regional transportation hubs (Trip Advisor, 2017). Of course, as numerous contributor comments reveal, tourism in this particular site is still impeded by Lake Xolotlán itself. Despite numerous clean-up attempts – from water-cleansing aquatic plants to large-scale water treatment plants designed to extract sewage sludge and use the dried runoff as fertilizer – visitors are assailed by the wafting scent of a dead lake, the current state of which is the complex product of infrastructural failures and years of concessions to capital interests.

Notwithstanding such shortcomings, tourism development efforts like those in Puerto Salvador Allende and elsewhere around the Managua are clearly outgrowths of larger political projects in pursuit under the heading of Vivir Bonito, Vivir Bien. Yessenia elaborates on how her own office understands the campaign:

> At INTUR, we've come to understand our work as complementary to what MARENA [Ministry of Environment] and MINED [Ministry of Education] are doing. MARENA plants the trees and cleans up the streets and cauces [city waterways]. MINED is teaching children not to cut down the trees, but to take care of them, and not to throw plastic bottles in the street, but to recycle them. We need them because their work in beautifying the city, and making a healthier place to live, also helps us bring people here. We help build solidarity. The Nicaragua that we desire cannot exist without the international community. Living well [vivir bien] for Nicaraguans cannot exist without tourism... We help show the world what we are, and what we can be.

Yessenia's comments mirrored those of other government and tourism industry representatives, who also connected tourism development with the broader principles of Vivir Bonito, Vivir Bien. They pointed to recent treatments of tourism development and the environment in the state-issued publication *Visión Sandinista*. The July 2016 edition, for example, provides some background for why Nicaragua initially joined Syria in rejecting the Paris Climate Agreement. (A later edition would explain why Nicaragua signed on, once again, when the US dropped out.) Here, Nicaragua's commitment to sustainability – framed as simultaneous investments in alternative energy, education and economic development – is on full display. At the same time, sustainability alone appears to fall short. According to the headline article, climate change is a direct manifestation of "yankee imperialism" (*el imperio yanqui*). Thus, the Paris Climate Agreement fails when it neglects to hold wealthy countries responsible for what they have wrought in the historical process of making themselves wealthy. In rectifying this situation, tourism development plays a key role. In the August 2017 edition of *Visión Sandinista*, for example, the FSLN announced their "Master Plan for Managua in the Next 24 Years" (*Plan maestro para Managua en el próximo 24 años*), which sketched their long-term approach to urban planning. In tandem with the Japanese Agency for International Cooperation (JICA), the Ortega administration intends to modernize Nicaragua's transportation system and controls horizontal growth while making the capital into a more "urban, attractive, sustainable, equitable, accessible, and active city" (Empresa Portuaria Nacional [EPN], 2017). On the radar of several of my interviewees was Japanese Prime Minister Shinzo Abe's plan to increase tourism to 20 million visitors per year in preparation for the 2020 Tokyo Olympics – a number that Japan actually surpassed twofold before the end of 2016. In Nicaragua, they conceded, tourism development would probably take a different course. As one government official put it, "The difference is that Nicaragua has something else to offer the world: We live simply in order to live well."

Sustainable tourism development and the Grand Canal

Since 2013, the quiet volcanic island of Ometepe has undergone a boom in tourist traffic. The once remote and rustic site accessible to only the most dedicated of travelers is now home to a new airport, dozens of new hotels and countless ecotourism operations (El Nuevo Diario, 2015; Vidaurre Arias, 2015). Though promising on its face, for local experts, challenges nevertheless loom on the horizon. To wit, just south of the island in the southernmost portions of Lake Nicaragua – which is, incidentally, the only freshwater lake in the world with sharks – the Ortega administration plans to construct the Interoceanic Grand Canal to rival that of neighboring Panama.

The project was announced in June 2012, as was its sole concessionaire, a Hong Kong-based company run by Chinese billionaire Wang Jing called the HKND Group.[11] The planned route runs through Lake Nicaragua and into the heretofore undeveloped rainforest region of the Río San Juan basin. Critics argue the canal will come with a significant human and ecological price tag: Each supertanker generates as much pollution as approximately 50 million cars; substantial amounts of fresh water must be drawn from surrounding rainforests to keep the lock system flowing; and the required land will likely be expropriated from already marginalized Indigenous populations (Baltodano, 2014; Carse, 2012; Hunt, 2016; Meyer & Huete-Pérez, 2014). That being said, the specific costs of Nicaragua's pending canal are unclear because the state's own environmental assessment has not been released publicly and independent academic inquiry has been met with swift retribution – including, in the case of numerous foreign scholars, deportation.

On the flip side, supporters argue that the canal will generate historic capital investment in Nicaragua. Government estimates indicate that the construction phase alone – arguably the largest civil earthmoving operation in history, with a proposed excavation of 5 billion cubic meters – would generate upwards of 25,000 jobs. The permanent infrastructure of the Grand Canal itself would also create long-term sources of employment, not only in terms of the work necessary to administer to the canal's operations but also through a number of lucrative subject-projects (Acevedo, 2015; El Nuevo

Diario, 2014; Torres, 2016). In making that case, the HKND Group has played an important role. Their homepage provides a vision of Nicaragua's future, cycling between overlaid images of Hong Kong and Managua (HKND, 2017a). Below those images, text spells out the project's "Profound Signifi-cance" as no less than a major step in the history of global trade since the fifteenth century, begin-ning with the conquest of the Americas. Separate entries outline the sub-projects at which government estimates hint. According to the company's website, the project will give rise to a world-class free trade zone system along the canal route, so that global business interests will be able to import and re-export commodities freely, without tariffs or duties, and swiftly ship them off to other parts of the Europe, Africa, Asia and elsewhere in the Americas. At the same time, the canal will pro-mote sustainable tourism development by opening up the heretofore unreachable interior of Nicara-gua's rainforests to ecotourism. As the website says:

> HKND Group intends to develop in Nicaragua several eco-tourism zones that embody the harmony between man and nature, modernity and antiquity. During the development process, HKND Group will attach great importance to and preserve local eco-environment and historical heritage, in a bid to promote the tourism, service industry, forestry, fishing and eco-environment along the Canal. Such eco-tourism zones will be a source of attraction for visitors around the world to appreciate the splendid Indian history and to enjoy the inviting views of the Carib-bean Sea and Nicaragua. (HKND, 2017b)

The text does not elaborate on how, exactly, the HKND Group or Ortega government will put the canal project to work to preserve "Indian history" or to restore balance between man/nature, moder-nity/antiquity. Insofar as the latter two distinctions are important elements of anthropocenic dis-course, however, what is instead clear is that sustainable development in general, and sustainable tourism in particular, are key elements of the campaign to sell the canal. Though mass tourism devel-opment has been proposed in the region since the late 1990s (see Hunt, 2011), new plans have been both scaled up and refigured to reflect the global demands of sustainability. Prospective investments include large-scale resorts for leisure and recreation as well as cultural, ecological and adventure tour-ism along the entirety of the canal route, from Ometepe to the Atlantic coast (Hunt, 2016, p. 172).

So far, it might appear that tourism development in Managua and the canal follow parallel paths. But the two projects met in the capital city of Managua, where the HKND Group has also coordinated efforts with the Vivir Bonito, Vivir Bien campaign. In June 2017, for example HKND sponsored the third annual "Ecological Parade." More than 6000 people from various environmental organizations, schools, churches and government institutions – including the Nicaraguan Army and National Police – marched along the Avenida Bolivar, lined by Trees of Life, on the way to the boardwalk in order to raise awareness of Nicaragua's untapped natural resources(HKND, 2017c). In late November 2017, HKND also co-sponsored the Fifth National Recycling Forum (FONARE), which also took place in Man-agua and was thematically organized around "The Environmental Management of Cities and Towns" (HKND, 2017d). While government officials raised questions about environmental citizenship, HKND representatives framed their own argument in terms of sustainable development. Despite their differ-ent outlooks – or at least discursive strategies – coordination was evident. Forum president Kamilo Lara emphasized the important role of private companies, in addition to public institutions, in "the responsible consumption of natural resources" such as water, which is one of Nicaragua's key com-petitive advantages. He also endorsed the canal as a significant opportunity for sustainable develop-ment. Meanwhile, HKND's representative articulated the many challenges facing Managua with the many opportunities provided by the canal in the face of global climatic change. As their subsequent press statement put it, "the Nicaragua canal will help Lake Nicaragua reverse its deterioration and thus will represent an opportunity for its sustainable development. The Canal will provide water sup-ply to compensate for the loss of lake water caused by climate change" (HKND, 2017d). As if to also gesture to the immediate social, cultural and infrastructural issues that provided the original impetus for Vivir Bonito, Vivir Bien, the event concluded with a fashion show in which a group of young, female Nicaraguan students modeled clothes made of recycled materials.

As coordination between state and capital interests suggests, the various differences between Vivir Bonito, Vivir Bien and sustainable development in terms of their priorities, goals and socio-ecological imaginaries are neither incommensurable or unsurmountable. Though contradictory on its face, the two appear to amplify one another, to extend one another's reach: While sustainable development allows government interests to tap into global capitalist streams, Vivir Bonito, Vivir Bien delves into the habits and attitudes of citizens. That point is perhaps most evident in Nicaragua's ongoing tourism development plans, which appear to be animated by both, simultaneously, on different registers and scales. In their complex conjuncture, it is also clear that the ascension of Buen Vivir is not necessarily the decline of sustainable development or vice versa. The contradictions between the two may actually invigorate already existing power (Ettlinger & Hartmann, 2015; Tsing, 2015).

Conclusion: Nicaraguan varieties of tourism development

The notion that tourism development projects are complex and contradictory assemblages, rather than pre-made frameworks, is not particularly novel or, for that matter, interesting. That these contradictions may actually invigorate such initiatives, however, suggests that it may be time to rethink sustainable development as well as Buen Vivir. To be sure, the latter category – Buen Vivir – signals a sea change currently underway in development thought in Latin America, one that is perhaps best positioned to cast tourism development in a different light. The high cost of progressive social and environmental policies has forced Latin American governments to reinvest in past strategies, including tourism. But under Buen Vivir, those strategies have also taken new form, invigorating an agenda that is at once more socially inclusive and ecologically sensitive as well as less so.

If tourism development is at least in part about generating the social, economic, infrastructural and ecological conditions for relationships between hosts and guests, then thinking solely in terms of markets – the dominant institution of neoliberalism – is no longer sufficient, especially as post-neoliberal proposals abound. As the changing face of Managua under Vivir Bonito, Vivir Bien suggests, tourism development refers to a dynamic system of relationships between habits and attitudes, citizens and states, communities and their environments. At the same time, as events not far away in the Río San Juan region illustrate, it would also be folly to ignore the enduring influence of capital interests and their ability to marshal discourses of sustainable development to their cause. Future research will need to avoid the temptation to place tourism either inside or outside of these logics. It will need to examine how these contradictions work. It will need to detail how projects shuttle between different socio-cultural and geographical locales as well as political contexts. And it will need to be open to the idea that tourism might serve as an interface not only between distant groups of people but also between different political and socio-ecological worlds.

In Nicaragua, Vivir Bonito, Vivir Bien is not merely a cynical Buen Vivir, fashioned into a strategy for tourism development. It is also not merely about tourism, although tourism is a central part. Rather, Nicaragua's Vivir Bonito, Vivir Bien is a hopeful, practical, and – yes – ideological response to a cluster of challenges that are at once endogenous, hemispheric and global. Its central concern is the many social and ecological pathways that comprise and erode collective well-being. That tourism plays a key part should not be surprising because, for better or worse, relationships between hosts and guests have long played an important political role. It is also not at odds with conversations about sustainable development, although their concerns are still very different. In contemporary Nicaragua, as that conversation moves away from the narrow focus on markets and makes space for Buen Vivir, it will be even more important to consider how those relationships between hosts and guests, citizens and states, communities and environments, are figured and refigured, as well as how different histories, political ecologies and socio-ecological imaginaries give rise to new varieties of tourism development.

Notes

1. I use the expansive term *tourism development*, here, to refer to the plurality of strategies oriented toward cultivating and maintaining the structures necessary for the operation of different kinds of tourism economies, including eco-tourism, community-based tourism, sustainable tourism and mainstream tourism.
2. For simplicity, I use the terms *buen vivir* and *vivir bien* interchangeably, as they are used in Nicaragua, while acknowledging their cultural and historical specificity in countries like Ecuador, Bolivia and elsewhere.
3. See Thomson (2011) and Gudynas and Acosta (2011) for brief surveys of differences that fall within the the the Buen Vivir conceptual framework, broadly speaking, including *vivir bien* ("living well"), *buen convivir* ("living well together"), *vida armoniosa* ("harmonious life"), *vida buena* ("good life"), *tierra sin mal* ("Earth without evil") and *camino noble* ("the noble way").
4. "Socialismo del Siglo XXI" was coined by Dietrich, but popularized by Hugo Chávez at his 2005 World Social Forum address.
5. In that sense, following Escobar (2010), the "post" in post-neoliberalism and post-development refers to a process of de-centering and de-naturalizing, that is, highlighting diversity and heterogeneity rather than signaling a pristine future unencumbered by the past.
6. The Morales government codified Buen Vivir first in the 2010 "Law of the Rights of Mother Earth" (*Ley de Derechos de la Madre Tierra*) and later in 2012 "Framework Law of Mother Earth and Integral Development to Live Well" (*Ley Marco de la Madre Tierra y Desarrollo Integral para Vivir Bien*).
7. FMLN refers to *Frente Farabundo Martí para la Liberación Nacional*, or "Farabundo Martí National Liberation Front.".
8. By semi-private, I mean that these roads, while constructed at public expense, were designed with occasional overhanging bridges that disallowed large public buses at the same time that private cars could pass underneath.
9. In Nicaraguan law, the argument for a cabinet position in Offices of Family, Community and Life was framed as revisions to the 1981 Family Law as well as Citizen Power Councils (CPCs) – partisan, grassroots structures similar to Venezuela's Community Councils (*Consejos Comunales*). The new Councils of Family, Community, and Life, as Murillo herself put it, sought to better reflect "Christian values, socialist ideals and solidarity practices" and to encourage "a community to reflect and work together, promoting family values and unity, self-esteem and dignity, responsibility, rights and duties, communication, co-existence, understanding and a spirit of community so as to achieve coherence of being, thinking and action" (quoted in Equipo Envío, 2013).
10. Perhaps the most publicized offender to date was the Organization of American States (OAS), headquartered in Washington DC, which was discovered to be disposing of their garbage in a local ravine. However, informal sector waste and recycling collectors have also felt the sting. Police issue them fines that they will never be able to pay or, in many cases, confiscate their collection carts.
11. HKND (Hong Kong Nicaragua Canal Development) Group is formally based in Hong Kong but registered in the Cayman Islands. It was founded in 2012 with the sole purpose of building the Nicaraguan canal. Wang Jing, HKND's controlling member, famously lost $10 billion (or 85% of his network) in the stock market in 2015, which led many Nicaraguan scholars that I talked to speculate that the project was better understood as having the backing of the Chinese state.

Disclosure statement

No potential conflict of interest was reported by the author.

Funding

This material is based on work supported by the National Science Foundation under Grant No. 1648667.

References

Acevedo, A. (2015). El Gran Canal y las 'grandes' expectativas de empleo [The Grand Canal and the 'grand job expectations]. *Envío*, 34(394), 21–24 .

Acosta, A. (2009). El Buen Vivir, una oportunidad por Construir [Living Well, an opportunity to build]. *Ecuador Debate, 75*, 33–48.

Acosta, A. (2010). *El Buen Vivir en el Camino del Post- Desarrollo: Una Lectura desde la Constituci ó n de Montecristi* [Living Well and the Path of Post-Development: A reading from the Constitution of Montecristi]. Policy Paper, 9(5), 1-36 .

Artaraz, K., & Calestani, M. (2015). Suma qamaña in Bolivia: Indigenous understandings of well-being and their contribution to a post-neoliberal paradigm. *Latin American Perspectives, 42*(5), 5–18.

Babb, F. (1999). Managua is Nicaragua: The making of a neoliberal city. *City & Society, 11*(1–2), 27–48.

Babb, F. (2010). *The tourism encounter: Fashioning Latin American nations and histories*. Stanford: Stanford University Press.

Baltodano, M. L. (2014). Canal Interoceánico: 25 verdades, 40 violaciones a la Constitución [The Interoceanic Canal: 25 truths, 40 violations of the Constitution]. *Envío*, 33(382), 24–26 .

Barkin, D. (2002). Indigenous ecotourism in Mexico: An opportunity under construction. In D. McLaren (Ed.), *Rethinking tourism and ecotravel* (pp. 125–135). West Hartford, CT: Kumarian Press.

Barreto, P. E. (2001). *Ciudad Sandino: 31 años [Ciudad Sandino: 31 years]*. Managua: Alcaldía de Managua.

Bautista Lara, F. J. (2017, June 29). El Buen Vivir [Living Well]. *El Nuevo Diario*.

Bebbington, A., & Bebbington, D. H. (2011). An Andean avatar: Post-neoliberal and neoliberal strategies for securing the unobtainable. *New Political Economy, 16*, 131–145.

Bebbington, D., & Bebbington, A. (2012). Post-what? Extractive industries, narratives of development, and socio-environmental disputes across the (ostensibly changing) Andean region. In H. Haarstad (Ed.), *New political spaces in Latin American natural resource governance* (pp. 17–37). New York, NY: Palgrave.

Becker, M. (2011). Correa, indigenous movements, and the writing of a new constitution in Ecuador. *Latin American Perspectives, 38*(1), 47–62.

Berger, D. (2006). *The development of Mexico's tourism industry: Pyramids by day, martinis by night*. New York, NY: Palgrave.

Borges, T. (1985). This revolution was to create a 'New Society.' In B. Marcus (Ed.), *Nicaragua: The Sandinista people's revolution* (pp. 22–38), New York, NY: Pathfinder.

Burdick, J., Oxhorn, P., & Roberts, K. (Eds.). (2009). *Beyond neoliberalism in Latin America? Societies and politics at the crossroads*. New York, NY: Palgrave Macmillan.

Cabrera, K. (2007). Los horrores del mundo [The horrors of the world]. *Interviu*. Retrieved March 29, 2017, from http://www.interviu.es/reportajes/articulos/los-horrores-del-mundo

Calisto Friant, M., & Landmore, J. (2015). The Buen Vivir: A policy to survive the Anthropocene ? *Global Policy, 6*(1), 64–71.

Calligaris, G., & Trevini Bellini, R. (2015). The end of the Yasuní-ITT initiative: Considerations in a Buen Vivir perspective. *International Journal of Environmental Policy and Decision Making, 1*(3), 240–260.

Caria, S., & Domínguez, R. (2015). Ecuador's Buen Vivir: A new ideology for development. *Latin American Perspectives, 43* (1), 18-33.

Carse, A. (2012). Nature as infrastructure: Making and managing the Panama Canal watershed. *Social Studies of Science, 42*(4), 539–563.

De La Cadena, M. (2015). *Earth beings: Ecologies of practice across Andean worlds*. Durham: Duke University Press.

El Nuevo Diario. (2014, July 28). El Gran Canal Interoceánico de Nicaragua y su necesidad. *El Nuevo Diario*.

El Nuevo Diario. (2015, January 12). Cargada agenda de promoción turística, según plan del gobierno [Charged tourism promotion agenda, according to government plan]. *El Nuevo Diario*.

Empresa Portuaria Nacional. (2017). *Visión Sandinista* (Vol. 20) [Sandinista Vision, Vol. 20]. Managua: Empresa Portuaria Nacional.

Equal Exchange. (2015). *Equal exchange – our co-op*. Retrieved April 1, 2017, from http://www.equalexchange.coop/our-co-op

Equipo Envío. (2013). Vivir bonito: ¿una 'revolución cultural' ? [Live beautiful: a cultural revolution?]*Envío*, 32(372.

Escobar, A. (2010). Latin America at a crossroads: Alternative modernizations, post-liberalism, or post-development ? *Cultural Studies, 24*(1), 1–65.

Escobar, A. (2011). Sustainability: Design for the pluriverse. *Development, 54*(2), 137–140.

Escobar, A. (2012). *Encountering development: The making and unmaking of the third world* (2nd ed.). Princeton: Princeton University Press.

Ettlinger, N., & Hartmann, C. D. (2015). Post/neo/liberalism in relational perspective. *Political Geography, 48*, 37–48.

Fisher, J. (2016). Cleaning up the streets, Sandinista-style: The aesthetics of garbage and the urban political ecology of tourism development in Nicaragua. In M. Mostafanezhad, R. Norum, & E. Sheldon (Eds.), *Political ecology of tourism: Community, power and the environment* (pp. 231-250). London: Routledge.

Fletcher, R. (2011). Sustaining tourism, sustaining capitalism? The tourism industry's role in global capitalist expansion. *Tourism Geographies, 13*, 443–461.

French, J. D. (2009). Understanding the politics of Latin America's plural lefts (Chávez/Lula): Social democracy, populism and convergence on the path to a post-neoliberal world. *Third World Quarterly, 30*(2), 349–370.

Galinier, J., & Molinié, A. (2006). *Les néo-Indiens: Une religion du IIIe millénaire* [The New Indians: A religion of the third millenium]. Paris: Odile Jacob.

Gudynas, E. (2009). La ecología política del giro biocéntrico en la nueva constitución de Ecuador [The political ecology of the biocentric turn in the new constitution of Ecuador]. *Revista Estudios Sociales, 32*, 34–47.

Gudynas, E. (2011). Buen Vivir: Today's tomorrow. *Development, 54*(4), 441–447.

Gudynas, E. (2016). Beyond varieties of development: Disputes and alternatives. *Third World Quarterly, 37*(4), 721–732.

Gudynas, E., & Acosta, A. (2011). La renovación de la crítica al desarrollo y el buen vivir como alternativa [The renewal of the critique of development and living well as an alternative]. *Utopía y praxis latinoamerica, 16*(53), 71–83.

Hartmann, C. D. (2016). *Public health, environment, and d e velopment in Nicaragua and Latin America: A post/neoliberal perspective* (Doctoral dissertation). The Ohio State University, Columbus, OH.

HKND Group. (2017a). *Nicaragua canal: The century old dream will come true.* Retrieved April 14, 2017, from http://www.hknd-group.com

HKND Group. (2017b). Ecotourism zones. Retrieved April 14, 2017, from http://hknd-group.com/portal.php?mod=view&aid=50

HKND Group. (2017c). HKND group supports Nicaragua's third ecological parade. Retrieved December 31, 2017, from http://hknd-group.com/portal.php?mod=view&aid=443

HKND Group. (2017d). The Nicaragua interoceanic canal, a boost to the sustainable development of the country's environment. Retrieved December 31, 2017, from http://hknd-group.com/portal.php?mod=view&aid=450

Hunt, C. (2011). Passport to development? Local perceptions of the outcomes of post-socialist tourism policy and growth in Nicaragua. *Tourism Planning & Development, 8*(3), 265–279.

Hunt, C. (2016). A political ecology of tourism in the shadow of an inter-oceanic canal in Nicaragua displacing poverty or displacing social and environmental welfare ? In S. Nepal & J. Saarinen (Eds.), *A political ecology of tourism* (pp. 163–178). London: Routledge.

Jarquín, E. (2015, May 15). El buen vivir. *La Prensa*, p. 11A.

Lancaster, R. (1992). *Life is hard: Machismo, danger and the intimacy of power in Nicaragua.* Berkeley: University of California Press.

Luna, J. P., & Filgueira, F. (2009). The Left turns as multiple paradigmatic crises. *Third World Quarterly, 30*(2), 371–395.

Meyer, A., & Huete-Pérez, J. A. (2014). Nicaragua Canal could wreak environmental ruin. *Nature, 506*(7488), 287–289.

Montoya, R. (2007). Socialist scenarios, power, and state formation in Sandinista Nicaragua. *American Ethnologist, 34*(1), 71–90.

Morales, A., & Cruz, E. (2016, January 31). El renacer de Managua. *La Prensa*, p. 1A.

Mosse, D. (2005). *Cultivating development: An ethnography of aid policy and practice.* London: Pluto Press.

Mostafanezhad, M., Norum, R., & Sheldon, E. (Eds.). (2016). *Political ecology of tourism: Community, power and the environment.* London: Routledge.

Murillo, R. (2013). *Vivir Limpio, Vivir Sano, Vivir Bonito, Vivir Bien* ! [Live Clean, Live Healthy, Live Beautiful, Live Well!] Managua, Nicaragua: Gobierno de Reconciliación y Unidad Nacional.

Nading, A. (2014). *Mosquito trails: Ecology, health, and the politics of entanglement.* Berkeley: University of California Press.

Nading, A., & Fisher, J. (2018). Zopilotes, Alacranes, y Hormigas (vultures, scorpions, and ants): Animal metaphors as organizational politics in a nicaraguan garbage crisis. *Antipode*. Advance online publication. DOI:10.1111/anti.12376

Nepal, S., & Saarinen, J. (2016). *Political ecology and tourism.* London: Routledge.

Oviedo, A. (2014). El Buen Vivir posmoderno y el Sumakawsay ancestral [Postmodern Living Well and ancestral Sumakawsay]. In A.-L. Hidalgo-Capitán, A. Guillén García, & N. Deleg Guazha (Eds.), *Antología del Pensamiento Indigenista Ecuatoriano sobre Sumak Kawsay* (pp. 267–297). Huelva: FIUCUHU.

Parés Barberena, M. I. (2006). *Estrategia municipal para la intervención integral de asentamientos humanos espontáneos de Managua, Nicaragua* [Municipal strategy for the comprehensive intervention of spontaneous human settlements in Managua, Nicaragua]. Managua, Nicaragua: CEPAL (Comisión Económica para América Latina y el Caribe).

Postero, N. (2007). *Now we are citizens: Indigenous politics in Postmulticultural Bolivia.* Stanford: Stanford University Press.

Prado Reyes, Y. (2014, March 5). Chávez vive en el corazón de los nicaragüenses [Ch áves lives in the heart of Nicaraguans]. *El 19 Digital*. Retrieved April 20, 2017, from https://www.el19digital.com/articulos/ver/titulo:16860-chavez-vive-en-el-corazon-de-los-nicaraguenses

Quijano, A. (2000). Coloniality of power, eurocentrism, and Latin America. *Nepantla: Views from the South, 1*(3), 533–580.

Quijano, A. (2011). Buen Vivir: Entre El Desarrollo Y la Des/colonialidad Del Poder [Living Well: Between Development and De/Coloniality of Power]. *Ecuador Debate, 84*, 77–88.

Radcliffe, S. (2012). Development for a postneoliberal era? Sumak kawsay, living well and the limits to decolonization in Ecuador. *Geoforum, 43*, 240–249.

Revels, C. S. (2014). Placing Managua: A landscape narrative in post-earthquake Nicaragua. *Journal of Cultural Geography, 31*(1), 81–105.

Rodgers, D. (2004). Disembedding the city: Crime, insecurity, and spatial organization in Managua, Nicaragua. *Environment and Urbanization, 16*(2), 113–124.

Rodgers, D. (2008). A symptom called Managua. *New Left Review,* 49(January–February), 103-120 .

Salazar, N. B. (2012). Tourism imaginaries: A conceptual approach. *Annals of Tourism Research, 39*(2), 863–882.

Salazar, N. B., & Graburn, N. H. (Eds.). (2014). *Tourism imaginaries: Anthropological approaches.* Berghahn: London.

SENPLADES. (2009). *Plan Nacional para el Buen Vivir 2009-2013: Construyendo un Estado Plurinacional e Intercultural* [National Plan for Living Well 2009-2013: Building a Plurinational and Intercultural State]. Quito, Ecuador: Author.

SETEPLAN. (2015). *Plan Quinquenal de Desarrollo 2014-2019: El Salvador Productivo, Educado, y Seguro* [Five Year Development Plan 2014-2019: A Productive, Educated, and Safe El Salvador]. Santa Tecla, El Salvador: Author.

Silva, E. (2009). *Challenging neoliberalism in Latin America.* New York, NY: Cambridge University Press.

Smith, T. (2013). Crude desires and 'green' initiatives: Indigenous development and oil extraction in Amazonian Ecuador. In B. Buscher & V. Davidov (Eds.), *The ecotourism-extraction nexus: Political economies and rural realities of (un)comfortable bedfellows* (pp. 149–170). New York, NY: Routledge.

Torres, E. (2016, March 3). Nicaragua, la nueva joya de los negocios en Centroamérica. *El 19 Digital.* Retrieved April 13, 2017, from

Trip Advisor. (2017). Puerto Salvador Allende. *Trip Advisor* [online]. Retrieved April 13, 2017, from https://www.tripadvisor.com/Attraction_Review-g294478-d4776907-Reviews-Puerto_Salvador_Allende-Managua_Managua_Department.html

Vanden, H. E. (2007). Social movements, hegemony, and new forms of resistance. *Latin American Perspectives, 34,* 17–30.

Vanhulst, J. (2015). El laberinto de los discursos del Buen Vivir: Entre Sumak Kawsay y Socialismo del siglo XXI [The laberynth of discourses of Living Well: Between Sumak Kawsay and Socialism of the 21st century]. *Polis, 14*(40), 233–261.

Vidaurre Arias, A. (2015, January 24). Nicaragua rica en turismo sostenible. *El Nuevo Diario.*

Villalba, R. C., Lastra Vélez, J., & Yánez Velásquez, J. (2014). La Marca País: Su origen y evolución [The National Brand: Its origin and evolution]. *Retos, 4*(8), 174–187.

Walker, T., & Wade, C. J. (2016). *Nicaragua: Emerging from the shadow of the eagle.* Boulder: Avalon.

Walsh, C. (2009). *Interculturidad, Estado, Sociedad* [Interculturality, the State, and Society]. Quito: Abya-Yala.

Whisnant, D. E. (1995). *Rascally signs in sacred places: The politics of culture in Nicaragua.* Chapel Hill: University of North Carolina Press.

Wilk, R. (1995). Learning to be local in Belize: Global systems of common difference. In D. Miller (Ed.), *Worlds apart: Modernity through the prism of the local* (pp. 110–133). New York, NY: Routledge.

Zapata Campos, M. J. (2010, August). *Branding poverty: La Chureca , the ' Slum Project Millionare ' : How a project becomes a project.* Paper presented at the workshop Exploring Spaces and Linkages between Services, Markets, and Society, Lund University, Lund, Sweden.

Zapata Campos, M. J., & Zapata, P. (2013). Switching Managua On! Connecting informal settlements to the formal city through household waste collection. *Environment and Urbanization, 25*(1), 225–242.

Li, T. M., (2007). The Will to Improve: Governmentality, Development, and the Practice of Politics. Durham: Duke University Press.

Thomson, B. (2011). Pachakuti: Indigenous perspectives, buen vivir, sumaq kawsay, and degrowth. Development, *54*(4), 448–454.

Villalba, U. (2013). Buen Vivir vs Development: a paradigm shift in the Andes? Third World Quarterly, *34*(8), 1427–1442.

Latour, B. (2013). An Inquiry into Modes of Existence: An Anthropology of the Moderns. Cambridge: Harvard University Press.

Walsh, C. (2010). Development as Buen Vivir: Institutional arrangements and (de)colonial entanglements. Development, *53*(1), 15–21.

Hidalgo-Capitán, A.L, A. Arias and Ávila J. (2014). Antología del Pensamiento Indigenista Ecuatoriano sobre Sumak Kawsay [Anthology of Indigenous Ecuadorian Though Concerning Sumak Kawsay. Cuenca, Ecuador: FIUCUHU

SENPLADES (2007). Plan Nacional de Desarrollo 2007-2010 [National Development Plan 2007-2010]. Quito, Ecuador: SENPLADES.

Stonich, S. (1998). Political Ecology of Tourism. Annals of Tourism Research, *25*(1), 25–54.

Strauss, C. (2006). The imaginary. Anthropological Theory, *6*(3), 322–344.

MINSA (2012). Plan de Acción de Manejo de Basura [Waste Management Action Plan]. Managua, Nicaragua: Gobierno Nacional de Nicaragua.

Tsing, A. (2015). The Mushroom at the End of the World: On the Possibility of Life in Capitalist Ruins. Princeton: Princeton University Press.

What are wilderness areas for? Tourism and political ecologies of wilderness uses and management in the Anthropocene

Jarkko Saarinen

ABSTRACT

In global imaginaries, wilderness areas are considered to represent the last parts of "original" nature, untouched by civilization and modernization. In most cases, this is misleading as wilderness environments have been exploited, explored and also converted into administrative units in various protected area networks. Indeed, most wilderness areas have been a part of human–environment interactions for a long time and they have been influenced and modified in that interaction. As a result, wilderness is constitutively a cultural and politically loaded idea. While the Western notion of wilderness as a place where "man himself is a visitor who doesn't remain" represents the global hegemonic conservation thinking, it does not necessarily work with different local realities, meanings and use values of "the wild". In addition, in recent decades, the tourism industry has placed an increasing interest on nature-based and adventure tourism products creating new kinds of ideas and use needs for the remaining wilderness environments. This paper analyzes empirically how wilderness environments and their roles are seen in the context of new and traditional anthropogenic uses and meanings of wilderness areas. More specifically, the paper uses a political ecology approach to evaluate the use and management priorities in the Finnish Wilderness system.

Introduction

In the contemporary global imaginary, wilderness environments are considered to represent the last existing fragments of untrammeled "real" nature, untouched by human systems and societies. As a manifestation of this, the International Union for Conserving Nature (IUCN) has labelled the "highest" protection area Category 1b as "wilderness" (Strict Nature Reserves, as a comparison, are labelled Category 1a). Such areas are considered to be large unmodified or slightly modified areas, retaining their natural character without permanent or significant human habitation. Indeed, in Western thought, there is a long tradition of perceiving wilderness as a space external, and in opposition, to people, culture and civilization (see Nash, 1967; Short, 1991; Tuan, 1974). In this sense, then, wilderness becomes an emblematic anthropocenic space – one that people endeavour to define, create and maintain as removed and distinct from *anthropogenic* spaces – this speaks to the complex politics involved in the culturally constructed separation of people and nature, and in its reckoning in contemporary situations.

However, the existence of any firm borderline between an organized society and wilderness (as a non-human and "unorganized" environment) has become increasingly unclear and problematic.

Following the era of industrialization, wilderness areas have become heavily utilized in modern economic development projects and programmes, leading to a growing and cumulative human presence in environments previously considered as "wild" and "pristine". For some, this increasing anthropogenic influence and/or presence in (former) natural systems has progressed to a state in which humans are now one of the "great forces of/in nature" (Morton, 2012; see Robbins & Moore, 2013). In current environmental discourses, this increasing and visible domination by humans is said to represent the Anthropocene (Latour, 2015), the new geological epoch in which human activity has become the dominant force and influence on climate and the environment, in general (Rockström & Klum, 2012). In addition, the idea of Anthropocene has also emerged as an epistemic system influencing practices and imaginaries that increasingly govern contemporary environmental management, which issues this paper focuses on the Finnish wilderness area context.

The idea of the Anthropocene was coined and popularized in the early 2000 by the Nobel laureate Paul Crutzen (see Crutzen, 2002; Crutzen & Stoermer, 2000) and since that time the concept has come in to popular use (Brondizio et al., 2016) but also critically debated, especially in the social sciences and humanities (see Lorimer, 2012; Robbins & Moore, 2013; Veland & Lynch, 2016). The concept of the Anthropocene is used as an indication of global change and environmental crisis and the Earth's limits to growth (Rockström et al., 2009). It is seen as "a result of population growth and resource use, humans are now a geological force in and of themselves, driving planetary change at an unprecedented rate" (Moore, 2015, p. 32). As an idea, it firmly connects global to local: due to the global processes linked with the Anthropocene, distant wilderness environments are perceived to have disappeared or they are increasingly fragmented and, thus, threatened and unable support the integrity of "wild" ecosystems. Indeed, the Anthropocene is characterized by powerful global imaginaries of threats and disasters operating in global–local nexus, which are estimated to have very serious consequences not only to the natural environment but also to human civilization and even the future existence of humans (Crutzen, 2002). Here, the question is not whether the Anthropocene is a geological fact and an epoch that follows the Holocene with the estimated consequences to humans and to the planet. Instead, the Anthropocene can be interpreted as a powerful global imagination and a *social* fact, although strongly dominated by natural science arguments and logics (see Castree, 2014, 2015, 2017), that is, at present, a prominent discourse influencing policy discussions and related practices, including environmental and conservation management issues.

As a result of these increased human impacts and pressures on the natural environment, since the nineteenth century, many of the planet's remaining wild environments have been converted into administrative units in conservation and protected area networks (see Hendee, Stankey, & Lucas, 1990; Hovik, Sandström, & Zachrassion, 2010; Nash, 1967) that are "managed so as to preserve their natural condition" (IUCN, 2013). Instead of existing external to "us", however, these protected wilderness spaces are integral parts of organized societies characterized by various governance structures, laws and nationally and internationally defined management procedures (Saarinen, 2016). While legally protected, however, wilderness areas are still considered endangered due to multiple processes linked to the Anthropocene. In addition to the impacts of global warming on biodiversity and certain species of flora and fauna, in general, various localized human activities are also seen to be increasingly "filtering in" to protected areas (see Duffy, 2015; Hall, 1992). In this respect, organized commercial tourism is regarded as one of the major threats and activities that utilizes wilderness and other natural areas (Buckley, 1999; Duffy, 2002). For example, Higham (1998, p. 27) has argued that "wilderness areas are arguably the most sensitive physical resources for tourism".

However, although contemporarily growing forms of nature-based tourism pose a potential threat to wilderness areas, there is a long symbiotic relationship between tourism and conservation (see Bukowski, 1977; Frost & Hall, 2010). From a political ecology perspective, this relationship has often led to an uneven development by marginalizing local and Indigenous communities and their traditional livelihoods in the use of natural resources (see Mostafanezhad, Norum, Shelton, & Thompson-Carr, 2016; Nepal & Saarinen, 2016). In current neoliberal governance of conservation areas, tourism is increasingly used for the commodification of nature by turning intrinsic or local use values into exchange values for the purposes of non-local touristic consumption (Büscher, 2013; Duffy, 2002, 2015). The exchange value built into tourism production and consumption systems provides a rhetoric and tool to create socio-economic importance for wildernesses and, thus, a promise for local and regional development (see Büscher & Dressler, 2012). Interestingly, in recent discussions, the growing and geographically evolving tourism industry has been noticeably connected to the processes of Anthropocene: as indicated by Gren and Huijbens (2014, p. 6), tourism has become "a geophysical force censoriously interrelated with the capacity of the Earth to sustain the human species" (see also Hall & Saarinen, 2010).

Indeed, tourism is a global activity with a substantial capacity to change the environment, particularly in the destination regions. According to Buckley (1999, p. 191), the emerging trend of commercial tourism "is important for wilderness management, because tourism is a large and powerful industry with considerable political power". Congruently, Büscher (2013, p. 57) has labelled tourism as a "Holy Grail" that holds a magical power to integrate remote conservation areas to global capitalistic markets. All this makes it challenging for wilderness managers to cope with changing use pressures and to limit or control the tourism industry, especially in the peripheral regions (where most wilderness areas are typically located) which offer limited alternative economic development options besides tourism (Butler & Boyd, 2000; Saarinen, 2007, 2013a). Specifically, the different forms of fast-growing commercial adventure tourism rely on wilderness environments, conditions and related images (see Buckley, 2006; Duffy, 2002).

Based on these multi-scalar threats, it seems that wildernesses are doomed with boundaries that are too porous or weak for them to cope with human-induced invasion, activities and pollutants operating in global and local scales. This distinction, imagined opposition and "fencing" between civilization and wilderness (Nash, 1967), however, is based on a certain kind of understanding of what nature is and what it should be. In contrast to this opposition perspective, most wilderness areas can be seen as having been part of human–environment interaction (i.e. human systems) for a long time; they have been influenced and modified by such relations, but still remain something we generally refer to as "wild" (see Dahlberg, Rohde, & Sandell, 2010; Hall, 1992; Sæþórsdóttir, Hall, & Saarinen, 2011). Clearly, in historical or pre-industrial times, the human-induced changes were not global in scale, but localized human–wilderness interactions have defined wildernesses in place-specific ways: wilderness is constitutively a cultural idea. While the Western notion of wilderness as a place where "man himself is a visitor who doesn't remain" (Public Law, 1964, p. 1) represents the current global hegemonic conservation thinking, the so-called fortress model (Spinage, 1998), it does not necessarily relate consistently with all local realities, traditional meanings and use values of "the wild" (see Bertolas, 1998; Neumann, 1998; Saarinen, 2016; Watson, Matt, Knotek, Williams, & Yung, 2011). This underscores the previous notion that wildernesses are not indisputably outside of societies and in many places these traditional uses and ideas of wilderness are still evident and active, although often marginalized and/or under a constant heated discussion and need for justification (Connor, 2014). It also draws attention to the notion that the ways in which the Anthropocene is understood, imagined and instrumentalized is and should be very much dependent on local contextual (e.g. social, cultural, political) factors.

Interestingly, the idea of Anthropocene is seen as challenging the principle duality between society and wilderness or culture and nature. As noted by Lorimer (2012, p. 593), "the modern understanding of Nature as a pure, singular and stable domain removed from and defined in relation to urban, industrial society (…) has been central to western environmental thought and practice". For him, the Anthropocene discourse represents a firm challenge to such thinking and the modern politics of nature or wilderness as distant from us. In this respect, and as I suggested above, the idea of the Anthropocene can empower us to see that there has never been a nature without us (see Castree, 2012; Mann, 2005). Therefore, and as I will argue in this paper, the idea of an apolitical and acultural wilderness may have reached an impasse. Indeed, for some, there has never been a symbolic or physical fence between culture and nature. In this respect, a political ecology perspective raises interesting questions about how relatively recent localized socio-ecological threats associated with the idea of Anthropocene, such as the evolving global tourism industry, relate to other existing human–nature interactions in the wilderness environments. Basically, are changing and growing tourism activities and traditional uses of wilderness areas seen and represented differently or pooled together in wilderness management thinking and, thus, seen as common threats to biodiversity and pristine characteristics of the wild? Do these different uses collide and exclude each other? Are they valued and governed differently? In addressing these questions, this paper seeks to contribute to the ongoing governance and management discussions of wilderness areas in the current context of environmental change and touristic uses of the wild.

Through an empirical analysis of the management of 12 wilderness areas legally established in Finnish Lapland in 1991, this paper aims to discuss how wilderness environments and their roles are seen in the context of new and traditional anthropogenic uses and meanings of wilderness areas. These new uses and meanings refer to evolving commercial tourism based on global imaginaries of the wild which is termed here as an *adventurescape* that serves for various evolving touristic activities in natural settings. These activities are increasingly linked to the above-described discourse of the Anthropocene on a local scale. Here, traditional views refer to Indigenous and other local uses of wilderness areas, namely reindeer herding and subsistence hunting, which have been practiced for a long time in these places and well before the establishment of the present-day wilderness area network that exists in northern Finland. However, the focus of this paper is not new or traditional uses of wilderness, or their characteristics or volumes and related material struggles per se, but rather how Finnish wilderness area managers (WAMs) perceive what Finland's designated wilderness areas are for and, equally as telling, what they are not for.

These are typical questions for a political ecology approach involving processes of exclusion, inclusion and conflict in natural resource management, as well as how nature – wilderness – is valued, perceived and controlled (see Blaikie & Brookfield, 1987; Forsyth, 2008; Mostafanezhad et al., 2016; Peet, Robbins, & Watts, 2011; Saarinen & Nepal, 2016). As an approach, rather than a theoretical perspective (Robbins, 2012, p. 5), political ecology provides fruitful lens and avenues to analyze and understand how different uses and meanings of wilderness are framed and practiced in natural resource management and what kind of potential power issues and discourses are evident (Dahlberg et al., 2010; Douglas, 2014; Wall-Reinius, 2012). Specifically, in this paper, I use a discursive political ecology (see Bryant, 2000; Bryant & Bailey, 1997) to evaluate the uses and management priorities and frameworks in the Finnish wilderness system. I do so by offering an analysis of how WAMs perceive wilderness and its role in the context of traditional uses, emerging tourism and, eventually, in the global biodiversity conservation agenda.

"Through the wilderness": wilderness as a cultural and contested idea

In the early 1980s, American singer Madonna recorded a popular song "Like a Virgin" in which she used a metaphor "through the wilderness". The rhetoric links the title phrase of the song and the idea of wilderness as a pristine and untouched object. This connection is based on the very context from which the song originates: western – and specifically an American – culture. In his seminal book "Wilderness and the American Mind", Roderick Nash (1967, pp. 1–2) indicates that the English term *wilderness* is composed of two root words ("wild" and "deor") to denote a place inhabited by wild beasts which is beyond human influence and control (see Short, 1991, p. 6). For Nash, the wilderness received its meaning by being located outside a culture and society in the Anglo-American context. However, at the same time, Nash (1967) underlines that wilderness is also a culturally defined concept, created by a civilization encountering and conquering the wild and constantly fencing it out. Similarly, Cronon (1998) has emphasized that wilderness is ultimately a social construction, while for Tuan (1974, p. 112), wilderness is "a state of mind" as, he argues, it is impossible to define a wilderness in an objective manner.

It is easy to agree with these philosophical views, underlining the culturally and socially defined nature of wilderness. However, they do not necessarily lead us far on a path to what wilderness environments *are* (or what they *mean*) for governance and management. In this respect, probably, the most well-known and influential technical definition of the wilderness originates from the world's first wilderness legislation, prescribed in the United States in 1964. According to the US Wilderness Act (Public Law, 1964, p. 1):

> A wilderness, in contrast with those areas where man and his own works dominate the landscape, is hereby recognized as an area where the earth and its community of life are untrammeled by man, where man himself is a visitor who does not remain.

The core idea of the Act is based on the historically constructed Anglo-American view on wild nature and conservation of natural and pristine areas (Shields, 1991; Short, 1991). For example, the designation of Yellowstone National Park in 1872, representing the world's first large-scale unit aiming to conserve wilderness environments, was influenced by increasing anthropogenic impacts on nature in frontier lands. Specifically, it aimed to reserve and withdraw the area "from settlement, occupancy, or sale" (Nash, 1967, p. 108), with a further emphasis on public use in tourism and recreation.

Indeed, in North America, the increasing, progressive encountering and disappearing of wilderness resulted in the need to protect any and all remaining wild places. Although this process was originally situated in North America, the resulting fortress model fencing both ideologically and practically "wild in and humans out" became a benchmark for global conservation. Currently, it represents the hegemonic conservation train of thought in wilderness management globally (see Hendee et al., 1990; Spinage, 1998). It involves a level of "anthrophobia" in conservation thinking in which "recovery, restrain, and control are imperative" (Robbins & Moore, 2013, p. 8). This *conserved wilderness* functions as a "biodiversity container" with minimizing or strictly limiting human presence and impacts (Saarinen, 2005). In contrast to Indigenous and other local cultures, it challenges the traditional (non-extractive) economic and cultural uses and meanings of wilderness (see Hall, 1992; Hallikainen, 1998). In Finland, for example, wilderness (erämaa) historically signified an area outside permanent settlement but which had a crucial economic importance in terms of hunting and fishing (Lehtinen, 1991), fashioning these areas into integral parts of the human system and organized society, including elements such as taxation (Hallikainen, 1998; Saarinen, 1998). This kind of *traditional wilderness* was and still is an economic resource and subject of usage. In spite (or perhaps in part because) of its somewhat "relic nature" in a modernized society, the traditional wilderness idea has remained and is manifested in the aims of the Finnish Wilderness Act (Erämaalaki, 1991). According to the Act, designated wilderness areas are founded for: (1) maintaining the wilderness character of the areas; (2) securing Sámi (Indigenous) culture and related traditional livelihoods; and (3) developing a versatile use of nature and possibilities for (economic) uses. Apart from the first target, the other goals clearly refer to wilderness that is a resource inside culture and economy, and which receives its meanings mainly through its local traditional uses, roles and values.

Despite these major differences, both conserved and traditional wilderness ideas situate those environments as integral parts of contemporary societies and their institutional structures in the globalized world (Saarinen, 2005). In addition, in recent decades, the tourism industry has placed an increasing interest on nature-based tourism products creating new kinds of ideas and use needs for the remaining wilderness environments, such as adventure tourism activities and organized snowmobile and sledge dog safaris (see Cloke & Perkins, 2002; Vail & Heldt, 2004). According to Hall and Page (2002, p. 236), a growing demand for wilderness experiences in tourism is a result of the changed and more positive attitudes towards the environment which have come about since the nineteenth century. Indeed, as noted, the symbiotic link between protected wild areas and tourism can be seen as old as the history of protected areas (Frost & Hall, 2010; Hall, 1992), but recently, the scale and nature of contemporary global scale tourism industry to utilize and impact wilderness environments is dramatically different than before (see Buckley, 1999; Fennell, 1999). The better access to natural areas has also integrated wilderness areas more closely to global and national tourism markets.

New touristic impacts and meanings are not only directly physical or based on site use alone, since the tourism industry and its marketing and commodification processes also construct certain kinds of images of wild places (see Bryant, 2000), which may in turn affect the practices related to wilderness use and management (for example, Budowski, 1977; Puhakka & Saarinen, 2013; Rutherford, 2011). Thus, this *touristic wilderness* is constructed through direct and indirect impacts based on consumption, marketing and visualizing natural environments and staging wilderness settings for touristic purposes, creating and using global imaginaries of wildernesses as commercialized wild spaces where a consumer is a visitor who does not remain (Saarinen, 2005). Indeed, nowadays, wild natural environments are "universally regarded as a source of pleasure" (Wang, 2000, p. 80), a certain type of adventurescape where the tourism industry has become a significant user, stakeholder and element of change and power (Buckley, 2007; Butler & Boyd, 2000; Saethorsdóttir, 2010). Thus, by defining wilderness as a specific site for tourist activities to take place, there is a danger to marginalize other uses and meanings of wilderness. As noted by Wall-Reinius (2012, p. 622), "the idea of wilderness as pristine and empty nature [for tourist consumption and/or conservation] can have negative implications for people living in or using these areas as they become silenced and ignored" (see also Reimerson, 2016; Saarinen, 2013b; Wilson, 1999).

Wilderness area governance and managers' perceptions

Finnish wilderness areas: governance and management

While, as may be inferred from what I have presented above, it is difficult to find a consistent or fixed definition for the idea of wilderness – particularly across different national and cultural contexts – several attempts have been made to do so in laws and in connection with various agreements to establish wilderness areas. In Finland, the Wilderness Committee (Erämaakomitean mietintö, 1988, p. 23) preparing the Wilderness Law (1991) defined the basic characteristics of wilderness areas as follows:

(1) A wilderness area should comprise a minimum of 15,000 hectares and usually be more than 10 km in width.
(2) The area should be ecologically as diverse as possible… and all human action should be adjusted to nature so as not to spoil the wilderness character of the area.
(3) The area should as a rule have no roads.
(4) The landscape should be in a natural state condition and unspoiled. Any structures connected with human activity should merge with the natural landscape.

This initial framework was relatively consistent with the global hegemonic idea of conserved wilderness. Due to conflicts and compromising political process in establishing the area network, however, the resulting Wilderness Law visibly emphasized the traditional and also new versatile use possibilities of the wilderness areas in the final legal text as referred to earlier. Thus, the Law does not conflict with every person's rights (i.e. common access to nature), for example, in Finland or previous economic and culturally defined activities, such as reindeer herding, in these areas. Based on the Law, 12 designated wilderness areas (total size 14,903 sq.km) were established in northern Finnish Lapland (Figure 1). A notable governance-related issue in the Wilderness Law is that it is separated from the legal framework of nature conservation areas. Simply put, legally, wilderness areas are not conservation areas, i.e. they are not separated from the surrounding spaces by the Nature Conservation Law but based on the Wilderness Act which allows human elements and versatile uses. These areas are managed by the state enterprise Metsähallitus (Parks & Wildlife Finland, a department funded by a public budget) which initially prepared a management and land-use plan for each site.

In its recently revised management plans, the Metsähallitus has utilized the *Akwé: Kon* Guidelines in its general management and land-use plans (see Juntunen & Stolt, 2013). These voluntary

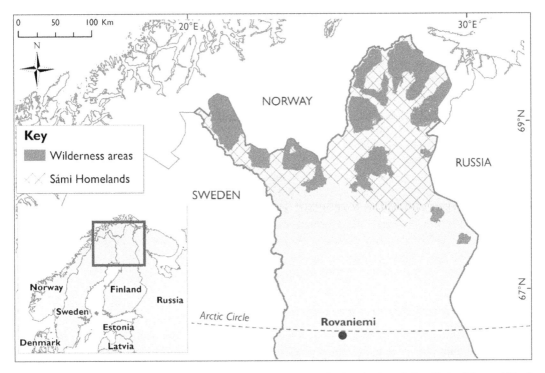

Figure 1. Location of Finnish wilderness area network in Lapland, Northern Finland (Author's map). Data: National Geospatial-Intelligence Agency (2017), SYKE (2017), MML (2017).

guidelines are derived from Article 8(j) of the United Nations Convention on Biological Diversity (1992). For the Metsähallitus, as the administering authority, a rationality to utilize the guidelines is based on a need "to further improve on securing the opportunities of Europe's only Indigenous people to influence the conditions for practicing their culture in the area" (Väisänen, 2013, p. 7). According to the Director of Natural Heritage Services, Metsähallitus, the *Akwé: Kon* work has "significantly improved the engagement of the Sámi people in the planning process" (Väisänen, 2013, p. 7) and the model is nowadays widely implemented in various areas of Sámi homeland in Finnish Lapland. As an Indigenous people forming an ethnic minority in Finland, Sweden, Norway and the Kola Peninsula of Northwest Russia, the Sámi people have their own languages, cultures, identities and histories (Aikio, Aikio-Puoskari, & Helander, 1994; Seurajärvi-Kari et al., 1995). Although the Sámi population does not comprise a homogeneous cultural unit, the various Sámi groups do share many common cultural and traditional features as well as economic livelihoods (see Ingold, 1976; Lehtola, 2004), such as reindeer herding (although in Finland, unlike in Norway and Sweden, non-Sámi can also by law practice herding).

Research materials

The research which follows is based on personal interviews conducted with wilderness managers (WAMs) by telephone between January and February, 2015. The WAMs are public officers employed by the Metsähallitus with a key responsibility to follow and guide the use of wilderness areas based on legal acts and existing management plans. The interviews were relatively short (approximately 20–35 min) and structured around open-ended questions with a specific focus on potential user conflicts and related issues in wilderness thinking and management, management priorities and finally conflict management. The interviews covered all 12 wilderness areas and their responsible managers (maximum one per area), and are represented here to preserve confidentiality. Thus, although the unit of analysis is a wilderness area, any distinguishing traits of the specific area in question are not reported here. Similarly, as the number of interviewees is limited and they represent a specific and relatively easily identifiable group managing those 12 areas (and some other conservation units), the following results do not reveal any details germane to specific informants. Therefore, no identity/ identifying information is used, nor are any concrete numerical references to the level of manager opinions and perceptions, since some of the issues raised may be considered delicate by the informants and/or stakeholders. In cases in which all or most of the informants share the same view, expressions such as "majority" or "clear majority" are used. In principle, this leaves room for individual WAMs to be considered as not sharing a given mentioned opinion, preference or argument.

Based on the following results, there are no noticeable differences between the informants and their personal information, such as gender and age, but obviously slightly different management issues, conflicts and challenges do exist across different wilderness areas. Still, because the focus of the interviews was based more on the principal-level opinions and perceptions concerning the rationale and priorities in wilderness management and related use and management issues and challenges, particular differences and details of specific wilderness areas are not in the scope of this article. Secondary materials (e.g. legal documents) were used in the analysis only if the interviewees specifically referred to such sources in their argumentation.

Manager views: what wilderness areas are for?

Based on the interviews conducted, a clear majority of the WAMs identified conflicting elements between traditional Sámi livelihoods, namely reindeer herding, and tourism. Specifically, new emerging organized forms of nature-based tourism in wilderness, such as snowmobiling and dog sledges, and related development and infrastructure needs were seen to be problematic. The issues raised were mainly focused on the negative feedback from the reindeer herders and disturbance that tourism was said to be causing for the animals and the herding practices. In addition, tourism activities were seen as influencing and limiting the (free) movement of reindeer to winter pastures critical for their condition and survival during a long cold snow period. This issue has also been noted in previous rangeland research (Helle & Särkelä, 1992; see Akujärvi et al., 2014). There were, furthermore, issues raised with respect to occurrences outside the wilderness areas, such as increased reindeer road-kills due to seasonally intensive tourism traffic, but were potentially linked to the growing touristic use of the wild. However, although growing, the visitation numbers in wilderness areas were estimated to be relatively low compared with national parks, which have better accessibility and infrastructures. However, due to common and open access (i.e. no entry fees or official entry points), there is no statistical ground for these estimates. Still, the number of nature-based tourism companies in wilderness areas was known to be growing as the businesses need a permit to operate in the areas.

Another management conflict also related to reindeer herding. Almost half of the WAMs mentioned the issue of over-grazing and resulting environmental degradation in wilderness areas, which is a widely debated issue in Finland and other Nordic counties having reindeer herding activities in protected areas (see Akujärvi et al., 2014; Benjaminsen, Reinert, Sjaastad, & Sara, 2015; Johnsen, Benjamisen, & Eira, 2015). While it was perceived to be a problem, the issue was also seen as being under control and manageable: "Better range use planning will reduce the problems in future" (Interviewed WAM). The over-grazing issue was also linked to tourism as the existing and evolving tourist activities were partly seen as causing a lack of pasture rotation and congestion of reindeer to certain places (see Helle & Särkelä, 1992). While some WAMs noted that the number of reindeer may be too large for the current pastures, a clear majority of them understood the economic pressures behind the issue of maintaining the high numbers of reindeer: modern herding with various equipment and motorized vehicles, such as snowmobiles, ATVs (all-terrain vehicles) and even helicopters, is compelling the reindeer owners to have a maximum amount of animals. At the same time, state subsidies have been reduced and the price the herders receive from selling reindeer meat has not increased at the same pace as petrol and other related operational costs. Thus, an increasingly market-driven economy in the utilization of wilderness areas is not serving the reindeer herders but is potentially serving tourist operations.

In respect the research aims, the key interview issue for the WAMs was the question: what are the wilderness areas for? And more specifically, what is or are their priorities in the management of wilderness areas? While the question focusing on what the areas are for was introduced by the researcher, the (hierarchical) priority aspect was initiated by the interviewees during discussions. Although the term Anthropocene was not used in the interview questions, a rationale behind the questions was to learn how changing and intensifying human presence and impacts were understood and seen in relation to different uses of the wilderness areas, i.e. how localized notions of the Anthropocene are imagined and operationalized by the WAMs.

The indication of management priorities was unproblematic for the WAMs to express: a very clear majority referred to the Wilderness Act (Erämaalaki, 1991) and the need to safeguard Sámi culture and related livelihoods. In addition to the Wilderness Act, the WAMs supported their priority preference by mentioning the Metsähallitus Act), which they considered not only giving a mandate but rather a demand that the Metsähallitus and, thus, the wilderness area management takes the Indigenous people and their socio-cultural needs and values seriously. Indeed, the recently revised Metsähallitus Act (2016) states that:

> The use and protection of natural resources managed by the Metsähallitus must be reconciled in the Sámi Homeland as referred to in the Sámi Law (974/1995), in order to safeguard the conditions for practicing the Sámi culture and in the way that the obligations of reindeer herding law will be fulfilled.

In addition to the strong need to safeguard Indigenous culture and conditions for practicing such culture – namely reindeer herding – some of the WAMs mentioned that small-scale tourism is suitable for the wilderness areas. Thus, small-scale tourism operations were also supported in management decisions and practices so long as they were not in conflict with the core priority. Basically, this "small scale" referred to individual and/or unorganized recreational tourism modes, such as hiking, that were practiced in the areas before they were designated as wilderness areas.

The support for tourism was also based on the Wilderness Act (Erämaalaki, 1991) by the WAMs and its specific aim to develop versatile use of nature and possibilities for its economic uses. That legal frame was interpreted to refer to small-scale, unorganized activities. In cases where there was to be organized tourism, the WAMs expressed a specific emphasis on "community-based" operations, i.e. locally driven, small-scale (sustainable) tourism development. According to the WAMs, this would also require official agreements between the Metsähallitus and an operator, based on the "principles of sustainable tourism" (see Metsähallitus, 2017a), which provide an administrative tool for the management to guide touristic activities.

What about conserved wilderness?

There was no clearly emphasized priority towards nature conservation expressed by the interviewed WAMs. However, they may well take conservation as a granted target and, thus, have no need to consider its justification in wilderness management. This probably relates to the hegemonic conservation thinking in which the other potential uses of nature are evaluated based on how they can support or conflict with conservation goals. Still, this silence about conservation in priority discussions is somewhat surprising as the conservation element is also included in the above-mentioned legal acts: the same legal frameworks provided major support for Sámi culture and related practices in wilderness, and on a limited scale to small-scale tourism development as well.

It seems that the WAMs perceive wilderness areas as locally embedded natural resource management and land-use planning units. This may represent "a necessity as virtue" thinking: the management institution is legally obligated to provide support for the Sámi and their cultural uses and practices of the wild. In addition, the Metsähallitus needs to be open to versatile uses of nature in the wilderness areas. Interestingly, in some designated wilderness areas, such versatile uses even include forestry activities, but as the WAMs interviewed stated, the Metsähallitus took the internal decision to stop logging many years ago. This internal decision indicates that conservation thinking and related priorities do indeed exist, although they were latent in the interviews. Still, there were emphasized conflicts related to wilderness characteristics: for example, some of the WAMs expressed challenges between reindeer herding and the natural "pristine" conditions of wilderness areas.

However, as noted, these challenges were considered to be manageable (i.e. tolerable) in the current governance context. Indeed, due to the legal frameworks of the wilderness areas and their management, there is not much the Metsähallitus can actually do in terms of limiting the cultural and related economic uses of the wild. In contrast to this, the case with evolving and transforming tourism and its impacts on wilderness and wilderness uses is a different one. The legal framework provides support for activities such as tourism. There is also major pressure for supporting tourism development in Finnish Lapland. In the regional economy, the tourism industry is a major actor and it provides more employment than any other sector utilizing natural resources in Lapland, including forestry and mining (see Regional Council of Lapland, 2015, 2016). In addition, in the peripheral parts of the province, where the wilderness areas are located, there are no substantial alternatives for tourism. Thus, the regional developers and local politicians see and promote the industry and related investments as a last resort for creating development that contributes to the regional economy and employment in local communities.

All this creates pressures for the WAMs to consider different kinds of tourism development initiatives. The general governance model of the wilderness areas also frames them as "more open" for socio-economic needs and uses: unlike national parks, the legal wilderness areas are not conservation areas and, thus, more strict conservation law cannot be used as a key guiding argument in the management priorities and decisions. As noted, however, the WAMs were not highly supportive of tourism development, especially not for non-locally driven organized and commercialized forms of nature-based and adventure tourism. In Lapland, these forms include activities such as snowmobile and sledge dog safaris which have grown considerably in the last two decades (Regional Council of Lapland, 2016; Tyrväinen, Uusitalo, Silvennoinen, & Hasu, 2014). These forms of commercial tourism did not really exist in wilderness areas that existed when the Act was set in 1991, and they are the very same activities that the reindeer owners often regard as disturbing to their herding practices and pasture use.

The WAMs resolve this existing (or potential future) wilderness use conflict between reindeer herding and changing forms of organized tourism in a convenient way by positioning themselves as facilitators of stakeholder relations. These conflicting uses and meanings of the wilderness are positioned as external to management; in other words, the Metsähallitus is not part of the conflict and problem but rather a solution to these that aims to deliver all the necessary targets based on their legal obligations. Sámi culture and Sámi cultural practices in the wilderness are clearly prioritized by the WAMs which is a logical outcome in the current governance context of the wilderness areas. First, contrasting a general notion of "versatile use of nature", the cultural needs and uses are explicitly mentioned in the legal frameworks and obligations with a specific reference to Sámi Law from 1995, and thus, they need to be specifically safeguarded. Second, this minimizes the use pressures and impacts of tourism representing a new force in the wilderness use with unknown future impacts and risks.

Conclusions: political ecologies of wilderness management

> Wilderness areas are large, uninhabited, road less and almost pristine natural areas. They are meant to be maintained as such. Therefore, only traditional livelihoods such as hunting, fishing and reindeer husbandry are possible. (Metsähallitus, 2017b)

There is nothing apolitical in wilderness, particularly in wilderness management. Different ideas of the wild have an influence on and are guided by different priorities, meanings, understandings, uses and values of wildernesses. These meanings and use priorities are not static, nor are they historically or spatially unconditional. In traditional societies, the role and uses of wilderness environments may not have been problematic for local people (Nash, 1967) but, due to modernization and the recent scale of globalization and anthropogenic impacts, the uses and meanings of wilderness have become a contested landscape of various interests characterized by power relations and concomitant inclusions and exclusions.

This is particularly evident in the Finnish wilderness area management case in which the interviewed WAMs prioritized traditional cultural uses over evolving organized commercial forms of tourism. This distinction and hierarchy between different understandings of what wilderness areas are for was anchored in the legal governance frameworks of wilderness management in the country. These emphasized practices and meanings refer to the traditional idea of wilderness as a local economic resource and an integral part of socio-ecological systems of everyday life. For a long time, this traditional wilderness thinking has constituted the hegemonic view of wilderness environments and their utilization in the peripheral parts of the country, especially in the northern and eastern regions of Finland. In these parts of the country, even national parks, which have been established under nature conservation law, are open for various local uses, including hunting by local people. In the context of wilderness areas, the small-scale recreational tourism emphasized by the WAMs also has connections to traditional wilderness thinking: hiking as an activity involves historical links to fishing and hunting practices in Finland (see Hallikainen, 1998). In addition, both elements – the cultural practices of the Sámi and other local people, and individual hiking activities – existed in the areas prior to 1991 when they were designated to become part of the legal wilderness area network.

For these traditional uses, wilderness is a place with specific cultural identities and socio-spatial characteristics. In contrast to this, commercialized and organized forms of tourism can be seen differently. For tourism, wilderness is often an empty space (see Edwards, 1996; Wels, 2004) where various activities can take place, and where tourists are clearly visitors who do not remain (Saarinen, 2005). From the WAMs' perspective, these activities are new and represent initiatives or influences that are "global", i.e. something driven and initiated non-locally. Thus, they are also potentially beyond local control. Indeed, the tourism industry is increasingly seen as a global force with a capacity to effect significant impacts on the environment (see Gössling & Hall, 2006; Higham, 1998; Stonich, 1998). This massive capacity of commercial tourism is deeply embedded in imaginaries of the Anthropocene and its localized impacts for wilderness environments and identities. It can also lend us insights into how people involved in the tourism industry and its land-use strategies and practices conceive of the role that tourism might be able to play in the Anthropocene. Based on current discourses in the tourism sector, this self-awareness in which the industry becomes conscious of itself as a planetary force is unclear, not least as UNWTO (2016, p. 10) celebrates the recent achievement in global tourist numbers: "One Billion Tourists, One Billion Opportunities". In relation to the sustainable tourism and use of natural resources, however, "opportunities for whom" is a critical question.

Global imaginaries of the wild in touristic production and consumption are based on and reproduce images of pristine and untrammeled nature. The potential impacts of these kinds of images and organized tourism are difficult to demonstrate and analyze, but they are quite likely challenging for wilderness area management as they can be gradually evolving, indirect but still profound. This assumed high capacity of the tourism industry is based on structural power relations (see Harvey, 1996) between different stakeholders across a local–global nexus. A dominance by the tourism industry, supported by neoliberal conservation emphasis (Büscher & Dressler, 2012), can lead symbolically and practically to an exclusion of culture and people – often people other than tourists(!) – from nature and its utilization (see Neumann, 1998; Wall-Reinius, 2012). In this respect, a common characteristic of colonial depictions of distant natural places and wildernesses has been based on the descriptions of wildernesses as "no man's land" or as an "Eden" (see Mann, 2005). These kinds of depictions are still widely used in the touristic representations of "exotic" destinations (see, for example, Edwards, 1996; Gibson, 2010; Saarinen & Niskala, 2009), where natural landscapes are marketed without people present, or if there are people they are most likely "fellow tourists" admiring a view or doing specific tourist activities (e.g. quad biking, white-water rafting, mountain biking, four-wheel driving, snowmobiling, sledge dog rides) in the wild. In the Finnish context, the sledge dog safaris, for example, are culturally out of place. The activity effectively evolved with the emergence of French winter season tourism to Finnish Lapland in the late 1980s; in this context, the Arctic receives meanings and images from French historical and cultural connections with (present-day) Canada, where dogs have been traditionally used for transportation. But this has not been the case in the Finnish Arctic, where tourist resorts located next to wilderness areas and reindeer pastures nowadays often have hundreds of huskies on hand, ready to serve visiting tourists.

Commodified wilderness can exclude local resource uses and Indigenous communities, and include other stakeholders and activities in local and regional development (see Ramutsindela, 2007), resulting in unwanted inequalities and exclusions/inclusions in development (see Mowforth & Munt, 1998) in order to meet what is considered as "the wild" in the West. However, the key point of this paper has not been to focus on whether the indicated priorities and practiced inclusions and exclusions in wilderness management are "good or bad". This is obviously an ideological question in which the answer partly depends on the specific wilderness idea and selected stakeholder view. Taking a political ecology perspective, however, the purpose was rather to discuss, understand and demonstrate that there are different ideas and discourses of wilderness (see Bryant, 2000) which do not abide by the globally hegemonic fortress model alone. As I have aimed to show, these distinct ideas of what wilderness is and what it is for guide and structure the uses, meanings and values of natural resources as well as their control, management mechanism and conflicts in various ways (see Robbins, 2012). This involves a struggle for power, in principle, where some images and ways of using the wilderness succeed in the political process of denoting meanings better than others (Saarinen, 1998, 2005). Despite the different, often competitive and contradictory ideas and related uses of wilderness, there exists an understanding – a discourse – that is, in a certain moment of time and socio-spatial context, a more dominant way of thinking and acting than others.

In the current Finnish context, the traditional wilderness idea is the hegemonic discourse serving the needs of the Indigenous people and locally defined and valued uses. This is not a conflict-free situation, but in respect to academic critical notions on the domination of neoliberal conservation with integral links to transnational tourism industry, the emphasis on traditional and cultural uses in wilderness areas is an unexpected finding. And due to the present governance structures of Finnish wilderness areas, it is difficult to envision that the situation would dramatically change in the near future. Interestingly, local traditional uses have been recently "reintroduced" to some wilderness environments elsewhere. For example, in Krüger National Park (South Africa) established in 1892 based on the fortress model "fencing people out", local communities have been allowed to engage in seasonal natural resource harvesting (see Ramutsindela, 2014). But more importantly than individual cases, with their historically contingent issues and challenges (see Ramutsindela, 2002), there has also been structural-level transformations towards more socially sustainable and inclusive governance models in global wilderness conservation thinking. IUCN, for example, has recently renewed its management guidelines for wilderness protected areas, strongly acknowledging partnerships and local and Indigenous values and uses of these areas (Casson, Martin, Watson, Stringer, & Kormos, 2016, see also Jamal & Stronza, 2009; Reimerson, 2016). This is also what is intended by the Metsähallitus with the utilization of *Akwé: Kon* Guidelines (United Nations Convention on Biological Diversity, 1992) in Wilderness Area and National Park management in Sámi homelands. Although the hegemonic conserved wilderness idea influencing the management of a global network of areas is still largely based on a nature-culture/people distinction, these are promising initiatives and signs of alternative and changing governance models acknowledging the cultural and political nature of wilderness and their management.

Obviously, a universal wilderness governance model may not be realistic nor desirable as socio-cultural, political and ecological conditions vary. This calls for further research that would focus on the contextual and place-specific discourses and (also material) formations of wilderness values, socio-ecological situations and uses. The Finnish wilderness governance context may be highly specific, and commonalities even with the neighboring Nordic countries may be limited (see Hovik et al., 2010; Johnsen et al., 2015; Wall-Reinius, 2012). Therefore, international comparative and context-sensitive studies would be valuable to conduct in future. In addition, there are clearly issues and conflicts between wilderness management and local views and uses in the Finnish context too, which would probably be much more evident and problematic were they to be approached from the local – not the management – perspective. For example, the government's unwillingness to ratify the International Labor Organization's Convention on Indigenous and Tribal Peoples (No. 169), specifically the Articles 13–19 regarding rights to land (ILO, 1989), is highly disturbing for the Sámi and creates a ground for distrust, disagreements and conflicts in land-use planning in and outside the wilderness areas of the Sámi Homelands.

Thus, there is a critical need for local and Indigenous perspectives on the wilderness values and management (see Watson et al., 2011). The role of growing tourism "as the key driver of neoliberalism" (Duffy, 2015, p. 529; see also Bryant, 2000; Mosedale, 2011) in the changing management ideals and practices of wilderness areas would also require more nuanced research. Local communities are not homogenous concerning their views and needs in wilderness governance, and the same complexity most probably applies to the tourism industry having different kinds of business orientations and modes. Thus, in order to fully understand the use needs, values, priorities and discourses linked to wilderness environments and their governance, a plurality of perspectives and scales in research is needed. To sum up, in future research and governance discourses, it is important to evaluate critically the ways in which wildernesses and their use priorities are argued and justified from different perspectives amidst the constantly transforming contexts of natural resource management and tourism.

Acknowlegements

The author would like to thank the Guest Editors Mary Mostafanezhad and Roger Norum and the anonymous peer-reviewers for their constructive comments. M.Sc. Outi Kivelä prepared the map. The study is part of the Academy Finland's RELATE Centre of Excellence and Resource Extraction and Sustainable Arctic Communities (REXSAC) Nordic Centre of Excellence in Arctic research, funded by Nordforsk.

Funding

Academy of Finland, RELATE Centre of Excellence [project 307348].

References

Aikio, S., Aikio-Puoskari, U., & Helander, J. (1994). *The Sámi culture in Finland*. Helsinki: Lapin Sivistysseura.

Akujärvi, A., Hallikainen, V., Hyppönen, M., Mattila, E., Mikkola, K., & Rautio, P. (2014). Effects of reindeer grazing and forestry on ground lichens in Finnish Lapland. *Silva Fennica, 48*(3), 1–18.

Benjaminsen, T. A., Reinert, H., Sjaastad, E., & Sara, M. N. (2015). Misreading the Arctic landscape: A political ecology of reindeer, carrying capacities, and overstocking in Finnmark, Norway. *Norwegian Journal of Geography, 69*(4), 219–229.

Bertolas, R. J. (1998). Cross-cultural environmental perception of wilderness. *Professional Geographer, 50*(1), 98–111.

Blaikie, P., & Brookfield, H. (1987). *Land degradation and society*. London: Methuen.

Brondizio, E., O'Brien, K., Bai, X., Biermann, F., Steffen, W., Berkhout, F., … Chen, C.-T. A. (2016). Re-conceptualizing the Anthropocene: A call for collaboration. *Global Environmental Change, 39*, 318–327.

Bryant, R. L. (2000). Politicized moral geographies: Debating biodiversity conservation and ancestral domain in the Philippines. *Political Geography, 19*, 673–705.

Bryant, R. L., & Bailey, S. (1997). *Third world political ecology*. London: Routledge.

Buckley, R. (1999). Wilderness in Australia. In S. McCool, D. Cole, B. Borrie, & J. O'Loughlin (Eds.), *Wilderness science in a time of change conference* (Vol. 2, pp. 190–193). Odgen: Department of Agriculture, Forest Service, Rocky Mountain Research Station.

Buckley, R. (2006). *Adventure tourism*. Oxfordshire: CABI.

Buckley, R. (2007). Adventure tourism products: Price, duration, size, skill, remoteness. *Tourism Management, 28*, 1428–1433.

Budowski, G. (1977). Tourism and conservation: Conflict, coexistence, or symbiosis. *Parks,* 3–7.

Büscher, B. (2013). *Transforming the frontier: Peace parks and the politics of neoliberal conservation in southern Africa*. Durham: Duke University Press.

Büscher, B., & Dressler, W. (2012). Commodity conservation. The restructuring of community conservation in South Africa and the Philippines. *Geoforum, 43*, 367–376.

Butler, R., & Boyd, S. (Eds.). (2000). *Tourism and national parks: Issues and implications*. Chichester: Wiley.

Casson, S. A., Martin, V. G., Watson, A., Stringer, A., & Kormos, C. F. (Eds.). (2016). *Wilderness protected areas: Management guidelines for IUCN Category 1b protected areas*. Gland: IUCN.

Castree, N. (2012). *Making sense of nature*. London: Routledge.

Castree, N. (2014). The Anthropocene and geography I: The back story. *Geography Compass, 8*(7), 436–449.

Castree, N. (2015). Geography and global change science: Relationships necessary, absent, and possible. *Geographical Research, 53*(1), 1–15.

Castree, N. (2017). Unfree radicals: Geoscientists, the Anthropocene, and left politics. *Antipode, 49*(1), 52–74.

Cloke, P., & Perkins, H. (2002). Commodification and adventure in New Zealand. *Current Issues in Tourism, 5*(6), 521–549.

Connor, T. (2014). *Conserved spaces, ancestral places*. Pietermaritzburg: University of KwaZulu-Natal Press.

Convention on Biological Diversity. (1992). United Nations: New York. Retrieved from https://www.cbd.int/doc/legal/cbd-en.pdf

Cronon, W. (1998). The trouble with wilderness, or getting back to the wrong nature. In J. Callicott & M. Nelson (Eds.), *The great new wilderness debate* (pp. 471–499). Athens: University of Georgia Press.

Crutzen, P. J. (2002). Geology of mankind. *Nature, 415*, 23.

Crutzen, P. J., & Stoermer, E. F. (2000). The Anthropocene. *Global Change Newsletter, 41*, 17–18.

Dahlberg, A., Rohde, R., & Sandell, K. (2010). National parks and environmental justice: Comparing access rights and ideological legacies in three countries. *Conservation and Society, 8*(3), 209–224.

Douglas, J. A. (2014). What's political ecology got to do with tourism? *Tourism Geographies, 16*, 8–13.

Duffy, R. (2002). *A trip too far: Ecotourism, politics and exploitation*. London: Earthscan.

Duffy, R. (2015). Nature-based tourism and neoliberalism: Concealing contradictions. *Tourism Geographies, 17*, 529–543.

Edwards, A. (1996). Postcards – greetings from another world. In T. Selwyn (Ed.), *The tourist image: Myths and myth making in tourism* (pp. 198–221). Chichester: Wiley.

Erämaalaki (1991). Suomen säädöskokoelma no. 62 [Finnish Act collection no. 62]. (pp. 129–131). Helsinki: Valtion painatuskeskus.

Erämaakomitean mietintö. (1988). *Komiteamietintö* [Committee report] (p. 39). Helsinki: Valtion painatuskeskus.

Fennell, D. (1999). *Ecotourism: An introduction*. New York, NY: Routledge.

Forsyth, T. (2008). Political ecology and the epistemology of social justice. *Geoforum, 39*, 756–764.

Frost, W., & Hall, C. M. (Eds.). (2010). *Tourism and national parks*. London: Routledge.

Gibson, C. (2010). Geographies of tourism: (Un)ethical encounters. *Progress in Human Geography, 34*, 521–527.

Gössling, S., & Hall, C. M. (Eds.). (2006). *Tourism and global environmental change*. London: Routledge.

Gren, M., & Huijbens, E. (2014). Tourism and the Anthropocene. *Scandinavian Journal of Hospitality and Tourism, 14*(1), 6–22.

Hall, C. M. (1992). *Wasteland to world heritage: Preserving Australia's wilderness*. Melbourne: Melbourne University Press.

Hall, C. M., & Page, S. (2002). *The geography of tourism: Environment, place and space*. London: Routledge.

Hall, C. M., & Saarinen, J. (2010). Geotourism and climate change: Paradoxes and promises of geotourism in polar regions. *Téoros, 29*(2), 77–86.

Hallikainen, V. (1998). The Finnish wilderness experience. *Metsäntutkimuslaitoksen tiedonantoja, 711*, 1–288.

Harvey, D. (1996). *Justice, nature, and the geography of difference*. Oxford: Blackwell.

Helle, T., & Särkelä, M. (1992). The effects of outdoor recreation on range use by semi–domesticated reindeer. *Scandinavian Journal of Forest Research, 8*(1–4), 123–133.

Hendee, J., Stankey, G., & Lucas, P. (1990). *Wilderness management*. Golden, CO: Fulcrum.

Higham, J. (1998). Sustaining the physical and social dimensions of wilderness tourism: The perceptual approach to wilderness management in New Zealand. *Journal of Sustainable Tourism, 6*(1), 26–51.

Hovik, S., Sandström, C., & Zachrassion, A. (2010). Management of protected areas in Norway and Sweden: Challenges in combining central governance and local participation. *Journal of Environmental Policy & Planning, 12*, 159–177.

International Labour Organization (ILO). (1989). *Indigenous and tribal peoples convention No. 169*. Geneva: Author.

Ingold, T. (1976). *The Skolt Lapps today*. Cambridge, UK: Cambridge University Press.

IUCN (2013). IUCN protected areas categories system. Retrieved from http://www.iucn.org/about/work/programmes/gpa p_home/gpap_quality/gpap_pacategories/

Jamal, T., & Stronza, A. (2009). Collaboration theory and tourism practice in protected areas: Stakeholders, structuring and sustainability. *Journal of Sustainable Tourism, 17*, 169–189.

Johnsen, K. I., Benjamisen, T. A., & Eira, I. M.G. (2015). Seeing like the state or like pastoralists? Conflicting narratives on the governance of Sámi reindeer husbandry in Finnmark, Norway. *Norwegian Journal of Geography, 69*(4), 230–241.

Juntunen, S., & Stolt, E. (2013). *Application of Akwé: Kon guidelines in the management and land use plan for the Hammas-tunturi wilderness area*. Helsinki: Metsähallitus.

Latour, B. (2015). Telling friends from foes in the time of the Anthropocene. In C. Hamilton, F. Gemenne, & C. Bonneuil (Eds.), *The Anthropocene and the global environmental crisis: Rethinking modernity in a new epoch* (pp. 145–155). Abingdon: Routledge.

Lehtinen, A. (1991). Northern natures: A study of the forest question emerging within the timber-line conflict in Finland. *Fennia, 169*, 57–169.

Lehtola, V. (2004). *The Sámi people: Traditions in transition*. Inari: Kustannus-Puntsi.

Lorimer, J. (2012). Multinatural geographies for the Anthropocene. *Progress in Human Geography, 36*, 593–612.

Mann, C. (2005). *1491: New revelations of the Americas before Columbus*. New York, NY: Knopf.

Metsähallitus. (2017a). Principles of sustainable tourism. Retrieved from http://www.metsa.fi/web/en/sustainablenature tourism

Metsähallitus. (2017b). Lapin erämaa-alueet - kauas kaikesta [Lapland wilderness areas – faraway from everything]. Retrieved from http://www.luontoon.fi/eramaa-alueet

Metsähallitus Act (2016). Act on Metsähallitus (234/2016). Helsinki: Ministry of Agriculture and Forestry.

Moore, A. (2015). Tourism in the Anthropocene park? New analytic possibilities. *International Journal of Tourism Anthro-pology, 4*(2), 186–200.

Morton, T. (2012). From modernity to the Anthropocene: Ecology and art in the age of asymmetry. *International Social Science Journal, 63*(207–208), 39–51.

Mosedale, J. (Ed.). (2011). *Political economy of tourism*. Abingdon: Routledge.

Mostafanezhad, M., Norum, R., Shelton, E. J., & Thompson-Carr, A. (Eds.). (2016). *Political ecology of tourism: Community, power and the environment*. London: Routledge.

Mowforth, M., & Munt, I. (1998). *Tourism and sustainability: A new tourism in the Third World*. London: Routledge.

Nash, R. (1967). *Wilderness and the American mind*. London: Yale University Press.

Nepal, J., & Saarinen, J. (Eds.). (2016). *Political ecology and tourism*. London: Routledge.

Neumann, R. (1998). *Imposing wilderness: Struggles over livelihood and nature preservation in Africa*. Berkley: University of California Press.

Peet, R., Robbins, P., & Watts, M. J. (Eds.). (2011). *Global political ecology*. London: Routledge.

Public Law (1964). *Public law 88-577*. 88th Congress. Washington D.C.: The Senate and House of Representatives of the United States of America.

Puhakka, R., & Saarinen, J. (2013). New role of tourism in national park planning in Finland. *Journal of Environment and Development, 22*(4), 412–435.

Ramutsindela, M. (2002). The perfect way to ending a painful past? Makuleke land deal in South Africa. *Geoforum, 33*, 15–24.

Ramutsindela, M. (2007). *Transfrontier conservation in Africa: At the confluence of capital, politics and nature*. Wallingford: CABI.

Ramutsidela, M. (Ed.). (2014). *Cartographies of nature: How nature conservation animates borders*. New Castle upon Tyne: Cambridge Scholars Publishing.

Regional Council of Lapland. (2015). Lapin matkailustrategia 2015–2018 [Lapland tourism strategy 2015–2018]. Retrieved from http://www.lappi.fi/lapinliitto/c/document_library/get_file?folderId=2265071&name=DLFE-25498.pdf

Regional Council of Lapland. (2016). Lapin matkailutilastot 2016 [Tourism statistics in Lapland 2016]. Retrieved from http://www.lappi.fi/lapinliitto/julkaisut_ja_tilastot/matkailu

Reimerson, E. (2016). Sámi space for agency in the management of the Laponia World Heritage Site. *Local Environment, 21*(7), 808–826.

Robbins, P. (2012). *Political ecology: A critical introduction*. Malden, MA: Wiley-Blackwell.

Robbins, P., & Moore, S. A. (2013). Ecological anxiety disorder: Diagnosing the politics of the Anthropocene. *Cultural Geographies, 20*(1), 3–19.

Rockström, J., & Klum, M. (2012). *The human quest – prospering within planetary boundaries*. Stockholm: Bokförlaget Langenskiöld.

Rockström, J., Steffen, W., Noone, K., Persson, Å., Stuart Chapin, F., III, Lambin, E. F., ... Foley, J. A. (2009). A safe operating space for humanity. *Nature, 461*, 472–475.

Rutherford, S. (2011). *Governing the wild: Ecotours of power*. Minneapolis: University of Minnesota Press.

Saarinen, J. (1998). Wilderness, tourism development and sustainability: Wilderness attitudes and place ethics. In A. E. Watson & G. Aplet (Eds.), *Personal, societal, and ecological values of wilderness: Sixth world wilderness congress proceedings on research, management, and allocation* (Vol. I, pp. 29–34) (General Technical Report). Ogden, UT: USDA Forest Service, Rocky Mountain Research Station.

Saarinen, J. (2005). Tourism in northern wildernesses: Nature-based tourism development in northern Finland. In C. M. Hall & S. Boyd (Eds.), *Nature-based tourism in peripheral areas: Development or disaster* (pp. 36–49). Clevedon: Channelview Publications.

Saarinen, J. (2007). Tourism in peripheries: The role of tourism in regional development in Northern Finland. In B. Jansson & D. Muller (Eds.), *Tourism in high latitude peripheries: Space, place, and environment* (pp. 41–52). Oxon: CABI.

Saarinen, J. (2013a). "Tourism into the wild": The limits of tourism in wilderness. In A. Holden & D. Fennell (Eds.), *The Routledge handbook of tourism and the environment* (pp. 145–154). London, Routledge.

Saarinen, J. (2013b). Indigenous tourism and the challenge of sustainability. In M. Smith & G. Richards (Eds.), *Routledge handbook of cultural tourism* (pp. 220–226). London: Routledge.

Saarinen, J. (2016). Wilderness conservation and tourism: What do we protect for and from whom ? *Tourism Geographies, 18*(1), 1–8.

Saarinen, J., & Nepal, S. K. (2016). Conclusions: Towards a political ecology of tourism - key issues and research prospects. In S. K. Nepal & J. Saarinen (Eds.), *Political ecology and tourism* (pp. 253–264). London, Routledge.

Saarinen, J., & Niskala, M. (2009). Local culture and regional development: The role of OvaHimba in Namibian tourism. In P. Hottola (Ed.), *Tourism strategies and local responses in Southern African* (pp. 61–72). Wallingford: CABI.

Seurajärvi-Kari, I., Aikio-Puoskari, U., Morottaja, M., Saressalo, L., Pentikäinen, J., & Hirvonen, V. (1995). The Sámi people in Finland. In J. Pentikäinen & M. Hiltunen (Eds.), *Cultural minorities in Finland. An Overview towards cultural policy* (pp. 101–146). Helsinki: Finnish National Commission for UNESCO.

Shields, R. (1991). *Places on the margin: Alternative geographies of modernity.* London: Routledge.

Short, R. (1991). *Imagined country.* London: Routledge.

Spinage, C. (1998). Social change and conservation misrepresentation in Africa. *Oryx, 32*(4), 265–276.

Stonich, S. C. (1998). Political ecology of tourism. *Annals of Tourism Research, 25,* 25–54.

SæÞórsdóttir, A.-D. (2010). Planning nature tourism in Iceland based on tourist attitudes. *Tourism Geographies, 12*(1), 25–52.

SæÞórsdóttir, A.-D., Hall, C. M., & Saarinen, J. (2011). Making wilderness: Tourism and the history of the wilderness idea in Iceland. *Polar Geography, 34,* 249–273.

Tuan, Y.-F. (1974). *Topophilia.* New Jersey, NJ: Prentice-Hall.

Tyrväinen, L., Uusitalo, M., Silvennoinen, H., & Hasu, E. (2014). Towards sustainable growth in nature-based tourism destinations: Clients' views of land use options in Finnish Lapland. *Landscape and Urban Planning, 122,* 1–15.

Vail, D., & Heldt, T. (2004). Governing snowmobilers in multiple-use landscapes: Swedish and Maine (USA) cases. *Ecological Economics, 48,* 469–483.

Väisänen, R. (2013). Foreword. In S. Juntunen & E. Stolt (Eds.), *Application of Akwé: Kon Guidelines in the management and land use plan for the Hammastunturi wilderness area* (pp. 7–8). Helsinki: Metsähallitus.

Veland, S., & Lynch, A. H. (2016). Scaling the Anthropocene: How the stories we tell matter. *Geoforum, 72,* 1–5.

Wall-Reinius, S. (2012). Wilderness and culture: Tourist views and experiences in the Laponian World Heritage Area. *Society and Natural Resources, 25,* 621–632.

Wang, N. (2000). *Tourism and modernity.* Amsterdam: Pergamon.

Watson, A., Matt, R., Knotek, K., Williams, D. & Yung, L. (2011). Traditional wisdom: Protecting relationship with wilderness as a cultural landscape. *Ecology and Society, 16*(1), 36.

Wels, H. (2004). About romance and reality: Popular European imagery in postcolonial tourism in Southern Africa. In C. M. Hall & H. Tucker (Eds.), *Tourism and postcolonialism: Contested discourses, identities and representation* (pp. 76–94). London: Routledge.

Wilson, R. K. (1999). "Placing nature": The politics of collaboration and representation in the struggle for La Sierra in San Luis, Colorado. *Ecumene, 6,* 1–28.

World Tourism Organization (UNWTO). (2016). *UNWTO annual report 2015.* Madrid: Author.

Tourism and environmental subjectivities in the Anthropocene: observations from Niru Village, Southwest China

Jundan Zhang

ABSTRACT

The increasingly popular notion of Anthropocene urges us to reflect and review the role of the human, the Anthropos, as part of the planet earth. In this context, tourism has been singled out as a global industry that is driven by neoliberal economic principles and is inevitably intertwined in the production of the Anthropocene. At the same time, tourism has been adopted also as part of environmental governance and management, aiming for a more sustainable economy. Based on the idea that ecotourism contributes to the discourse of "nature" (and Anthropocene) disruptively as well as productively in unsettling the normative ideas of "nature" and "culture", in this article I attempt to understand more specifically how ecotourism may enable individuals' subject formation in relation to the broader environmental discourse. Drawn on fieldwork in Niru Village, Shangri-La, Southwest China, I employ a political ecology approach and examine the ways individuals relate themselves to "nature", through a process of negotiation and exchange with others engaged in ecotourism activities. The tourism encounters in Niru Village, therefore are also embodied encounters of different environmental subjectivities.

Introduction

In this paper I look at how, as recent studies discuss (Gren & Huijbens, 2014, 2016; Huijbens & Gren, 2016; Moore, 2015), the notion of Anthropocene may help us make sense of the Earth through tourism practices and development. Derived from the natural sciences, the notion of Anthropocene is proposed for naming the new geological epoch in which "the Anthropos", humanity, is the main operating force (Crutzen, 2002; Crutzen & Stoermer, 2000; Zalasiewicz et al., 2011). Condensing all disparate environmental crises into a single word (Davies, 2016), the Anthropocene once again evokes familiar themes such as sustainability, conservation and resilience within the studies of sustainable tourism (Hall, 2016; Svensson, Sörlin, & Wormbs, 2016). Immediate responses regarding the relationship between tourism and the Anthropocene are often directed at the most apparently relevant elements in the tourism industry, such as energy and resource consumption (Eijgelaar, Amelung, & Peeters, 2016) and the ethical behaviours of tourists and tourism operators (Baruchello, 2016; Kristoffersen, Norum, & Kramvig, 2016). What is more relevant, yet less discussed, is how the Anthropocene on local levels is shaping our perception, decision-making processes and actions in tourism studies and practices.

This article therefore draws attention to local villagers and tourists who took part in my 2013 ethnographic fieldwork in Niru Village, Shangri-La City, Southwest China. Remote from the global centres where discourses on the Anthropocene are forming, Niru Village nevertheless offers local accounts of it in how the villagers show their awareness regarding recent natural disasters and environmental degradations. However, as I will discuss below, this awareness is illustrated through the villagers' increasing interactions with tourists and their other involvements in the local tourism development. I will then elaborate on individuals' ideas and attitudes towards "nature" and how they form certain sets of environmental subjectivities, or social formations, mediated through ecotourism practices. Fifty-three kilometres from the central town of Shangri-La, Niru villagers have historically depended on the forest resource and have developed a semi-farming, semi-grazing, structured economy; although, in recent years, this economy has taken a turn towards tourism. In contrast to the well-established ecotourism products in Shangri-La City (thanks to its name-changing strategy), such as Pudacuo National Park and Dukezong Old Town, Niru Village has struggled and finally come up with a niche with which to brand itself, as "Shangri-La's backyard" (xianggelila de houhuayuan) (Zhang & Tucker, 2018). During the process of developing ecotourism products, individuals are brought together in Niru Village as tourists and hosts, who may articulate differently on the subject of "nature" yet are somehow able to exchange and negotiate these differences through their tourism encounters.

I have argued, from a poststructuralist political ecology perspective (Zhang, 2016b), that people as carriers of ideas and ideologies are crucial players in the hybridization of different regimes of "nature". This means that tourism (and tourism destinations) should always be viewed within a vibrant environmental discourse, since individuals constantly mingle and negotiate with their biological and social others through tourism products at tourism destinations (Zhang, 2016a). Bearing this in mind, in this article I see and use the notion of Anthropocene as further adding to the ever-expanding forms of discourse on "nature", both deriving from and contributing to these discourses. In other words, I believe the significance of the term Anthropocene lies more in opening new ways of understanding and responding to the modern ecological catastrophe than in accurately defining the geological epoch.

Thus, this study seeks to contribute not simply to the growing literature on tourism and the Anthropocene, but more generally to the long-existing discussions on tourism and environmental discourse. This is done in three aspects. First, with the intention of bringing forward the individual perceptions of and engagements with the broader environmental discourses, I look at environmental subjectivities from a political ecology perspective. Environmental subjectivities address the formation process of how individuals construct and reconstruct a set of discursive relationships with "nature". Second, the focus on environmental subjectivities provides local contexts within which the metanarratives of environmental discourses, such as Anthropocene, operate. This will help us understand how new possibilities of subjectivities appear when people's livelihood and lifestyle are challenged in a context of large-scale modern practices and development. In the case of Niru Village, increasing social mobilization and the introduction of new conservation policies and technologies offer an opportunity for a re-examination of people's ideas about the role of "nature" that closely interrogates the binaries of nature/culture and Indigenous/modern. Finally, studying environmental subjectivities may shed light on new collaborations that emerge from individual encounters. This in turn will elucidate the diversity that has always existed and contributed to environmental discourses.

Political ecology of tourism in the Anthropocene

In Castree's (2014a, 2014b, 2014c) trio of Anthropocene papers, he presents a comprehensive picture of the concept's past, present and future. In a nutshell, Anthropocene was coined by Eugene Stoermer in the 1980s, and then gained prominence when Nobel Prize-winning atmospheric scientist Paul Crutzen proposed that, after the Holocene epoch, the Anthropocene is the new era, with humans being the predominant force shaping the Earth's surface (Crutzen, 2002; Crutzen & Stoermer, 2000; Zalasiewicz et al., 2011). While the term remains to be verified within geological and environmental sciences, its symbolic charge and popular appeal have long since left behind the verification processes of natural science, and the term has gained increasing credence both within and outside academia. As Castree (2014a) notes, the term can be seen as "a new, more graphic way to frame an existing idea, namely, that of 'global environmental change' caused by human activities and extending in causes and effect beyond climate change" (p. 441). Purdy (2015) rightly points out that the term Anthropocene is similar to "the environment", which gained prominence in the 1960s, and is a call to take responsibility for a changing planet (p. 4). In a sense, Anthropocene urges us to reflect on and review the role of the human, the Anthropos, as part of planet Earth, from local community policies to global industrial franchises. In this context, tourism has been singled out as a global industry that is driven by neoliberal economic principles and is inevitably intertwined in the making of the Anthropocene (Eijgelaar et al., 2016; Hall, 2016). At the same time, tourism has also been adopted as part of environmental governance and management, aiming for a more sustainable economy (there are many examples of this in the *Journal of Sustainable Tourism*). Haraway (2015) proposes that our job as world citizens is to "make the Anthropocene as short as possible and to cultivate with each other in every way imaginable epochs to come that can replenish refuge" (p. 160). What tourism can contribute in fulfilling this task is intriguing.

As mentioned above, recent works within tourism studies have begun to outline tourism's planetary impacts (Eijgelaar et al., 2016; Hall, 2016) and how an Anthropocene ethics can be mediated through nature-based tourism practices (Kristoffersen et al., 2016) and tourism destinations that provide different understandings of landscape (Svensson et al., 2016). To problematize the seemingly straightforward role tourism plays in the Anthropocene, several scholars have explored the question through different approaches in order to address the responsibilities tourism may take on in the proposed epoch. Jóhannesson, Ren, and Van Der Duim (2016) employ Actor Network Theory, regarding tourism encounters as being produced, and producing, within the complex and fragmented notion of the Earth. In a similar vein, Jönsson (2016) argues that the abstraction and generality of the idea of Anthropocene can be problematic when simply added as a background in tourism studies, as it is the particular processes in environmental transformations that tell stories about power relations and inequalities within tourism development. It is through a focus on the particularities that we, as tourism researchers, can address tourism's responsibilities in the Anthropocene and what comes after.

Following a line of argument similar to Jönsson (2016) and Jóhannesson et al. (2016), in this paper I employ political ecology to scrutinize some particular encounters among individuals and their environments, enabled by ecotourism, in Niru Village, Shangri-La, Southwest China. Political ecology frameworks have been used to conceptualize and contextualize tourism within the discourse of environment (Mostafanezhad, Norum, Shelton, & Thompson-Carr, 2016; Nepal & Saarinen, 2016; Stonich, 1998). Historically situated and place-based, political ecology frameworks are useful in examining relationships between and across actors at different scales within tourism practices, and can help us tackle the intricate dilemmas existing within sustainable tourism as an alternative form of development (Mostafanezhad et al., 2016). A growing body of literature investigates the position of nature-based tourism in green economies and green governmentality (Law, De Lacy, Lipman, & Jiang, 2016; Reddy & Wilkes, 2015). However, I have suggested that it is important to also recognize the poststructuralist elements in political ecology and to rethink ecotourism within the local-global and nature-culture dialectics (Zhang, 2016b); that is, to consider ecotourism as a constant hybridization of different regimes and ideologies of "nature" and to recognize tourism's contribution to the discourse of "nature" (and Anthropocene) as disruptive, but also productive, in unsettling the normative ideas of "nature" and "culture" (Zhang, 2016b). One way to understand how this process is realized is to examine the ways individuals relate to "nature", through negotiation and exchange with others in tourism activities. Therefore, in this paper, I specifically explore the process of negotiating, and interaction between, the assemblages of ideas, attitudes and actions, which, in political ecology studies, is contained within the concept of environmental subjectivity.

Tourism and environmental subjects in Niru Village, Shangri-La City, Southwest China

Here, my analysis is based on stories from a village located in the rural area of Shangri-La City, Southwest China. Since its name change (and tourism branding) in 2001, Shangri-La City has attracted a great deal of attention from tourists and researchers alike. Since the establishment of the People's Republic, Shangri-La and its surrounding regions, together with other peripherals, have acted as raw material suppliers for infrastructure development in the coastal areas. The exploitation of natural resources led to increasing numbers of natural disasters and eventually a devastating flood of the Yangtze River in 1998, claiming thousands of people's lives and causing tremendous economic loss (Zhao & Shao, 2002). Following this was a logging ban by the central government, and a series of policies and plans to install a "green economy", of which tourism is a crucial part, to provide substitute income (Mu, Yang, & Zhou, 2007). Now, millions of tourists visit in order to admire and worship the alpine grassland, Snow Mountains and grand gorges.

The historical background of (eco)tourism development in Shangri-La, in combination with a top-down approach of conservation, reflects what Xu, Cui, Sofield, and Li (2014) maintain in their paper, that ecotourism development in China is closely associated with the country's progression towards modernity and is strongly influenced by global environmental movements. In this context, ecotourism products are often packaged with introduced Western forms of conservation, such as national parks, which also introduces a dilemma as it is weaved in the building of an "ecological state" in China (Zinda, 2014). It has been increasingly recognized that the current ecotourism management model in China is based on Western ideas of human–nature relations and could be more efficient if it integrated traditional Chinese culture, or sacred ecological knowledge (that is, beyond ethnic Han Chinese), into conservational tools (Grumbine & Xu, 2011; Xu et al., 2005, 2014). Such discussions are again part of the environmental discourses in which a hybridization of different regimes of "nature" is constantly happening. It is important to remember here that economic and ecological transformations are carried out *through* tourism.

Located within this context of tourism development, Niru Village caught my attention during my 2013 fieldwork in Pudacuo National Park in Shangri-La. As a consequence of a shift from the logging industry towards a tourism-oriented economy, Pudacuo National Park, a 50-minute drive from the central town of Shangri-La, was established in 2007 as the first national park in Mainland China. Because of its geographic proximity and the ecological compensation agreement the park made with nearby villages, some villagers from Niru Village were given jobs as bus drivers and bus interpreters in Pudacuo. I got to know one of the bus drivers, and he and his family introduced me to the village. While Niru Village has also experienced the ecological and economic transformation as a general result of China's environmental governance, the desire for and interest in tourism development have been greatly influenced by the opening of Pudacuo National Park. First and foremost, the park demonstrates a shift in the regional economy, which makes nearby villages like Niru Village acutely aware of the possibilities offered by tourism. Second, the economic success of the park affects the incomes of the Niru villagers in both direct and indirect ways; and as I will discuss later, this generates both hope and frustration in terms of how much the villagers can participate in their own "tourism dream" (Zhang & Tucker, in press). Last but not least, combining a protected area and ecotourism, Pudacuo National Park exemplifies that "nature" can make money in a less destructive way. Together with other events, this brings the metanarrative of environmental changes closer to local villagers' everyday life.

Fifty-three kilometres from the central town of Shangri-La (Compilation Committee of Luoji Township Annals, 2007), Niru Village covers 406 km²; however, only 0.16% of this area is arable (Lu, 2005). All villagers in Niru Village are ethnic Tibetans, making it the only solely Tibetan-populated village in Luoji town, and the largest Tibetan-populated village in Shangri-La. Six hundred and forty-three villagers live in 108 households in three village groups, namely Nizhong, Baizhong and Pula (Compilation Committee of Luoji Township Annals, 2007). Historically reliant on the forest resource, farming and husbandry, the villagers in Niru Village have built their knowledge, beliefs and everyday life around their interconnections with the environment. As mentioned above, the recent tourism development, such as Pudacuo National Park in Shangri-La, has nevertheless permeated Niru Village; however, it is a process not without its hitches.

While the tourism in Shangri-La has developed rather dramatically (Hillman, 2003; Kolås, 2008), Niru Village struggles to attract tourists even while packaging its remoteness and wilderness with other types of place-branding. Before 2002, there was no road to the village. The inaccessibility, and thus the undisturbed originality, caused the village to be selected by the regional government in order to showcase the universal value of an ecological environment to the UNESCO investigation team in the process of applying for the region to be named a natural World Heritage Site (Yang, 2012). The UNESCO team's visit to Niru Village played a major role in the successful listing of the Three Parallel Rivers of Yunnan-Protected Areas, but the tourism development promised in the local government's rhetoric has not been realized to the extent of the villagers' anticipation (Zhang & Tucker, in press). Indeed, the villagers were not only disappointed by the fact that fewer tourists than expected travel to Niru Village, but were also confused about whether or not the application for the UNESCO listing had been approved. Zhang and Tucker (2017) argue that the UNESCO outstanding universal values criteria have singularly prioritized the physical and biological aspects of the local environment, which is a dualistic view of "nature" the local villagers do not share. Consequently, this dualism influenced the selection process and the construction of identities at the local level. Although Niru villagers were proud that their village was selected for the UNESCO visit, thus allowing them to identify themselves as "Niru Tibetans", they soon realized that UNESCO does not see "culture" and "intangible heritage" as part of the celebrated universal values of natural heritage. As I observed in my fieldwork, identity construction is thus shifted in an environmental direction, through which the local villagers can negotiate with the dominant discourse on "nature" as illustrated in the outstanding universal values.

The very move of changing the direction of self-identification, together with the will to negotiate with the broader environmental discourse, can be formulated using one of the themes within political ecology, namely the "environmental subject" (Robbins, 2012). Broadly speaking, the topics of environmental subject explore "the way that people's behaviours and livelihoods (their actions) within ecologies influence what they think about the environment (their ideas), which in turn influences who they think they are (their identities)" (Robbins, 2012, p. 216). There is only one space and one world that all people are dwelling in, but the way of engaging with the world is never fixed and singular. Working in a poststructuralist direction, Agarwal (2005) suggests that, although practices are always situated in the context of a construct of expectations and obligation and political relations, "to point to the situatedness of practices and beliefs is not to grant social context a deterministic influence on practice and subjectivity. Rather, it is to ground the relationship among structure, practice, and subjectivity on evidence and investigative possibilities" (p. 172). Such an interpretation opens up the endless circle of the producing and the produced. Robbins (2012) suggests that engaging with theories of subject is crucial for a better understanding of how certain environmental actions and identities fit together as the products of power. In the process of production, "the subject" often becomes a vague entity, for there is never a point at which we can regard "the subject" as a fixed status; rather, it is always in the process of formation. Fletcher (2014, pp. 61–62) implies that Foucault's theories on subjectivation (Foucault, 1983) may be read in a dialectical way: the internalized subjectivity can be exercised and transformed into a set of future subjectivities that either consent to or diverge from the formal "pre-existing subjectivity."

To better understand and articulate this process, I will examine the narratives from my informants in relation to recent discussion of the notion of "the Green Tibetan." Originally emerging in the 1980s (Huber, 1997, 2001), "the Green Tibetan" re-emerged in the late 1990s and early 2000s as a "a rapid mobilization of international and domestic environmentalists and transnational investors is accompanied by intensive Chinese and foreign scholarship, media coverage, and explosive growth in tourism" in northwest Yunnan (Coggins & Zeren, 2014, p. 211; Litzinger, 2004; Moseley & Mullen, 2014; Yeh, 2014). In short, "the Green Tibetan" is a set of representations in which Tibetans are perceived as resourceful in their ecological knowledge and their harmonious ways of living with nature (Huber, 1997). However, the recent trend of using the term addresses more the connection between environmental awareness and the local Tibetan ethnic identity.

Some methodological notes should be made here. Although the bulk of my ethnographic fieldwork was conducted between April and October 2013, other field trips both before and after have also contributed to my understanding of the context of tourism development in Shangri-La and Niru Village. Pursuing ethnographic research in Niru Village, engaging with people and their environment, meant I had to situate myself, and others, in a culturally, socially and politically constructed setting. However, situating does not mean the ethnography is framed within a particular period of time or location. Keeping in mind the ever-present ethical dilemmas, the ethnography started even before I was physically in Shangri-La and Niru Village, and continued to demand my attention and devotion well after I had physically left these places. My belief that human experiences and meaning-making are rooted in people's interpretations, activities and interactions involving themselves and others has resulted in the intersubjective nature of this study. In Niru Village I was a "Jia", which means outsider, visitor, tourist and ethnic Han all at once. In order to make myself familiar to my informants I lived with a local family, engaged in daily chores, and worked with them in the fields. This familiarity eventually helped me gain access to other families/individuals in Niru Village. Once I was more familiar with the villagers, I was able to openly engage in participant observation and interviews.

Tourism encounters in Niru village

After the series of natural disasters mentioned above, and due to overexploitation, the central government started adopting tourism as a way to turn the economy green. Starting in the late 1990s, the ecotourism development gained increasing support and investment from the state, as well as from international environmental organizations, often working closely with nature reserves and later through a new model of national parks (Zhang, 2013). The idea of tourism as a sustainable alternative economic activity has now been transformed into the everyday discourse of Shangri-La's tourism development (Zhang, 2016a). In Niru Village, the changing ways of utilizing natural resources, and the pending worries and concerns about the natural environment, continue to shape the villagers' daily discussion topics and practices. For example, unusual natural events are often mentioned to express their worries, and nearby mining constructions have always been a concern because of their negative impacts on the environment. As noted in the notion of "the Green Tibetan", local villagers in Niru Village have closely coexisted with nonhuman beings for generations and the local religion, Bon, still spiritually influences their ideas of "nature"; thus, disturbing nature is both spiritually and physically unacceptable. However, as Woodhouse et al. (2015) suggest, there is a more complex picture available concerning the potential collaboration between local geopiety and environmental protection, and "local people should be able to negotiate and engage in the terms for conservation and natural resource management themselves" (p. 305). Below are three sets of events, narratives and scenarios of tourism encounters around Niru Village. Through retelling these narratives, and reading them alongside the notion of "the Green Tibetan", I attempt to shed light on how tourism can be a productive space for the villagers to represent and remake their environmental subjectivities and, consequently, contribute back to the environmental discourse.

The checkpoint

My first visit to Niru Village was with Wu, a key informant in my fieldwork in Pudacuo National Park and later my host in Niru Village. After several hours traveling on the road, Wu stopped his minivan in front of a checkpoint, made from a spruce trunk. A tent was set up beside the checkpoint and two villagers came out, smiling and talking to Wu in Tibetan. Following Wu, I stepped out of the minivan. The two villagers nodded at me and asked Wu who I was. Wu said "*Jia!*" to his village fellows. As we entered the tent, Wu pointed at the spruce trunk and explained to me:

This is the protection checkpoint we set up ourselves. Because in Niru we still have so many endangered species, like animals, plants and medicinal vegetation, there are many people from outside who come in for poaching and illegal logging. So we set up this tent to scan every vehicle coming and going, every day, day and night. Every household sends one person, and three people take turns staying here for three days. We've done this for many years. The villagers had a meeting to discuss the system and voted to pass it. It's all voluntary.

I listened, and was intrigued. Inside the tent it was dark and simple, with a small fire and some sleeping mats lying around. I asked if they really did this every day. Wu replied quickly "Of course! Those thieves have stolen too much from us so we had to do something. Ever since we built this checkpoint the poaching is much less." I nodded. At this point, he asked me meaningfully:

What do you think of this? Shouldn't you say that we have very good environmental awareness?

Pondering the purpose of his questions, I responded with positivity, "Yes! It's very good, very pioneering, and very strong environmental awareness!" Wu seemed satisfied, but still wore a bitter smile on his face, saying:

It won't last very long if there's no further development.

At first glance, Wu presented a voluntary and genuine stewardship of villagers regarding their "nature", just like the "Green Tibetan" notion would suggest. But the Tibetan Buddhism explanation is absent. In 2000, the Natural Forest Conservation Program was launched, as a 10-year programme to completely ban logging across 41.8 million hectares of natural forest, including the area where Niru Village is located. The villagers are allowed to take only limited amount of wood for their own house building, but are still allowed to harvest mushrooms and herbs in the forest (those which are economically valuable). Thus, from a localism logic, it is understandable that they want to guard the natural resources.

However, more layers are added by Wu's questions about environmental awareness and his comment that it would not last long, especially when put in the context of sustainable development. From what Wu presented, the natural resources in Niru Village's territory need to be guarded for their potential economic value; but based on what measurement? Having worked in Pudacuo National Park for many years, and interacted with many tourists from big cities, Wu is very familiar with the modern concept of conservation and its combination with ecotourism. Thus, it is very likely that it has become part of his perception that nature can be utilized not only as raw materials but also as aesthetic entities. Wu's concern for no further development involved the low number of tourists coming to Niru Village. Having "environmental awareness", in his opinion, must be attractive to me as a tourist. When he subtly lamented that the self-organized conservation would not last long, he was actually expressing a will to realize the economic value of the natural resources through ecotourism. In this context, "the Green Tibetan" is presented as the conventional/normative environmentalist thinking. And, later, after talking more with the villagers about the checkpoint, I realized many of them, especially young people, do not share the wish to "do something with the natural resources." For example, I heard comments like:

When it was bad (outsiders' poaching) before, it was necessary to do something. Now that it's gotten better it's kind of a waste of time to do it. …better to use our time and energy to do something else.

These are doubts as to the necessity of guarding their forests and resources. But what is this "something else" that is more valuable? Many suggested it would be tourism. And when I posed the question "What is a good life for you?" to many villagers in different circumstances, one of the answers I heard many times was:

City people want to come here, and we want to go to the city.

Here, they are referring directly to tourists from big cities. Although only a few hundred tourists come to Niru Village every year, the Niru villagers want to be more engaged with tourism and to generate new ways of expressing themselves. Not only are they saying "what they want"; they are also juxtaposing this with "what other people want." The juxtaposition of "city people" and "we" creates a contrast, a tension and eventually an irony. The irony between city people and the villagers is reflected in the acute awareness of the villagers regarding the great paradox in ecotourism – that one person's desire to escape the urban reality is used to fulfil another's desire to obtain an urban lifestyle.

Yet, there is still a negotiation process between modern and Indigenous ideas of "nature". Later, it became clear that Wu holds an important position in the religious and spiritual practices in Niru Village. However, when the checkpoint episode happened I was merely a tourist and a visitor to the village, and Wu did not talk about the spiritual connections between the villagers and "nature". Rather, he chose to communicate in terms he assumed would be more familiar to me – talking about "environmental awareness", "ecological village" and so on. For him, "nature" is spiritual, economic, and social all at once; therefore, his subjectivities are much richer than what "the Green Tibetan" alone would represent. He is perhaps proud of his culture, but somehow feels shame at the lack of development in Niru Village. His beliefs, knowledge, experiences and assumptions regarding others' perceptions of "nature" were interrelated and displayed through his encounter with me, a tourist.

Against the background of the Anthropocene, environmental activism seems to be more logical than ever. However, the story of the checkpoint shows that different views may exist at the local level. In contrast to the protests in the Qinghai–Tibetan Plateau (Liu, 2012), reported in *chinadialogue*, whereby local residents refused to have tourism development on their sacred lake, Niru villagers look forward to a new method in order to upgrade their quality of life, possibly engaging with the "integration of Indigenous knowledge and conservation", or "the Green Tibetan."

"Constructing beauty"

If the story of the checkpoint reflects a process of negotiating and expressing ideas about "nature" (or values of nature) between Wu as a host and me as a tourist, my conversations with Uncle Denzeng in Niru Village present another understanding of this process, but among the villagers. Operating a family tourism business with his son and daughter-in-law, Uncle Denzeng was one of the first people in Niru Village to start developing tourism, as he describes:

> Many tourists came in 2002 and 2003. There was nowhere for them to live back then, so we arranged for them to camp in the schoolyard. I was still working in town, but I already knew I wanted to do this (tourism)…The most important thing is to protect the environment (because) that's what tourists want to see. But every time I talked about environmental issues at village meetings, people were upset and against me. They said, "We have to fetch firewood, build houses, feed cows!" I told them that protecting nature is good for them. But they didn't understand it. They cannot think. They don't have scientific knowledge.

Again, the environmental protection Uncle Denzeng advocated here can be interpreted as what "the Green Tibetan" entails, but here he openly criticized other Tibetan villagers and regarded them as backward. Knowing that many of the practices he criticized were in fact carried out in a sustainable manner before the 1950s, when the logging industry changed people's ways of living, I nevertheless asked Uncle Denzeng if there was an element of tradition at play there. He asserted that it should nevertheless not be excused, and that it was the village cadres who were to blame. According to him, none of the cadres were creative and innovative enough to lead the villagers to a better life. "What is a better life?" I asked him. He answered:

> A better life is to follow what the central government requires, for example, now there's a national project to promote "Beautiful China (měilìzhōngguó)" and "Beautify Villages (měilìxiāngcūn)." First, it has to be beautiful people. But people here can never be beautiful. So what can we do? We have to make nature beautiful. We need to do a lot of things to do that. Replanting those trees that were cut before, and not littering, etc. …we have to construct beauty (zàoměi). Otherwise, people come here and they've read on the Internet that Niru is "Heaven on Earth, the last paradise, the real Shangri-La" and then they see that the reality is like this – dusty, ugly and empty… In the winter, the whole village is grey and dusty. I've heard many tourists say that the reality was different from what they expected, and this is ridiculous.

It is obvious here that Uncle Denzeng was influenced by the official regime of romanticizing rural areas, as well as believing he knew tourists' views on what Niru Village should be like. This led to his emphasis of the phrase "construct beauty", particularly when related to "natural beauty." As a local-born Tibetan, Uncle Denzeng went to work in town at a young age and came home after he had retired. He seemed to have rather negative opinions of his fellow villagers regarding how they handle "nature", and the "traditional way of living", and felt the self-contented lifestyle is backward. Viewing traditional resource management as constraining people in the position of being regulated, Uncle Denzeng advocated a more active position entailing the villagers regulating their own environment. As an innovator in the village, he tried to appropriate not only his own home environment but also the village's surroundings. For instance, he made road signs to direct tourists to the right trek to a lake, on which he signed "made by Denzeng". He believed that improving the facilities is fundamental for ensuring that tourists are satisfied and that exploring the beauty of "nature" will, in turn, help the village become a better place and help the villagers have a better life.

However, other villagers expressed rather different views on Uncle Denzeng's business, which today attracts nearly a thousand tourists per year. The villagers claimed that local rubbish had been discarded by Uncle Denzeng's tourists rather than by them. One villager stated: "And they talk loudly and shout when they go to our sacred sites." "That's absolutely forbidden, the gods will be angry and our crops will grow poorly!" Interestingly, these villagers were also using the "nature" argument to negotiate what tourism should be like. For these villagers, tourists are welcome but must follow certain rules. We can thus see "the Green Tibetan" operating here as well, with an emphasis on the "Tibetan" part. The villagers' complaints also revealed that it was Uncle Denzeng's attitude that was the most provocative to them, as he spoke as if he knew best. Uncle Denzeng was not the only one trying to develop tourism entrepreneurship, but he was not approved of because he differentiated himself from the other villagers, portraying himself as "cultivated" and "modern" and the others as "uneducated" and "premodern". The other villagers also took offence at his appropriating the environment and turning it into what tourists wanted to perceive and experience. In other words, Uncle Denzeng was trying to "modernize" and "ecologize" the village into a metropolitan tourist's product for so-called green consumption, and the villagers' narratives in fact indicate a willingness to negotiate and take part in the place-making process.

It is useful to pause here for a moment; when Uncle Denzeng talked about "making beauty" and "constructing nature", was he acting as the agent that was going to construct nature, or had he been enticed by the subject position of "constructing nature"? Like Wu, Uncle Denzeng promoted the idea of ecotourism, but unlike Wu's ambiguous position between traditional and modern, Uncle Denzeng was explicit in expressing his willingness to change the villagers' current livelihoods, even the human–nature relationship in the traditional sense.

Uncle Denzeng's story is informative, and shows that within a seemingly self-regulated community dramatically different opinions can be generated. While Uncle Denzeng is aware of the external "reality" – that tourists desire a remote and pure "nature" experience – the villagers seem to have difficulty taking in the whole package of ecotourism, which entails some ideas implying that their way of life is backward and primitive. Also, it shows that through the process of negotiating with other villagers, Uncle Denzeng identifies more with the "scientific" way of relating to nature. When asked about the local deities residing in the mountains and rivers, he said "if throwing rubbish in the mountains didn't give them (the villagers) bad luck, perhaps there's no god then", at once rationalizing and deconstructing their beliefs. The story of Uncle Denzeng presents us with a complicated process of translating environmental discourse into local contexts, and identifies the tensions involved in this process.

Niru village, the real Shangri-La

During my stay in Niru Village I talked with a small number of tourists, and majority of them mentioned "the real Shangri-La." Some of the tourists who come to Shangri-La are disappointed at the "inauthentic" tourism experiences in the centre of town, where activities are commercially packaged and staged, and thus search elsewhere for "the real Shangri-La." Others, who had heard of or experienced the commercialization in Shangri-La and were thus searching for "the real Shangri-La", found Yading, a nearby village on the border of Sichuan Province and Yunnan Province. Yading Village is home to three holy Snow Mountains. Visitors have to cross several mountains to get there, and Niru is normally the starting point. In Niru Village I met a tourist from Guangzhou City, one of the biggest cities in China. He said:

> Actually I'd heard of Niru before I came to Shangri-La. I knew you could go to Yading by mountain treks, and to do that you have to start from Niru. I wasn't interested in touristic activities in Shangri-La. The real Shangri-La is in less exploited areas, for example Niru or Yading.

Interestingly, Niru has become known for both the inauthentic experiences in official "Shangri-La" as well as the imagined authentic experience in "the real Shangri-La." Since Yading has been imagined and promoted as "the real Shangri-La", after the renaming of Zhongdian, it has experienced growing visitation from international and domestic tourists. It is recorded elsewhere that villagers in Yading have considerably changed their lifestyle (Guo, 2012). It now seems that Niru is becoming another alternative to Yading; villagers know that city people appreciate Niru's beauty and charm, and are also highly aware of the air pollution and lack of natural aesthetics in the big cities, making tourists flee to places like Niru. I often heard villagers imitate tourists' exclamations, such as "I wish I could live here forever!" and "Let me die here drunk!" The villagers, on the one hand, find such expressions ridiculous. On the other hand, they understand tourists as having great demands as well as generating tremendous value from the very "being" of the village. Learning that I was going to stay in Niru Village for three months, a young villager commented:

> That's insane. I can't stand to be here even one more day. But I know you people are here to enjoy your leisure time, right? Many tourists come here for that. They do nothing, just sleep and stare into the air!

The sense of boredom expressed above needs to be read in context. Here, the young villager tried to make sense of the city tourists who would come all the way to Niru and "do nothing". For him it is at the same time both a peculiar idea and novel, perhaps even resembling modernity. I have also observed other villagers, very often young ones, copying tourists' ways of dressing, and sometimes also their ideas of traveling. One day, after having been in the village for about two months, I asked the younger son in my host family if I could accompany him to the family's pasture up in the mountains. He agreed. The next day he asked to borrow my 50 L backpack and filled it with snacks, Coca Cola and other mountaineer gear he had collected. Carrying my backpack and holding a walking stick he had gotten from an American tourist who had stayed in the family house, he looked like a mountaineer. Smiling, and posing to my camera, he joked, "I'm a tourist!"

My friend's temporarily enacting the performance of "the tourist" is more than mimicry; it also shows the process of his forming an identity through othering himself from his own local social context. This does not mean he feels he belongs in city life but rather, as he later expressed, that he feels pulled in different directions:

> Every time I get to town I feel tired after just a short while. There's nothing I can do there in town. I prefer to be in the mountains. I can stay up in the mountains several days. In the mountains I feel free. But whenever my friends ask me if I want to go to town, I can't resist.

In this reflective way, he expressed his sense of belonging to the mountains, as opposed to the city. But his performance of a hiker exposes another layer of his subjectivities – that to fully express his sense of belonging he sometimes needs to become someone else (the tourist). However, this is not to say he is simply influenced by another culture, but rather that his awareness of the contrast between city and village life enables him to represent himself in a different way. The way the villagers express themselves also influences outsiders, who sometimes hold stereotypical views on things.

Another young woman, who had studied in Shangri-La for several years and returned to Niru Village in 2012, said:

> Life is different in the city. Many people asked me if I could get used to the village life again. The first year (after I came back) was a bit hard. But now I'm totally used to life in the village. I even think it's quite good. We grow our own vegetables and grains, and feed our own animals. We have a house and live with our families. We live close to the holy mountains. I feel pretty happy here.

This woman and my friend both seem to be connected to the local environment in an emotional way, perhaps mirroring the Western image of sublime nature as well as the ecological thinking within Tibetan Buddhism. Esler (2016) observes that some Tibetans have internalized the image that Tibetan Buddhism and the traditional lifestyle in Tibetan areas are superstitious and backward; however, the recent promotion of Tibetan geopiety in environmental protection (such as in the rise of "the Green Tibetan") has made more Tibetans feel a sense of a superiority of their culture. I would add here that new environmental subjectivities are constantly in the making, and that the sense of belonging or pride can come from a mix of ideas, just as Yeh (2014b) suggests: "The Green Tibetan was…not merely a top-down imposition by transnational conservation organization; rather, it was forged out of the production of local and translocal Tibetan interests and identities arising out of radically different frameworks and communities" (p. 265).

Furthermore, the young woman's narrative above shows that the formation of new environmental subjectivities is a two-way process: after experiencing different environments and the interaction and exchange between radically different frameworks, she needs to find a new position in life for herself. Her feeling happy after going through a period of hard times reflects her adoption of a new understanding of herself and of her relationship with the environment. Although her reaction is the opposite of that of the other villager, who complained about the boring life in the village, the girl's representation of a good lifestyle in the village similarly presents how she internalized and made use of the imaginaries of the "ecological life" city people dream about. Meanwhile, these new understandings cast new influences and contributions back to the externally generated representations of the environment; as she talked to me, her multifaceted subjectivities challenged the earlier statement that "city people want to come here and we want to go to the city", which again indicates the complexity and dynamic nature of subject formation.

The process of subject formation and embodiment of the local residents in Niru Village in response to "the Green Tibetan" is similar to the "translocal assemblage" Kumar (2014) theorizes. This author describes how each of the different actors involved in protecting Niyamgiri holds diverse ideologies, beliefs and priorities and, through choosing the aspects which fit their own ideologies and contexts, transform each other and further develop each other's positions (Kumar, 2014). Lockwood and Davidson (2010) also employ the theory of assemblage to describe how Australian natural resource governance is a hybrid of three mentalities of government, namely neoliberalism, localism and ecocentrism. The inevitable overlaps into each other's domains may cause conflicts and tensions, but also provide opportunities for creating further emergent assemblages; an act which then allows institutional and discursive spaces to be inserted into and used for intervening in different situations.

Discussion and conclusion

Moore (2015) criticizes the use of the notion of Anthropocene to generalize humanity as a homogeneous acting unit, and suggests that the new epoch instead be named Capitalocene, as the capitalist way of dealing with nature has had more impact on the Earth than any set of actions has. Moore (2017) also stresses that the starting date of capitalism should be the beginning of the era of Columbus (1450), in order to include capitalism's early-modern origins, in which the first signs of landscape transformation can be seen. However, while discussions on terminology are perhaps useful in regard to what has happened in the past, they may not be as important concerning what will happen in the future. Our current goal is to recognize what we have been living with now and will live with in the future. Such a goal may allow us to see that, due to the precarity we all live in and the collaborations we all need for survival, capitalistic and alternative ways of dealing with nature are interdependent on each other now more than ever before. As the discourses of environmental crisis, global environmental change and, overall, the Anthropocene, permeate all corners of our lives, it is easy, by way of remedy, to construct and essentialize a fixed Other. In the situation studied here, this fixed Other is tourism, or "the Green Tibetan." What we often forget is the space between the one and the Other, where conflicts as well as positive possibilities may arise.

While the tourism development in Shangri-La can be seen as a result of organizing nature, it is also an important medium through which to challenge the essentialized idea of nature/culture, and can provide a space through which the villagers are able to present their environmental subjectivities as multilayered and ever-changing. As the basic ideas within ecotourism are to combine nature conservation and community-based development, brought together through a more general tourism industry sector ecotourism is often used as a substitute to replace other staple, and stable, economies ("Cheap Nature", in Moore's (2015) terms), especially when the cost of the staple economies is no longer cheap and they are increasingly destructive. However, neither conservation nor ecotourism can escape from the capitalist logic of utilizing "nature". Instead, both emerge from the reduction of the ecological surplus and, at the same time, still remain part of the capitalist system and thus continue to contribute to the same exploitation of natural resources. Yet, as this paper has shown, all is not clear. The exploration of what happens when individuals who have different environmental subjectivities but are still enmeshed within tourism is still open and evolving. Producing compelling readings of these encounters is challenging, but is also likely to be productive.

It is thus important to look at these encounters from new perspectives. Tsing (2015) proposes the notion of contaminated diversity for how we might move forward in our thinking and doing, as we face the global state of precarity: "We are contaminated by our encounters; they change who we are as we make way for others…Everyone carries a history of contamination, purity is not an option" (p. 27). Recently, in a tourism context, Tucker (2017) suggests that we, too, should reflect on the phenomenon of encounter-as-contamination, in order to recognize the open-endedness of indeterminate encounters in tourism. In this article I have tried to show some of these embodied encounters of different environmental subjectivities and the new meanings emerging from these encounters. Similar to the process of contamination leading to diversity and collaboration, these encounters play important roles for the constitution of future environmental subjects, through tourism.

In Niru Village, discussions relating to "nature" were often buried beneath other verbal exchanges about daily chores and were clearly formulated only in limited contexts, with tourism acknowledged as one of these day-to-day contexts. In relation to "nature", the tourism encounters in Niru Village present us with a complex and diverse process whereby the local villagers try to translate, appropriate and negotiate the broader discourse of environment into their own ideas, actions and identities, which is a process of (re)producing environmental subjectivities. In order to understand this process, the recent notion of "the Green Tibetan" was read alongside my informants' narratives. The villagers' narratives illuminate different ways of negotiating the notion of "the Green Tibetan" with their own previously existing and available subjectivities. While they were aware of this projection and its potential benefits for the local economy, the villagers also actively took part in the projection through their encounters with tourists as well as with their fellow villagers. What is happening in Niru Village thus responds to the question posed at the beginning, regarding how the concept of Anthropocene can help us make sense of the Earth through tourism (Gren & Huijbens, 2016). Although the grand idea of the Anthropocene, in the local context, appears only to be vaguely manifest through people's concerns over degrading nature, through tourism encounters the villagers may become more aware of what is going on and make local-global connections of environmental changes.

To conclude, I suggest that the Anthropocene is best considered within the broader environmental discourses, of which sociocultural justice is also a part. While the most dramatically changing environments are in the areas where the poorest populations live, the inhabitants do not always hold the same ideas about sustainable development. However, tourism development does make villagers in even a remote village in Southwest China aware of conservation, and thus provides social space for them to negotiate these terms for their own interests. Furthermore, different local contexts can provide more detailed and diversified examples of how the embodied encounters are enabled through tourism, and allow us to better understand an inevitable process of negotiation and appropriation. The political ecology approach will continue to be a useful tool for assisting the analysis in situated contexts, while we should also note that even in places in the so-called developed (or "First") world, environmental and social transformations are happening and are affecting individuals in similar ways as in less developed areas. Therefore, the intersecting field of ecotourism, environmental subjectivities and metanarratives emerging from environmental discourses (such as that of the Anthropocene) can be a fruitful arena for future studies.

Acknowledgement

I would like to than three reviewers for their insightful feedbacks and suggestions. Thanks also go to the research participants in Niru Village who took part in the fieldwork.

Disclosure statement

No potential conflict of interest was reported by the author.

Funding

This work was supported by the Swedish Research Council Formas under Grant "Mobilizing the rural: Post-productivism and the new economy [Grant Number: 2011-9018-18821-43]."

References

Agrawal, A. (2005). Environmentality: Community, intimate government, and the making of environmental subjects in Kumaon, India. *Current Anthropology, 46*(2), 161–190. doi:10.1086/427122
Baruchello, G. (2016). Good versus bad tourism: Homo viator's responsibility in light of life-value onto-axiology. In M. Gren & E. H. Huijbens (Eds.), *Tourism and the Anthropocene* (pp. 111–128). London: Routledge.
Castree, N. (2014a). The Anthropocene and geography I: The back story. *Geography Compass, 8*(7), 436–449. doi:10.1111/gec3.12141
Castree, N. (2014b). Geography and the Anthropocene II: Current contributions. *Geography Compass, 8*(7), 450–463. doi:10.1111/gec3.12140
Castree, N. (2014c). The Anthropocene and geography III: Future directions. *Geography Compass, 8*(7), 464–476. doi:10.1111/gec3.12139
Coggins, C., & Zeren, G. (2014). Animate landscapes: Nature conservation and the production of agropastoral sacred space in Shangrila. In E. Yeh & C. Coggins (Eds.), *Mapping Shangrila: Nature, personhood, and sovereignty in the Sino-Tibetan borderlands* (pp. 205–228). Seattle: University of Washington Press.
Compilation Committee of Luoji Township Annals. (2007). *Luoji township annals*. Luoji: Luoji Township Government.
Crutzen, P. (2002). Geology of mankind. *Nature, 415*, 23.
Crutzen, P., & Stoermer, E. F. (2000). The Anthropocene. *Global Change Newsletter, 41*, 17–18.
Davies, J. (2016). *The birth of the Anthropocene*. Oakland: University of California Press.
Eijgelaar, E., Amelung, B., & Peeters, P. (2016). Keeping tourism's future within a climatically safe operating space. In M. Gren & E. H. Huijbens (Eds.), *Tourism and the Anthropocene* (pp. 17–33). London: Routledge.

Esler, J. (2016). 'Green Tibetans' in China: Tibetan geopiety and environmental protection in a multilayered Tibetan landscape. *Asian Ethnicity, 18* (3), 1-12. doi:10.1080/14631369.2015.1090367

Fletcher, R. (2014). *Romancing the wild: Cultural dimensions of ecotourism.* Durham: Duke University Press.

Foucault, M. (1983). The subject and power. In H. Dreyfus & P. Rabinow (Eds.), *Michel Foucault: Beyond structuralism and hermeneutics* (pp.208–264). Chicago: University of Chicago Press.

Gren, M., & Huijbens, E. H. (2014). Tourism and the Anthropocene. *Scandinavian Journal of Hospitality and Tourism, 14* (1), 1–17. doi:10.1080/15022250.2014.886100

Gren, M., & Huijbens, E. H. (Eds.). (2016). *Tourism and the Anthropocene.* Abingdon: Routledge.

Grumbine, R. E., & Xu, J. (2011). Creating a 'conservation with Chinese characteristics'. *Biological Conservation, 144,* 1347–1355. doi:10.1016/j.biocon.2011.03.006

Guo, J. (2012). *Xue Shan Zhi Shu [Tales of Kha ba dkar po].* Kunming,Yunnan: Yunnan Renmin Chubanshe [Yunnan People Press].

Hall, C. M. (2016). Loving nature to death: Tourism consumption, biodiversity loss and the Anthropocene. In M. Gren & E. H. Huijbens (Eds.), *Tourism and the Anthropocene* (pp. 52–74). London: Routledge.

Haraway, D. (2015). Anthropocene, capitalocene, plantationocene, chthulucene: making Kin. *Environmental Humanities, 6,* 159–165.

Hillman, B. (2003). Paradise under construction: Minorities, myths and modernity in Northwest Yunnan. *Asian Ethnicity, 4*(2), 175–188.

Huber, T. (1997). Green Tibetans: A brief social history. In F. Korom (Ed.), *Tibetan culture in the Diaspora* (pp. 103–119). Vienna: Verlad der Österreichischen Akademie der Wissenschaften.

Huber, T. (2001). Shangri-La in exile: Representations of Tibetan identity and transnational culture. In T. Dodin & H. Räther (Eds.), *Imagining Tibet: Perceptions, projections, and fantasies* (pp. 357–371). Somerville, PA: Wisdom Publications.

Huijbens, E. H., & Gren, M. (2016). Tourism and the Anthropocene: An urgent and emerging encounter. In M. Gren & E. H. Huijbens (Eds.), *Tourism and the Anthropocene* (pp. 1–14). London: Routledge.

Jóhannesson, G. T., Ren, C., & Van Der Duim, R. (2016). ANT, tourism and situated globality: Looking down in the Anthropocene. In M. Gren & E. H. Huijbens (Eds.), *Tourism and the Anthropocene* (pp. 77–93). London: Routledge.

Jönsson, E. (2016). Anthropocene ambiguities: Upscale golf, analytical abstractions, and the particularities of environmental transformation. In M. Gren & E. H. Huijbens (Eds.), *Tourism and the Anthropocene* (pp. 152–170). London: Routledge.

Kolås, Å (2008). *Tourism and Tibetan culture in transition: A place called Shangrila.* New York: Routledge.

Kristoffersen, B., Norum, R., & Kramvig, B. (2016). Arctic whale watching and Anthropocene ethics. In M. Gren & E. H. Huijbens (Eds.), *Tourism and the anthropocene* (pp. 94–110). London: Routledge.

Kumar, K. (2014). The sacred mountain: Confronting global capital at Niyamgiri. *Geoforum, 54,* 196–206, doi:10.1016/j.geoforum.2013.11.008.

Law, A., De Lacy, T., Lipman, G., & Jiang, M. (2016). Transitioning to a green economy: The case of tourism in Bali, Indonesia. *Journal of Cleaner Production, 111,* 295–305.

Litzinger, R. (2004). The mobilization of "nature": Perspectives from north-west Yunnan. *The China Quarterly, 178,* 488–504.

Liu, J. (2012). Tibetans fight tourism on holy lakes. Retrieved from https://www.chinadialogue.net/article/show/single/en/5114-Tibetans-fight-tourism-on-holy-lakes, Accessed on 27 March, 2015.

Lockwood, M., & Davidson, J. (2010). Environmental governance and the hybrid regime of Australian natural resource management. *Geoforum, 41*(3), 388–398, doi:10.1016/j.geoforum.2009.12.001.

Lu, Y. (2005). *Meili Nizu: Laizi Xianggelila Zangzu Shengtai Wenhua Cun de Baodao* 魅力尼汝:来自香格里拉藏族生态文化村的报道 [Beautiful Niru: Report from a Tibetan Eco-cultural Village in Shangri-La]. Beijing: The Ethnic Publishing House.

Moore, A. (2015). Tourism in the Anthropocene Park? New analytic possibilities. *International Journal of Tourism Anthropology, 4*(2), 186. doi:10.1504/IJTA.2015.070067

Moore, J. W. (2015). *Capitalism in the web of life: Ecology and the accumulation of capital* (1st ed.). New York, NY: Verso.

Moore, J. W. (2017). The capitalocene, part I: On the nature and origins of our ecological crisis. *The Journal of Peasant Studies, 44*(3), 594–630, doi:10.1080/03066150.2016.1235036

Moseley, R. K., & Mullen, R. B. (2014). The nature conservancy in Shangri-La: Transnational conservation and its critiques. In E. T. Yeh & C. Coggins (Eds.), *Mapping Shangri-La: Contested landscapes in the Sino-Tibetan Borderlands* (pp. 129–152). Seatle: University of Washington Press.

Mostafanezhad, M., Norum, R., Shelton, E., & Thompson-Carr, A. (Eds.). (2016). *Political ecology of tourism: Community, power and the environment.* London: Routledge.

Mu, B., Yang, L., & Zhou, M. (2007). The progress of eco-tourism researches of nature reserves in China. *Journal of Fujian Forest Science and Technology, 34*(4), 241–247.

Nepal, S., & Saarinen, J. (2016). *Political ecology and tourism.* London: Routledge.

Purdy, J. (2015). *After nature: A politics for The Anthropocene.* Cambridge, MA: Harvard University Press.

Reddy, M. V., & Wilkes, K. (2015). *Tourism in the green economy.* London: Routledge.

Robbins, P. (2012). *Political ecology: A critical introduction.* Chichester, MA: Wiley.

Stonich, S. C. (1998). Political ecology of tourism. *Annals of Tourism Research, 25*(1), 25–54. doi:10.1016/S0160-7383(97)00037-6

Svensson, D., Sörlin, S., & Wormbs, N. (2016). The movement heritage: Scale, place, and pathscapes in Anthropocene tourism. In M. Gren & E. H. Huijbens (Eds.), *Tourism and the Anthropocene* (pp. 131–151). London: Routledge.

Tsing, A. L. (2015). *The mushroom at the end of the world: On the possibility of life in capitalist ruins.* Princeton: Princeton University Press.

Tucker, H. (2017). Contaminated tourism: On pissed off-ness, passion and hope. Paper presented at the Critical tourism studies VII conference, 25–29 June, Palma de Mallorca, Spain.

Woodhouse, E., Mills, M.A., McGowan, P.J., & Milmer-Gulland, E.J. (2015). Religious relationships with the environment in a Tibetan rural community: Interactions and contrasts with popular notions of indigenous environmentalism. *Human Ecology, 43*(2), 295–307.

Xu, H., Cui, Q., Sofield, T., & Li, F. M. S. (2014). Attaining harmony: Understanding the relationship between ecotourism and protected areas in China. *Journal of Sustainable Tourism, 22*(8), 1131–1150. doi:10.1080/09669582.2014.902064

Xu, J., Ma, E. T., Tashi, D., Fu, Y., Lu, Z., & Melick, D. (2005). Integrating sacred knowledge for conservation: Cultures and landscapes in southwest China. *Ecology and Society, 10*(2), 7. [online]. Retrieved from http://www.ecologyandsociety.org/vol10/iss2/art7/

Yang, S. (2012). *Connected with the three rivers: An account of the application process for the three parallel rivers of Yunnan protected area.* Kunming: Yunnan People Publishing.

Yeh, E. T. (2014). The rise and fall of the Green Tibetan: Contingent collaborations and the vicissitudes of harmony. In E. T. Yeh & C. Coggins (Eds.), *Mapping Shangri-La: Contested landscapes in the Sino-Tibetan Borderlands* (pp. 255–278). Seatle: University of Washington Press.

Zalasiewicz, J., Williams, M., Fortey, R., Smith, A., Barry, T. L., Coe, A. L., … Stone, Philip (2011). Stratigraphy of the Anthropocene. *Philosophical Transactions of the Royal Society A: Mathematical, Physical and Engineering Sciences, 369*(1938), 1036–1055. doi:10.1098/rsta.2010.0315

Zhang, J. (2013). *Political ecology of tourism worldmaking: A case of Shangri-La County*, Southwest China, in J. Sarmento & E. Brito-Henriques (Eds.), *Tourism in the Global South: landscapes, identities and development* (pp. 193–207). Lisbon: Center for Geographical Studies.

Zhang, J. (2016a). *Political ecology of Shangri-La: A study of environmental discourse, tourism development and environmental subjects* (Doctor of Philosophy), University of Otago, Dunedin. Retrieved from http://hdl.handle.net/10523/6361

Zhang, J. (2016b). Political ecology of tourism: Community, power and the environment. In M. Mostafanezhad, R. Norum, E. Shelton, & A. Thompson-Carr (Eds.), *Rethinking ecotourism in environmental discourse in Shangri-La: An antiessentialist political ecology perspective* (pp. 151–168). London: Routledge.

Zhang, J., & Tucker, H. (2018). Knowing subjects in an unknown place: Producing identity through tourism and heritage in Niru Village, Southwest China. In G. Hooper (Ed.), *Heritage at the interface: Interpretation and identity* (pp. 106-120). Gainesville: University Press of Florida.

Zhao, G., & Shao, G. (2002). Logging restrictions in China: A turning point for forest sustainability. *Journal of Forestry, 100*(4), 34–37.

Zinda, J. A. (2014). Making national parks in Yunnan: Shifts and struggles within the ecological state. In E. Yeh & C. Coggins (Eds.), *Mapping Shangrila: Nature, personhood, and sovereignty in the Sino-Tibetan Borderlands* (pp. 105–128). Seattle: University of Washington Press.

Fueling ecological neglect in a manufactured tourist city: planning, disaster mapping, and environmental art in Cancun, Mexico

Matilde Córdoba Azcárate

ABSTRACT

This article explores how tourism urban governance fuels patterns of ecological neglect. It turns a critical eye on Cancun, a leading Caribbean beach tourist destination and battered epicenter of anthropogenic climate change. First, the article contextualizes Cancun's design and construction as a state development project and manufactured tourist city. It describes the city's socio-spatial segregation and highlights the role of hurricanes in processes of beach enclosure. Second, it explores a series of risk maps elaborated as responses to international demands on coastal disaster mitigation and beach erosion. I show how local authorities, academics, and the Mexican state are bound to disregard risk maps to further enclose the Caribbean beach and keep the city productive for tourism. Finally, I look at the adoption of anthropogenic narratives on climate change as tourist attractions in Cancun's Underwater Museum of Art, a unique coalition between conservation, art and tour-operators in the city. I show that turning sea level rise and ocean acidification into tourist spectacles through copyrighted art, this attraction depoliticizes tourism's responsibility in patterns of environmental degradation. The article serves to reflect on the tacit paradoxes that plague efforts to imagine alternative environmental politics and sustainable tourism urbanisms outside neoliberal trends.

Introduction: the collapse of paradise

In June of 2015 Gina booked for her daughter, Lisa, a week in an upscale all-inclusive hotel in Cancun. Lisa was soon to begin her undergraduate degree at the University of California San Diego. Spending a week in a spa or luxury hotel in a foreign country before starting college has become an extended practice, Lisa told me, among some upper-middle class international students from Asia coming to the States. Cancun's all-inclusive hotel promised a sensory paradise and a balanced vacation in "a resort inspired by a yin-yang philosophy". For Gina, Cancun was an affordable vacation in a well-serviced, safe, and controlled leisure environment. For less than a thousand dollars Gina had a good package with an international direct flight and seven days' accommodation with all meals included. Cancun's all-inclusive hotel's marketing images of clear blue skies, sandy beaches, and large swimming pools descending directly into the shoreline, invited her "to relax and connect with nature". As Lisa told me in an interview,

My mum knew I'd always dreamt to go to a place like this (…) to run, to swim and to relax on the beach. The swimming pools going down to the beach in the hotel's promotion looked ideal for a morning run and a dip in the ocean without leaving the hotel (…) It was paradise (…) the beaches were paradise (…) solitary, quiet, inviting (…) so beautiful and peaceful.

In August, Lisa flew directly from her home in Shanghai to Cancun. A pre-paid hotel van was waiting for her and other visitors at the airport to transport them non-stop to the resort. Upon arrival, her room was ready, "looking exactly as in the pictures". On her first morning at the resort, Lisa went to the beach ready for jogging and that ocean swim that she had longed for, and then, as she put it, "paradise collapsed".

Not only was the beach smaller than in the marketing images she and her mother had seen, but the sand was "shellish and hurtful" to her feet. She did not jog because "it was too painful" and decided instead to get into the sea. In a first dip, the water felt "muddy". Then, as she vividly remembers, when the water reached only her knees she suddenly found a big hidden slope which, combined with a strong wave, made her fall and rolled her back to the shore. "It was so embarrassing" and "uncomfortable" she recalls. During the rest of her stay, she did not go back to the beach, but rather spent the time between her room and "the swimming pool (…) feeling sort of trapped there (…) not able to connect with nature at all" and "disappointed" with the experience.

Lisa's feeling of deception with Cancun's beaches is common among tourists in Cancun since hurricanes Ivan (2004) and Wilma (2005) left its beaches systematically sand-less and in constant need of re-filling. Dredging sand from the ocean and nearby coral reefs to refill the eroding beaches has become a recurring practice. Cranes, tow-trucks, pipes, large artificial rocks, and plastic containers filled with concrete have become regular elements of how tourists (and locals) experience Cancun's beaches. As a consequence, the once soft and powdery sands of the ideal Caribbean beach have been transformed into a wet, darkened rough, and shell filled seashore. The "peaceful" imagined beachscape is interrupted by the constant noise and strong smell of diesel coming from the heavy machinery that operates right next to the hotels' beach lounge chairs. Rosaline, a tourist in the area, described this landscape as "totally unexpected and surreal" in a conversation with me. Further, the destruction of the ocean's tidal sandbar has created large slopes and generated strong waves like the ones Lisa encountered. Turquoise sea waters often turn green over the course of a single day as a result of the presence of algae which regularly, and unpredictably, come ashore. Small storms can create tall sand cliffs in the beach, and when this happens tourists' pictures posing with them proliferate throughout the web. Aware of the dangers of being sued, oceanfront hotels have circulated posters warning tourists that if they swim in the ocean it is "at their own risk" since the beach is "a public space" and not part of the hotel's property.

This clash between imagined and lived tourist experiences of the Caribbean beach pose tremendous challenges for destinations like Cancun whose existence is predicated upon securing the popular and idealized representation of the place as a paradisiacal Caribbean beach destination for Western and Westernized audiences.

In this article, I explore in ethnographic detail, the management of ecological crisis through the lenses of tourism governance and practice. First, the article elaborates on Cancun's historical characteristics as a state development project and as a manufactured tourist city. I describe the spatial and social segregation upon which tourism was predicated, and highlight the role of hurricanes in furthering processes of beach enclosure. Second, I look at the systematic disregard for environmental expert knowledge tools in the management of extreme weather events, like hurricanes. I do so by discussing the elaboration and appropriation of international scientific measures on climate change research and sustainable urbanism. I focus in particular on disaster mitigation maps, or risk maps. I suggest that they are used to legitimate beach engineering projects and have as their main result to further enclose the beach for tourists. Finally, I look at the recent transformation of anthropogenic narratives about climate change as tourist attractions. As

I will highlight, Cancun's Underwater Museum of Art (MUSA), inaugurated in 2009, has formed a unique connection between a marine protected area, environmental art, and tour-operators in the city. I show that the contemporary arrangements of tourism in the museum are taking place in a marine protected area that lacks proper conservation regulations. I suggest that transforming human-led environmental degradation into copyrighted objects of consumption without paying attention to the city's urban and tourism dynamics contribute to de-politicize both tourists and the industry's responsibility in furthering contemporary patterns of ecological neglect.

While risk maps are examples of how internationalized scientific approaches to global ecological crisis and sustainable urban governance have permeated, and are used by local governance practices, MUSA exemplifies the commodification of anthropogenic narratives – those that locate human activity at the core of Earth's ecological problems – and in particular, ocean acidification and rising sea waters, as the city's latest natural tourist must-see. Considered together, the systematic disregard for expert knowledge tools and the transformation of human-led environmental degradation into tourist attractions, show how ecological crisis are being addressed, discussed, and manipulated at different institutional levels. In a city manufactured for tourism, these practices help to maintain the destination's appeal despite the evident decline of its beaches and amidst tourists' increasingly disappointing experiences.

Cancun's tourism governance of environmental hazards and crises, counter overly optimistic readings of alternative socio-environmental politics outside current neoliberal trends. This is the case for example of eco-modernists and proponents of the "good Anthropocene" who place the hope in expert scientific knowledge and planetary stewardship to resolve global environmental degradation (Hamilton, 2015; Steffen et al., 2011). Instead, I suggest, Cancun's tourism and ecological governance invite discussions on the eco-Marxist alternative of the Capitalocene (Moore, 2015, 2016); an alternative that calls to historicize patterns of ecological degradation side by side the development of mercantile and financial capitalism and hence, profit-oriented infrastructures.

The examples addressed in this research illuminate the difficulties of elaborating inclusive urban political ecologies when using standardized scientific canons and measures to understand and imagine local environmental patterns and solutions. They bring to light some of the paradoxes that plague efforts to imagine alternative environmental politics and sustainable urban tourism in manufactured cities that, like Cancun, are predicated upon exploiting their geographies for economic profit.

A note on methods

This article is the result of a larger multi-sited ethnographic research on tourism governance and sustainable development conducted in Cancun.[1] It is informed by a political ecology theoretical framework that urges a consideration of the role of political, economic, and environmental concerns in tourism practice and planning (Mostafanezhad, Norum, Shelton Eric, & Thompson Car, 2016).

Empirical evidence has been gathered through secondary sources, archival research, and ethnographic fieldwork with participant observation both in Cancun's city and hotels, as well as through qualitative interviews with urban planners, local newspaper leaders, municipal authorities, regional academics, service workers, hotel managers, local residents, migrant workers, tourists, and survivors of hurricanes Gilbert and Wilma during different one-month fieldwork stays in the city (2010, 2012, 2015). Secondary sources and archival research were used to trace the origins of the city and to locate and analyze official documents from the United Nations and Mexico's climate change and development and conservation reports. Ethnographic fieldwork was fundamental to gain access to all of the disaster maps not published as well as to capture the nuances of map production and implementation from the perspectives of scholars, environmental activists, politicians, inhabitants, and tourists. The arrangement of interviews was made possible due to my long-standing relationship with the region and social networks already in place from previous fieldwork. Inspired by Geertz' (1998) call for *deep hanging out*, I immersed myself in the everyday practices of knowledge production and tourism organization. Spending time physically in Cancun has helped me to grasp many of the ecological and moral dilemmas that scientists, urban planners, and Cancun's inhabitants and tourists face when participating in a state driven tourism model – a model they deem both socially and ecologically unsustainable but which they are still bound to nurture.

Since 2015, I have complemented this ethnographic data with content analysis of tourism marketing images and narratives of the city from governmental, real estate and travel agencies' documents and online platforms. I have also conducted semi-qualitative interviews with international tourists who have visited Cancun, like Lisa. I have carried out web content and image analysis from tourists' blogs, travel journals, travel sites, airlines, and social media platforms in an effort to trace the stability and disruptions of the city's popular tourist representations. As part of this effort, I have recorded tourists' comments and impressions on their uses of the beach, their selection of attractions to visit and their uses of city services (water and transportation) after tropical storms. My interest and research in MUSA emerges from this web content analysis and an interest in art as a form of environmental political practice. *Online deep hanging out* (Barendregt, 2017) has been central to trace the incipient coordinated displacement of Cancun's tourism imaginary and tourism governance from the beaches to underwater activities in the face of evident coastal erosion.

Manufactured tourist cities and the Capitalocene alternative

For many, Cancun would be the perfect example of the Anthropocene (Crutzen & Steffen, 2003), a geological epoch characterized by humans' impact on Earth. But for many others, myself included, Cancun is perhaps and more precisely, a textbook example of the Capitalocene. This is an alternative conceptual imaginary in which ecological crises are the recognized result of processes of uneven capital accumulation rather than the result of an abstract and homogenous humanity (Moore, 2016).

Cancun, like Dubai, Las Vegas, Palm Springs, or Benidorm, is a man-made manufactured city; a "fake city", like media regularly puts it, that has been built entirely for the leisure classes through deeply extractive and labor-intensive activities. These tourist cities are gateway service class cities (Karjanen, 2016) that attempt to look pretty much like modern versions of a Western Eden. Tropical beaches, floating private luxury islands, timeless casinos, and fountains in the desert have become iconic representations of what is, and should be, a leisure space in Western consumer cultural imaginaries. Recently, however, and like many of these manufactured tourist places, Cancun has become a battered epicenter of catastrophic events of global political and environmental relevance. Periodic toxic summers of green algae; deforestation; routine water-shortages; and contamination, extreme noise, and light pollution; coastal erosion and ocean acidification are just some of their visible effects.

Well before scientists attempted to consecrate this geological era as the Anthropocene, there was never a doubt that, in these manufactured cities, ecological degradation was caused entirely by human actions, driven by economic profits, and facilitated by vast land dispossessions. The construction of these artificial leisure places requires, in each instance, rapid and intense urbanization, the displacement of indigenous populations, the over-exploitation and commodification of natural and cultural resources, and the privatization of land and communal resources in the name of tourist infrastructural development.

As an artificially created and planned leisure city, the human–nature relationship in Cancun, was always an uneven and explicit relationship of domination – a relation predicated upon particular humans producing nature for market exchange through its spectacularization (Igoe, 2010). The force behind this marketization is not the abstract notion of humanity that figures prominently in anthropogenic narratives; here, the production and packaging of nature calls attention to specific collectives and institutions –like international development agencies and the estate– actively and obviously engaging in environmental destruction and largely motivated by consumerism. As such, these manufactured cities are best understood through a Capitalocenic framework. This framework enables us "to challenge the naturalized inequalities, alienation and violence inscribed in modernity's strategic relations of power and production" (Moore, 2015, p. 170), and hence, to historicize profit-oriented infrastructures like those made for tourism.

Designed, planned, and built as an economic device to attract foreign capital through ocean front tourist development, Cancun and its urban development serves as a paramount example of larger socio-ecological tensions. They demonstrate intimate connections with neoliberal and state driven extractive practices, as well as Capitalocenic narratives on the human uses and abuses of nature for profit.

To think ecological crisis through the lenses of tourism governance and practice reveals at large salient contradictions in contemporary human-nature relations. In Mexico, tourism has been used as a state modernization tool and it has been predicated upon social and economic exclusion. Looking at contemporary practices of tourism infrastructure development and eco-logical governance in cities manufactured for global leisure helps to show that not all Anthropos in the equation of the Anthropocene are at the center of this geological era in the same way. Some of them, in particular those closer to transnational capital, the state, international politics and mobility, arguably bear more responsibility and should be held more accountable than others in generating and reproducing patterns of ecological neglect (Higham and Miller, 2017; Moore, 2015, 2016; Wissenburg, 2016). It is also possible to observe that the Anthropocene is more than a geological epoch. It is a deeply political and politicized debate open to appropri-ation by many, from governments, to scientists, and artists, and therefore serves heterogeneous interests (Sayre, 2012; Swyngedouw, 2013).

Study of the governance of ecological crisis in man-made tourism destinations like Cancun also reveals a particular "politics of scale" that matters when articulating sustainable responses to Earth's environmental problems. In Cancun, globalized discourses and imaginaries about anthropogenic climate change, as well as the scientific production of knowledge about it, are strategically appropriated by regional and local structures of power in accordance with the logics of land enclosure and domination of nature for the market. After all, the city came into being through the global tourist gaze and exists primarily to offer travelers an escape to an idealized Caribbean of white, sandy beaches, and turquoise waters. Securing this popular tourist represen-tation of place in the global scale is crucial for Cancun as a tourist place as well as it is so for all tourist places (Sheller & Urry, 2004; Vicuña Gonzalez, 2013). In Cancun, this securing has meant the ongoing engineering of unsustainable and segregated spaces for consumption with the col-laboration of expert knowledge tools and populations. An engineering that is capable of sustain-ing the relationship between nature and society – beach and pleasure – that makes possible the inscription of the economic into the ecological (Escobar, 2012).

Building and securing a segregated tourism enclave in a hurricane prone area

"This was a land abandoned by our ancestors for a reason" remarked José, a migrant worker who lives in one of Cancun's irregular settlements. This is a common refrain among Cancun's migrants when speaking of the strong winds and flooding that shattered their makeshift houses and that preceded their forced relocation after hurricane Wilma hit the destination in 2005. Here, he pointed out, "there was nobody from whom to learn how to cope with a storm" as there were back in his hometown and neighboring villages. Unlike other tourist sites, there were no previous modes of inhabiting Cancun. There were no previous historical or cultural spatial mem-ories that tourism displaced. Indeed, Cancun, did not exist before 1974.

The city was designed and built from scratch by the Mexican state in 1974 upon a fragile natural barrier between the Caribbean Sea and the Laguna Nichupté. In Mexico, this was a moment of profound national crisis where tourism was conceived of as an export-oriented development industry just like *maquiladoras*[2] in the north of the country. Like the maquiladora export-system of electronics and textiles production, and similarly to other coastal resort areas in the Caribbean (Clancy, 2001), tourism was deployed as a modernization tool through the attraction of foreign investment and ocean front infrastructure expansion in a selected development pole (SEDESOL, 2010). To activate the region's Caribbean beaches within this extractive logic, or as they put it, "to take advantage of otherwise useless resources", Cancun was planned, designed, and built from scratch as an Integrally Planned Center (IPC) by a select group of Mexican government officials, bankers, and international development agencies. It was a state driven manufactured tourist city.

Cancun's Master Plan (1974), as the initiative was called, designed a spatially and socially segregated dual city. A city for tourists, Cancun Island or the Hotel Zone, and a city for workers, Downtown Cancun or Cancun City (Córdoba Ordoñez & Córdoba Azcárate, 2007; García de Fuentes, 1979; Torres & Momsen, 2005). Both areas were separated by differential routes of access and were designed to remain invisible to each other. A federal governmental institution, The Fondo Nacional de Fomento al Turismo (FONATUR), was created *ex profeso* to regulate the destination and it is still today the major federal governmental body in charge of the territory and urban planning in Cancun's Hotel Zone.

This original segregated design is effective to this date. The Hotel Zone is a highly homogenous space, a "fortified enclave" (Caldeira, 1996) designed and built as a single purpose, "enclavic tourist space" (Edensor, 2001). The Hotel Zone extends over 23 km of standardized architectures, gates, and walls where the Caribbean beach has been enclosed and domesticated as an attraction. According to the municipality's official data, in 2011 the Hotel Zone had over a hundred hotels and close to 25,000 hotel rooms. Local researchers from the Universidad del Caribe estimated that this number is really closer to 35,000 rooms, and well above the Hotel Zone's established carrying capacity of 30,000 rooms (Puls, de la Rosa, & Olivares Urbina, 2013).

Downtown Cancun or "the city", as it is called among its inhabitants, has grown meteorically through unregulated waves of labor migration. It has expanded from a population of less than a hundred inhabitants in the 1970s, to over half a million in 2015. During the decade of the 1990s, Cancun experienced one of the largest sustained demographic growths in the world (Pérez Villegas, 2000). The result is a "concrete jungle" (Castellanos, 2010), a large city that expands inland through a mix of consolidated housing, new residential housing units (*fraccionamientos*) and makeshift housing in irregular settlements (*colonias*) that burst with social life and economic informality.

In less than 40 years, Cancun has become a well-recognized symbol of global tourism, a major economic hub and the perfect example of a divided global city (Low, 1999). Its growth as a tourist city has been a tremendous success, at least if we judge it by the numbers. In four decades, Cancun has become the most visited Caribbean destination, attracting more than 5 million international tourists per year (SECTUR, 2015). The city is fundamental to the economic well-being of Mexico as well as of the Yucatan Peninsula where it is located, and where tourism is the major economic force.[3] Among locals, Cancun is conceived as "the engine" and tourism's spatial segregation almost accepted as second nature. Cancun is popularly considered a Gringolandia (Torres & Momsen, 2005); the major Spring Break destination for US college students. It is an example of translocality as a "city engaged in the circulation of foreign bodies, global commodities and transnational capital" (Castellanos, 2010, p. xxxii). Since the mid-2000, and possibly due to its segregated spatial design, Cancun's Hotel Zone has also become a favorite hub for international summits. Ironically, one of them was the 2010 United Nations Climate Change Conference, better known as Cancun Summit.

Interestingly, Cancun's apparent tourist success, has taken place despite being located in the North Atlantic hurricane path. In its short history, the city has been hit by three major category five hurricanes according to the Shaffir-Simpson Hurricane Scale,[4] Gilbert (1989), Ivan (2004), and

Wilma (2005). These storms have devastated the tourist infrastructure of the city, displaced populations, and irreversibly damaged its coastal ecology. Paradoxically, after each of these hurricanes, the city has not only multiplied its population but also its room capacity and tourist arrivals. The reason behind these post-hurricane speedy recoveries is that over the years, different international, national, state, and local governmental strategies have prioritized the enclosure of the beach for tourism use in an orchestrated effort to keep the city productive as a globalized Caribbean beach destination. As I have suggested elsewhere (Córdoba Azcárate, Baptista & Domínguez, 2014), hurricanes have been appropriated by the government and transnational real estate corporations, as creative destructive occasions (Harvey, 2003). They have been used to advance the logic of beach enclosure in the city's original plan. Flooding and material destruction have legitimated *ad hoc* land changes in the name of the public good, and prioritized tourism-oriented architectonic caprices in the Hotel Zone over housing programs in the city. The multiplication of all-inclusive hotels after hurricane Gilbert in 1989 and the shift towards high rise condominiums and time-share resorts after hurricane Wilma in 2005 are two examples of this trend (Córdoba Azcárate et al., 2014).

While the governmental appropriation of hurricanes has allowed the destination to update its tourist infrastructure and move towards mass tourism, the ecological and environmental effects of these meteorological and human-led interventions have become all too evident. Beaches are permanently eroded, the sand bar irreversibly broken, and coral reefs endangered by engineering equipment that regularly refurbishes the beach by dredging the sand from the ocean. Attempts to keep the beach stable for tourist consumption have not only altered the texture and feeling of the sand, as Lisa lamented in the opening vignette to this essay. They have also disrupted the function of natural coral reefs, a major natural protection against hurricanes in coastal cities. Mangrove forests are also dredged for new constructions, golf courses and artificial marinas. This interrupts the natural water currents that allow them to grow, spoiling the main natural protections against the strong winds and water surge that hurricanes generate (Spalding, Ruffo, & Lacambra, 2014).

Recent climate change research estimates consider that Cancun will suffer the impact of more virulent tropical storms as a result of increased heat waves and the rise in sea surface temperatures (UNFCCC, 2011). Storms will bring further erosion to its beaches, destruction to mangrove forests, devastation to the coral reef barriers as well as, of course, recurrent flooding and material devastation, forced relocations and deprivation to its population (Padilla, 2015). Mexico's 2007 *Third National Communication to the Parties* document to the *United Nations Framework Convention on Climate Change* (UN-FCCC: 112) explicitly said that Cancun "is highly vulnerable to extreme hydro-meteorological phenomena" and predicted an increase in the virulence of hurricanes affecting the destination.

When I was discussing these official documents with Violeta, a local environmental activist from Mexico City residing in Cancun, she lamented, "this is an all-well – too-known tragic scenario of the sinking to come". But far from sinking, Cancun's place as a leading tourism destination seems quite intact, at least if we judge it by the numbers. In the next sections, I examine two of the contemporary tourism urban governance practices that are contributing to keep the destination up and going while ironically, deepening its ecological problems.

Disaster mapping and the systematic disregard of environmental expert knowledge tools

The tension between the logic of enclosing the Caribbean beach for tourist consumption and its manifest erosion became salient with hurricane Ivan in 2004. Although Ivan did not have the same devastating effects in Cancun as it had in other areas of the Caribbean and the USA, it revealed how vulnerable the city had become after just thirty years of existence. Hurricane Ivan, as one expert put it, left Cancun's beaches too narrow for recreation and it was then when the federal authorities took prevention seriously (…) when they realized Cancun could disappear as a beach destination and hence risk the economic survival of the entire region.

Hurricane Ivan marks the beginning of disaster mitigation mapping strategies for the Mexican Caribbean. Alarmed by the destructive path of the hurricane, the United Nations International *Office* for Disaster and Risk Reduction (UNISDR) pressed the Mexican government as well as other Caribbean nations such as Grenada, to put in place a monitoring system. Governments were also urged to engage in risk knowledge production through the elaboration of assessment reports for their coasts. In Cancun, the agency in charge of assessing the city's coastline's risks was the Comisión Federal de Electricidad (CFE), a state productive entity in charge of distributing and marketing electric power. The first assessment of Cancun's coastline is called among analysts "the 2004 dossier" and was the result of an international mandate.

The 2004 dossier must be understood as a result of the growing International pressure towards measuring anthropogenic impacts on climate change through geoengineering practices such as risk assessments. These assessments are aimed at creating more sustainable urban governance agendas and cities, otherwise also known as regenerative urbanisms (Thomson & Newman, 2016). Risk and vulnerability maps are part of international initiatives to render national and municipal governments accountable by tying the allocation of development and/or tourism funding to their ability to meet international criteria, in this case, to have a coastal risk assessment plan as an urban planning and prevention tool. Yet, the knowledge produced has also become a tool to further empower uneven tourism infrastructure developments. This is evident in the case of the 2004 dossier.

While the stated objective of Cancun's 2004 dossier was to develop a plan to conserve and nourish Cancun's beaches and to assess hydro-meteorological risks on the city's coast, it was obvious from the start that these maps were political artifacts generating a partial production of scientific knowledge about the city. The maps only covered one part of the city, the one built for tourism consumption, not Cancun city itself. What this meant, in practical terms, is that only one part of the city, that embedded into the global circulation of capital, would enter the circuit of environmental knowledge production and risk prevention. One of the engineers who has been working at Cancun's Office of Civil Protection for over 25 years, lamented this compartmentalization. Assessment risks pointed to the fact that after any storm the lack of potable water and energy have always been the main risks for those living in the city. These issues he said, "have never and would never be addressed (for Cancun)".

Although partial, the CFE did propose a series of environmental and urban regulations to restore the coastal dunes as a main barrier against storm waves. However, they were never implemented.[5] As an expert involved in the project explained to me, "an agreement was never achieved among the CFE, the municipality and the hotel's association and the plan remained in its design phase". The main reason behind this was that for tourism developers, as well as for the federal and municipal government, the kind of risk mapping strategies and prevention tools proposed by the CFE were a threat that did not allow flexibility in the distribution of construction permits. As a result of the post-hurricane land changes, in 2004, Cancun's shoreline remained federal land ruled by FONATUR, but its coastal dunes were already a densely-populated area filled with all-inclusive hotels with mixed national and transnational, public and private, systems of ownership of land and buildings. The tools proposed by the CFE would had introduced the strict governance of a piece of land that needed to remain as flexible as possible for capital development. As a result, their recommendations were ignored and the maps stored in drawers.

Storing risk maps in drawers is an instance of what I call a *strategic and institutionalized disregard for expert knowledge tools*. Municipal authorities were able to obtain international funding by meeting international standards and then promptly disregarding them. This strategic disregard for environmental expert tools is not incidental, but must be understood as a logic of urban and tourism governance that allows man-made leisure coastal cities like Cancun to meet international demands and standards of sustainability and climate change knowledge production whilst persisting with unsustainable urban and economic models predicated upon ocean front infrastructure development.

In Cancun, this logic of urban tourism governance has become evident as the city has continued to grow and to be affected by hurricanes and storms. This has increased the international pressure to produce adequate risk assessments in order to mitigate their potential destructive outcomes and to develop more sustainable growth. Since hurricane Wilma hit Cancun in 2005, there have been three more risk mapping strategies in Cancun, which, like the 2004 dossier have been forced upon the city by external actors and have all ended up stored in drawers.

The first of these strategies was the Atlas of Risks for the City of Cancun, or the "2005 Atlas", which was created immediately after hurricane Wilma. The Atlas was funded under auspices of the United Nations, the Inter-American Development Bank and the Mexican Secretary of Social Development (SEDESOL) through its Habitat Program created two years earlier with the aim of alleviating urban poverty in Mexico. The 2005 Atlas was elaborated by a Mexican private consultancy from Monterrey, Outsourcing en Mercadotecnia S.A., C.V. The analysis was aimed at creating indexes to measure levels of risks and quantifying its perceptions among inhabitants. Unlike the 2004 dossier, the 2005 Atlas included the mapping of risks outside Cancun's Hotel Zone and elaborated a large sample of social variables which included for example, construction hazards. However, most local experts and environmentalists who had the opportunity to read the Atlas, found its data "unreliable". The main reason they highlight for this is the perceived risk of the vested interest that private consultancies brought to bear on the project. However, there was also another major reason for the 2005 Atlas to fail that points at a larger problem in the generation of reliable knowledge to measure human impacts on the environment through standardized scientific measures. All the social data generated by the Atlas were obtained through secondary sources, mainly censuses from the Instituto Nacional de Estadística y Geografía (INEGI). This data is not very reliable given the speed at which Cancun has grown and where more than 22% of the population lives in irregular settlements. As the director of the Municipal Institute of Urban Planning (IMPLAN) said, the maps only represented the official Cancun and did not account for Cancun's reality of irregular labor migration flows and informal urbanization.

The 2005 Atlas was never published. It was never distributed among local or regional scholars. It never arrived at the local Office of Civil Protection, nor was it publicized within the UN or shared in schools as has been done for example in the Caribbean state of Grenada (UN, 2007). In a tourist city that grew out of segregated planning, publicizing the vulnerability of informal settlements located in flooding areas could lead to protections that could hamper the ability of the municipal and federal governments to incorporate these lands into tourism ventures as well as to keep constructing housing for inhabitants and building real estate revenues. As a result, its recommendations were ignored.

Two years later, in 2007, another major mapping attempt took place in Cancun. The Oceanographic Risk Atlas of Cancun was conceived as the result of a scientific National Research Project co-funded by the largest Mexican national and regional scientific public institutions. This Atlas was produced by the *Grupo de Ingeniería de Costas y Puertos del Instituto de Ingeniería* from the *National Autonomous University of Mexico* (II-UNAM) with the support of the *Centro de Información Geográfica* from the *University of Quintana Roo* (CIG-Uqroo) in a pioneer interregional scientific effort to make local research institutions a visible part of the urban governance of the destination. Or to put it in other words, as an effort to bring geoengineering to urban planning. The 2007 Atlas was also the first attempt to provide a wider comprehensive and diachronical analysis of hurricanes and tropical storms affecting not only Cancun, but the Caribbean coastal region. As Luisa, a local researcher producing the Atlas, put it, "it is unthinkable why the previous risk maps never included a regional scale (...) the Oceanographic Atlas (was) pioneer in fostering this regional understanding of disaster prevention".

The 2007 Atlas developed a complex aero-spatial and Geographical Information System technology to analyze flooding risks resulting from storm surge. It directly pointed at informal urbanization as one of the main risks. It proposed risk zoning and the design of the first Integral Program of Disaster Prevention and Management intended to be used by the municipal government of Cancun as a prevention tool. However, in common with the previous maps, the 2007 Atlas was never implemented. Its "failure" demonstrated once again the need to keep risk maps out of sight, and especially those that highlighted areas vulnerable to flooding. These maps threatened to stifle the ability of federal and municipal governments to distribute land contracts in the Hotel Zone, as well as the ability of real estate developers to construct new condos for tourist workers in downtown Cancun. This was especially critical after the destruction caused by Wilma in 2005 and materialized in the shift in building trends at the hotel zone from all-inclusive hotels to high rise time-share condominiums (Córdoba Azcárate et al., 2014). The hurricane proved that the all-inclusive model was too expensive and risky in the face of recurrent natural disasters, which implied rising operating costs for vast oceanfront infrastructures and higher insurance premiums and precipitated the shift to the more profitable business model of high rise condos. Producing scientific maps and then ignoring them allowed Cancun's municipal government and the state to address international demands on sustainable environmental governance while allowing the city to reinvent itself in the face of a demanding tourism market.

The last disaster mapping attempt took place in 2010. It was called the *Atlas de Riesgos y Desastres Naturales, Municipio de Benito Juarez* and was commissioned by the federal government and the municipality of Cancun. It was elaborated by the *Centro de Información Geográfica* of the *Universidad de Quintana Roo* and found its justification and resources in the recently launched United Nations Climate Change Initiative. As with its predecessors, the underlying idea behind these risks maps was to generate more sustainable urban growth by transforming scientific knowledge about the Earth into active governance tools.

The Atlas produced up-to-date representations of the city detailing its most vulnerable areas as well as potentially 'dangerous' infrastructures (i.e. gas stations), and it identified urban areas according to their degree of vulnerability beyond the Hotel Zone. It generated 100 maps and identified 37 out of Cancun's 75 regions as "highly vulnerable areas to flooding".

All the materials were handed over to the Municipal Institute of Urban Planning in Cancun (IMPLAN) a year after they were done, in 2011. Interestingly, from 2009 to 2010 the Institute not only faced a 70% reduction of their personnel but its main office was relocated outside the city council. Its director recalls, "I knew that it (the 2010 Atlas) was born dead". To understand why this was the case, we have to focus, once again, in the particular political structures through which international demands to produce scientific knowledge about climate change are played out in practice.

First, the process of knowledge creation about human impacts in Cancun was constantly outpaced by the intense informal growth of the city. As one of the scholars involved in the 2010 Atlas put it, "we had to deal with the fact that the city is faster than us". According to the estimates of the experts, the maps they were elaborating would be useless in just three months. Local scholars were trapped between a rock and a hard place. While they knew the maps would not be used in planning, the international funding for their research on natural disasters depended on producing them.

Second, although this obsolescence could be avoided by turning maps into live models through Geographic Information Systems technologies, the municipal government and local planning institutions lacked the necessary infrastructures, manpower and expertise required to manage them. As Rafael pointed out,

we were never the executers of the Atlas (…) we generated the data but we never have had an executor's role (…) we handed them to the authorities along with the usernames and access instructions (…) but how to use them if their computers were not updated? (…) if most technicians working there did not even have the basic Internet knowledge to operate them? (…) it was obvious that all the data and the maps were just stored in drawers.

And third and most importantly, these problems could have been solved if there had been a political will to do so. But there was not. The reason for this is that, in Cancun, the production of these maps was not done to increase knowledge about human destructive effects on earth or to improve land occupation patterns and urbanism as the UN imagined maps to be. As one of the leading scholars in charge of the 2010 Atlas explained in an interview,

> A lot of times maps are just interpreted as a (governmental) requirement to be able to access funding of a very different nature, let's say tourist infrastructure or housing infrastructure as new condominiums (…) There is this perception within governments and institutions that maps generate information, and having information in such a direct way, in such a transparent way as the system created by the Atlas, is in itself a risk (…) there is always the doubt, the fear that the information could end with people or groups of people that they don't want them to have it. If everybody can access the information, everybody can act on the city.

Looking at the genealogy of risk mapping attempts in Cancun shows the difficulties in deploying the kind of formal scientific knowledge and climate change planning tools required by standardized scientific approaches to the environment. These difficulties do not stem from a lack of governance or lack of knowledge, nor in this case from competing understandings of nature. Rather, the issue is the strategic and institutionalized disregard for these mapping projects. And this disregard is not the result of an omission or negligence but, as I have argued, a distinctive form of urban tourism governance in a city built for leisure. In Cancun, the knowledge generated by risk maps needs to be stored so the whole "tourist machine", as locals put it, can keep going on and the city can remain stable in its popular geopolitical representation as a Caribbean beach paradise for the leisure classes. Here we have a city pushed to, but also maneuvering to, appropriate global environmental demands on disaster mitigation and sustainability for internal political reasons.

The production and disregard of disaster maps also indicates that caution should be exercised with regards to the culture of environmental bureaucracy, that is, the formal requirement to document, measure, map, and monitor environmental risks and land uses through standardized devices. These tools aim to hold social groups and societies accountable and responsible for human impacts upon nature or for certain anthropogenic hazards (Wissenburg, 2016; Richard, 2008). Ethnographic attention on how climate change data is generated, how it is ignored, stored, and disregarded once it has been produced, cast a new light onto how standardized measures – like risk maps – are understood and used at a local level. It shows how deeply scientific research and internationalized environmental categories – like that of risk – are entangled with local urban planning and environmental politics; and how sometimes, the cure can serve as a threat.

In this section, I have focused on how the generation of knowledge of anthropogenic climate change is trapped and de-activated in its claims towards more sustainable urbanisms as it is captured in and appropriated by existing geopolitical structures that force Cancun to operate under the global tourist gaze of escape to the beach. The next section shows how environmental concerns are once more absorbed in the logics of tourist consumption and a for profit industry. Exploring Cancun's Underwater Museum of Art, MUSA, I want to draw attention to the aestheticization of human-led environmental destruction and its transformation into yet another object of consumption under the guise of ecological consciousness and conservation planning.

Anthropocene©: the latest nature tourist attraction

During the last four years, Cancun's marketing in magazines, official campaigns, banners, and ads has been commonly associated with a submerged sculpture of a Volkswagen Beetle with a naked man curled up in its exterior front windshield. The car is the central piece of the Underwater Museum of Art, MUSA, inaugurated in 2009 as a joint initiative among the Director of Arrecifes de Cozumel's National Marine Park, Cancun's Nautical Association, and a renowned English-born sculptor who defines himself as an environmental artist and activist.

The Volkswagen Beetle sculpture is called The *Anthropocene* and whilst its image circulates widely, the sculpture can only be visited by scuba diving. The piece was submerged in 2011 in Sala Manchones, the second of the two reef bodies composing the Underwater Art Museum between Cancun and Isla Mujeres. As with the rest of the 500 sculptures in the museum, it is made of a pH neutral clay, marine grade cement. It is also engineered as an artificial reef intended to attract coral life, plant-life, and fish. According to the artist, the underwater sculptures are an example of "sculpture as a conservation method". They are made and submerged to provoke reflections on rising sea waters due to global warming and to subvert existing attitudes through "a new tourism" which is responsible with regards to the environment and local populations. His work aims to reveal the anthropogenic ocean acidification, or as his official website reads,

> to usher a new era for tourism, one of cultural and environmental awareness, with the hope that more tourists may begin to re-conceptualize our beaches as more than sunny paradises but living and breathing ecosystems

Building on the artist environmental discourses of awakening to a new era of sustainable tourism and "deep seeing" (McCormick & Scales, 2014), MUSA Cancun suggests that it "demonstrates the interaction between art and environmental science" and represents one "of the most ambitious underwater artificial art attractions of the world". The museum is not unique, however. It is part of a larger, worldwide group of six other underwater museums – in Grenada, the Bahamas, Lanzarote, and England. All of these sites draw on the work of the same artist. As a result, some critics say that he is exploiting sea level rise and his life long experience as a scuba diver, to play with and profit from artworks already conceptualized by others (Davis, 2017).

MUSA figures prominently in a multiplicity of internationally recognized media and always provides optimistic accounts of the city's developments.[6] In the larger context of Cancun's tourist industry and official discourses, *The Anthropocene* sculpture at MUSA speaks to carbon economies and ocean acidification. It talks to Mexico's industrial export economy and calls for an ecologically responsible experience of the destination. MUSA is advertised and talked about with pride by authorities, developers, journalists, mainstream conservationists, and tourists. In a mass tourism destination like Cancun, the submerged Beetle along with the rest of submerged sculptures symbolize for many that tourism can be planned otherwise and that tourism can be practiced otherwise. Nature tourism activities, like scuba diving or snorkeling are normally associated with ideas of balanced interaction among distinctive species and societies, and thought to serve to conserve the oceans that human activity has irreversibly acidified and altered (Mota & Frausto, 2014). As with many forms of tourism in protected areas, scuba diving in MUSA is then promoted along win–win discourses according to which nature tourism in conservation areas contributes to social, environmental, and economic sustainability (Duffy, 2010; Fletcher, 2011). Considered at close hand, however, the museum directly challenges the hypotheses of the good Anthropocene (Hamilton, 2015) as well as other optimistic readings of this era as a moment of renewed socio-environmental politics outside of market logics.[7]

MUSA's stated aim is to regenerate vegetation as well as fish fauna by building a reef replica for tourists to gaze upon and for researchers to build evidence from, while saving the real reefs from human traffic. The museum's widely reproduced sculptures – in promotional brochures and destination branding – attest to this claim showing in their dynamic transformation how reef algae and sponges, fishes, mollusks, crustaceans, worms, and echinoderms flourish in them.

Tellingly, however, MUSA is located in a marine protected area that lacks the proper legislation for diving activities and does not have an established limit on the number of tourists that can visit it. The Universidad del Caribe estimates that *Arrecifes National Park* receives 800,000 tourists per year, and that 100,000 of them visit MUSA. To do so, tourists can choose from three different experiences, that is, scuba diving, speed boats, and snorkeling. They can also avoid getting wet, by visiting the exhibition in a glass bottomed structure. All of these activities require entrance into the Marine Protected Area. Tours must be booked through one of the seven Museum's seven sponsors, all national and transnational diving corporations. Diving excursions must be booked and pre-paid in advance. They are accessible to novice and amateurs as well as to expert and professional divers. Multiple combinations have directly contributed to increase the traffic of tourists to the Cozumel barrier reef. The reef is part of the Mesoamerican Barrier Reef, also known as the Great Maya Reef. This is a unique marine region hosting the world's second largest coral reef system and whose species have been classified as critically endangered, endangered or have been under some sort of protection since the mid-1990s (The Nature Conservancy, 2016).

In the larger tourism dynamics of Cancun, the world's number one packaged tourism destination according to Expedia (2015), MUSA has become just one more stop on the road. The museum is promoted as "one of a kind" excursion into an "eco-project". It is advertised alongside non-environmentally friendly and crowded water activities such as Skyriding and Paragliding. Targeted at the environmentally aware and art lovers, it is characterized as a "top pick for every adventurer" that visits Cancun. MUSA has just added environmental art as the latest tourism niche and symbolic practice and aims to build on tourists' social distinction. The museum offers the "sophisticated natural package" for the "art tourist" and further opens the ocean, and specifically, the Cozumel reef system for mass tourism. As the artist put it, "installations bring environmental messages to a viewer already primed with an interest and passion in our oceans. These environments push land-goers to experience them through art". However, MUSA's showcase of the union between conservation knowledge and art also helps to rationalize anthropogenic marine destruction as inevitable. It is critic of anthropogenic narratives of environmental destruction and yet, it works alongside them and exists because of them. This alliance does not contribute to altering the production of nature for the market that is inherent in capitalist societies. Rather it contributes to further marine ecological deterioration. It capitalizes on it for further economic profit.

At the regional scale, MUSA's diving excursions are the city's *in situ* natural tourist attraction for Cancun's all-inclusive visitors and it is promoted alongside natural and cultural packages like the Biosphere Reserve Sian Kaan, the Maya village of Chan Kon, or the archeological site of Chichen Itza. And yet, in an urban context like Cancun, environmental attractions are packaged not in the form of wild encounters with nature, but instead, as niche and boutique experiences. Amanda, a tourist from New York whose husband Peter was diving when I talked to her, bluntly elaborated on this comparison when I asked her about why her husband chose this package in MUSA among the many others their hotel offered,

> diving in MUSA is different than let's say going to Chichén Itza because not everybody can do it. Not everybody can afford it, and not everybody is interested in diving with art. This is just for some and if you love diving as Peter does this is unique (…) it is one of the only places in the world where you can do it!

In their encounter with the ocean at Cancun's Underwater Museum, tourists like Peter do not deal with just any form of natural tourist attraction. They encounter artworks transformed into tourist commodities, and tourist commodities that are green and sit within preservationist frameworks. At MUSA, the voices of the artist, the Mexican National Commission on Natural Protected

Areas, in charge of environmental regulations, and Aquaworld, the major tour-operator in charge of the promotion of the museum, are indistinguishable. Their discourses cohere around narratives of awakening and the idea that tourism is an industry capable of minimizing environmental damage by transforming nature into an educational product of tourism consumption. In all excursions to MUSA, however, there is a 10USD governmental fee that tourists need to pay to enter the protected marine zone, and a 16% tax extra that they have to pay if they want to take photos. These practices directly contribute to privatizing the ocean as well as to perpetuating the inequalities and exclusions inherent in tourism.

Lacking an operational Management Plan[8] and relevant information on the tourist activities within the protected area, information on MUSA tours is barely available outside of institutionalized frameworks. This is not because of a direct attempt to discourage unregulated tourist flows to the area, but is mostly because of art copyright issues. Most promotional videos in Aquaworld depend on the artist's work dynamics and do not always fulfill tourists' demands for environmental education, or their desires to learn about an excursion before booking and paying for it. Some tourists complained that when it came to viewing the museum's scuba tours for example, they were sent a common screen message saying, "this video contains content from TiVI Media, who has blocked it on copyright grounds". The artist's own webpage offers an online shop where tourists can buy printed and signed photos of his sculptures. These photos are protected by copyright and do not allow replicas. The result is that a perfect homogenous and enclosed underwater ideological coalition has been formed. One that, disguised under a positive spin, normalizes environmental destruction, and rationalizes social exclusion.

The museum offers a perverse example of how to strategically appropriate discourses and imaginaries about humans' destruction of nature through the spectacularization and commoditization of the ocean. Despite its claims, interventions like MUSA seem to just reproduce nature as a passive object, open to be acted upon, dominated, admired, and rescued. A form of visibility that reproduces and legitimizes Enlightenment dichotomies between society and nature that lie at the problematic core of the good Anthropocene's awakening calls on benevolent scientists and expert knowledge tools as universal paths to save the Earth. For MUSA, like for nature conservation knowledge production that is reduced to inoperative management of resources, the ocean is objectified as resource, fragmented into collections of marine habitats and coastal assets. In this framing, the ocean legitimates the practices that operate upon it to preserve it, to protect it, to understand it, to contemplate it, but always through consumption. It is not the ocean as commons (Sayre, 2012) but the Anthropocenic ocean as the new capitalist "fix" (Fletcher, this volume).

It is not a coincidence that, as Brockington, Duffy, and Igoe (2008) suggest, the growth of natural protected areas parallels the expansion of neoliberal policies around the world. The Arrecifes de Cozumel Marine Protected Area where MUSA is located was created in 1996 under President Ernesto Zedillo. The president was one of the leading figures in the implementation of neoliberal policies in Mexico, which included manufacturing nature into reserves and implementing nature tourism and ecotourism as modernization vehicles to make nature profitable for the economy. In this line, MUSA exemplifies a neoliberal tourism logic in which a market-oriented understanding of nature as resource is privileged. This economic logic is furthered by the aestheticization of the ocean through environmental art and its conversion into a distinctive tourism attraction in a city created from scratch for beach leisure and that is slowly, but steadily, seeking to diversify tourists' attention from the sand and into the ocean.

The transformation of anthropogenic environmental destruction into a copyrighted artistic product is happening through institutionalized and corporate controlled forms of environmental art and conservation knowledge and planning. To do so in a manufactured and segregated tourism enclave like Cancun, provides yet another example of how fast eco-tourism relies on ego-tourism (Munt, 1994). Here, the consumption of nature as spectacle is a favorite route to achieve social distinction. However, this is also a practice that brings us back to basic questions of who is able to care for nature, what nature and at the expense of whom. MUSA and *the Anthropocene* sculpture cannot be understood if we lose sight of the historically and geographically specific reality of Cancun as a governmentally manufactured economic device under the global tourist gaze. The city's tourist predicament and organization as an internationally packaged Caribbean beach destination called for a segregated enclave. It has been the reproduction and securitization of the city's historical material and social segregation through tourism planning and infrastructure development what has maintained it competitive in the global market. Cancun is a city that draws tourists into the coasts, into the beach and into the water. To forget the destination's geopolitical strategic location as an affordable tourism beach destination for international travelers and well off citizens, or to forget its corporate governance as a top packaged destination in creating new nature attractions only contributes to the de-politicization of tourism. It does so because when embracing anthropogenic ecological crisis as aestheticized objects of tourist consumption, it overlooks that the tourism industry and tourism practices are major agents in the reproduction of environmentally and socially unsustainable human–nature relations that end up exploiting marine ecosystems for economic profit. Capitalist policies and values, noted Brockington et al. (2008, p. 3). "pervade conservation practice and in some places they infest it". Cancun is one of these places.

Conclusion

In the winter of 2010, just a year after MUSA's inauguration and a few months after the completion of the first maps of the 2010 Risk Atlas, Cancun hosted the United Nations Climate Change Conference (UNFCC COP 16), best known as the Cancun Summit. Climate change discussions took place in the Hotel Zone, in well-known, upscale all-inclusive hotels. It also made use of the city's newest Convention Center, the CancunMesse, opened for the occasion. This is a massive exhibition complex that claims to be the largest trade fair venue in Latin America as well as "a socially responsible and friendly to the environment" building in "perfect harmony with nature". Along with others in the Hotel Zone, some of these service infrastructures were immersed in ongoing legal battles for their questionable environmental practices. This included construction on areas of vulnerable coastal dunes that the authorities were trying to shore up amidst their inevitable erosion.

Cancun's Summit generated several environmental protests some of which went viral. The image of a polar bear lying on its beaches, has become an icon of environmental activism worldwide. Protests took place at the CancunMesse's doors as well as in the *de facto* privatized and refurbished sand-less beaches of the all-inclusive resorts hosting the conference. In the midst of international meetings and environmental protests, MUSA inaugurated its largest display submerging "The Silent Evolution", a sculpture group of 400 human life size replicas in a well-attended and media friendly public performance. The Silent Evolution aimed, in the artist's words, to bring awareness "for each million individuals living in coastal areas" in the "age of the Anthropocene". Summits, protests, and art performances though, all blatantly forgot about Cancun's own local environmental crises. In doing so, they also turned a blind eye to contemporary city residents' like José and their forms of spatial knowledge of the area. There was a reason why nobody was here before, José put it. His words deserve to be listened to.

And yet, in 2018, the Marine Protected Area where the museum is located still lacks an operative Management Conservation Plan. Moreover, the city still does not have an operative Atlas of Risks and Disaster Prevention for its citizens. Although revisions to the 2010 Atlas have been commissioned by the Mexican federal government, none have yet seen the public eye. Unsettlingly and ironically though, Cancun hosted the fourth meeting of the UN Disaster and Risk Reduction Platform in May 2017. Once again, experts and international delegates flocked to the city's all-inclusive hotels and sinking shorelines to discuss global environmental questions, all the while disregarding and contributing to the city's ongoing socio-ecological crises.

In this article, I have discussed three instances where global environmental concerns and crisis – beach erosion, sea level rise, and ocean acidification – are incorporated into the governance and tourist image of a man-made city manufactured by the Mexican state as a development tool through global leisure consumption. Firstly, I have traced the origins of the city as a state development tool. I have highlighted the processes of spatial enclosure of the beach and how hurricane's reconstruction efforts have helped to further keep the Caribbean beach enclosed for tourism consumption. Secondly, I have demonstrated how environmental concerns on coastal beach erosion are strategically appropriated and systematically disregarded by local governments, hoteliers and the Mexican state. As I have revealed, this practice has only contributed further to local ecological neglect. It does so by advancing uneven urban planning developments that foster the enclosure of the beach for tourism consumption. Paradoxically enough, these are the practices that assist business in the face of recurrent hazards such as hurricanes. Thirdly, I have shown how the transformation of anthropogenic environmental degradation into a tourist attraction contributes to the commodification of nature. This blurs the responsibility of the tourist industry for environmental degradation. By transforming human-led environmental destruction into an aesthetic device, and exploiting a marine protected area that lacks proper regulation, MUSA and its visitors are not helping but hurting the area's marine ecosystem and hence directly contributing to ecological neglect.

Tourism governance in Cancun can be taken as an example of how the language of the earth sciences has been coopted to strengthen a neoliberal project. The examples from Cancun force us to reconsider the social and political alignments of the human–nature relationships and the extractive logic that still informs much of the tourism industry worldwide. Planning for leisure, climate change research, the scientific assessment of environmental risks, and the artistic embrace of narratives about the Anthropocene need to be understood as part of a Capitalocene urban governance. Recognizing this is the starting point towards planning real sustainable tourism cities and imagining alternative socio-environmental politics outside the market.

Notes

1. This research is the basis of a book in progress on the Mexican state's use of mass tourism, ecotourism, indigenous luxury tourism, and domestic souvenir production as state development tools in the Yucatan Peninsula. Part of this research and fieldwork in Cancun was done under the auspices of two international and interdisciplinary research projects: "Understanding the Dynamics of Urban Flexibility and Reconstruction", Oxford Programme for the Future of Cities, University of Oxford (2010–2012) and "Tourism and Mobilities in Times of Crisis", Ministry of Science and Innovation of Spain (CSO2011-26527).

2. Maquiladoras are also referred to as maquila factories, factories under foreign ownership where products are processed for export under labor intensive practices. See Kopinak (1996) for a comprehensive history of the origins and organization of maquiladoras in Mexico and their relationship with the USA.

3. Tourism represents almost 9% of the gross domestic product in Mexico. Tourism ranks as the 4th largest source of foreign exchange in the country, 50% of which is estimated to come from the estate of Quintana Roo where Cancun located (The World Tourism and Trade Council, 2015). Since the dismantling of the henequen industry and Cancun's creation, tourism is the driving economic force of the Mexican Yucatan Peninsula.

4. The Saffir–Simpson Hurricane Scale estimates the potential property damage and life loss of hurricanes based on a scale of 1–5 rating hurricane's sustained winds speed (NOOA).

5. A comparison could be elaborated here with how disaster mapping took place in Grenada also through the UN mandates and the National Disaster Management Agency (NaDMA) right after the hurricane. Differently from Cancun where emphasis was placed on the reconstruction of the hotel zone despite the city, in Grenada authorities used reconstruction efforts to strengthen a social understanding of disasters. The state built a partnership with UNICEF and elaborated a teaching guide and workbook to raise preparedness awareness among children, parents and teachers in all public schools. The initiative has been summarized by the United Nations ISDR Towards a Culture of Prevention Manual (2007).

6. MUSA and the artist's environmental art has been featured in a multiplicity of media, including The New York Times, The National Geographic, CNN, BBC, Smithsonian.com and travel magazines like Caribbean travel and life, the Lonely Planet. The Anthropocene sculpture figures too in the front image of the *South Atlantic Quarterly Journal* special issue Autonomía on the Anthropocene (Nelson & Braun, 2017) although none of the articles studies the museum in question.
7. Another interesting comparison could be drawn here with Grenada where the artist's museum has been classified as one of the Top 25 Wonders of the World by National Geographic and the artist's work has been used by the local government to get approval and resources to establish a Marine Natural Protected Area.
8. The Management Plan of this marine protected area was published in August 2015 but it is still not operative. The Plan aims to control the number of diving authorizations and tourists that can access the reserve for diving and snorkeling.

Acknowledgements

I gratefully acknowledge the support from researchers at the Centro de Información Geográfica, Universidad de Quintana Roo (UqRoo), Cancun's Municipal Planning Institute and the Departments of Ecologia Humana and Recursos del Mar, Centro de Investigación y Estudios Avanzados del Instituto Politécnico Nacional de Merida (CINVESTAV). An earlier version of this article was presented at the 2017 American Geographers Association Meeting's session "The Geopolitics of Tourism" that the author co-organized with Mary Mostafanezhad and Roger Norum. Special thanks to Jennie Germann Molz, Liz Montegary, Javier Caletrío, Elana Zilberg, Kathryn Kopinak, Fernando D. Rubio, and the anonymous reviewers for their generosity in their comments to this article.

Disclosure statement

No potential conflict of interest was reported by the authors.

References

Barendregt, B. (2019). Deep hanging out in the age of the digital; contemporary ways of doing online and offline ethnography. *Asiascape: Digital Asia, 4*, 307–315. DOI: 10.1163/22142312-12340082

Brockington, D., Duffy, R., & Igoe, J. (2008). *Nature unbound. Conservation, capitalism and the future of protected areas*. London: Earthscan.

Caldeira, T. (1996). Fortified enclaves: The new urban segregation. In J. Holston (Ed.), *Cities and citizenships* (pp. 114–138). Durham: Duke University Press.

Castellanos, B. (2010). *A return to servitude. Maya migration and the tourist trade in Cancun*. Minneapolis, MN: University of Minnesota Press.

Córdoba Azcárate, M., Baptista, I., & Domínguez, F. (2014). Enclosures within enclosures and hurricane reconstruction in Cancun, Mexico. *City and Society, 26*(1), 96–119. DOI: 10.1111/ciso.12026

Córdoba Ordoñez, J., & Córdoba Azcárate, M. (2007). Turismo y Desarrollo: la Eterna Controversia desde el caso de Cancún. In A. García Ballesteros (Ed.), *Un mundo de ciudades* (pp. 180–210). Barcelona: GeoForum.

Clancy, M. (2001). Mexican tourism: Export growth and structural change since the 1970. *Latin American Research Review, 36* (1), 128–150. http://www.jstor.org/stable/2692077

Crutzen, P.J., & Steffen, W. (2003). How long have we been in the Anthropocene Era?. *Climatic Change, 61*, 251–257. DOI: 10.1023/B:CLIM.0000004708.74871.62

Davis, B. (2016). An "Unbelievable" coincidence? Damien Hirst's Venice show looks almost exactly like the Grenada Pavilion. *ArtNet News*. ArtNet News 16 May 2017 https://news.artnet.com/art-world/damien-hirsts-unbelievable-coincidence-with-the-granada-pavilion-962066

Duffy, R. (2010). Nature-based tourism and neoliberalism: Concealing contradictions. *Tourism Geographies, 17*, 529–543. DOI: 10.1080/14616688.2015.1053972

Edensor, T. (2001). Performing tourism, staging tourism: (Re)producing tourist space and practice. *Tourist Studies, 1*, 59–81. https://doi.org/10.1177/146879760100100104

Escobar, A. (2012). *Encountering development. The making and unmaking of the third World.* Princeton, NJ: Princeton University Press.

Fletcher, R. (2011). Sustaining tourism, sustaining capitalism? Tourism industry's role in global capitalist expansion. *Tourism Geographies, 13*, 443–461. DOI: 10.1080/14616688.2011.570372

Fletcher, R. (this volume). Ecotourism after nature: Anthropocene tourism as the latest capitalist fix. *Journal of Sustainable Tourism.*

García de Fuentes, A. (1979). *Cancún: Turismo y Subdesarrollo Regional.* México City: Universidad Autónoma de Mexico.

Geertz, C. (1998). Deep hanging out. *New York review of books*, October 22.

Hamilton, C. (2015). The theodicy of the "Good Anthropocene". *Environmental Humanities, 7*(1), 233–238. DOI: 10.1215/22011919-3616434

Harvey, D. (2003). *The new imperialism.* Oxford, UK: Oxford University Press.

Higham, J., & Miller, G. (2017). Transforming societies and transforming tourism; sustainable tourism in times of change. *Journal of Sustainable Tourism, 26*, 1–8. DOI: 10.1080/09669582.2018.1407519

Igoe, J. (2010). The spectacle of nature in the global economy of appearances. Anthropological engagements with the spectacular mediations of transnational conservation. *Critique of Anthropology, 30*, 375–397. https://doi.org/10.1177/0308275X10372468

Karjanen, D. (2016). *The Servant Class City. Urban revitalization versus the working poor in San Diego.* Minneapolis, MN: University of Minnesota Press.

Kopinak, K. (1996). *Desert capitalism.* Tucson, AZ: The University of Arizona Press.

Low, S. (1999). *Theorizing the city. The new urban anthropology reader.* New Brunswick, NJ: Rutgers University Press.

McCormick, C., & Scales, H. (2014). *The underwater museum.* San Francisco, CA: Chronicle Books.

Moore, J. (2015). *Capitalism in the web of life: Ecology and the accumulation of capital.* New York: Verso Press.

Moore, J. (Ed.). (2016). *Anthropocene or capitalocene?. Nature, history, and the crisis of capitalism.* Oakland: PM Press.

Mostafanezhad, M., Norum, R., Shelton Eric, J., & Thompson Car, A. (2016). *Political ecology of tourism: Community, power and the environment.* London: Routledge.

Mota, L., & Frausto, O. (2014). The use of scuba diving tourism for protected area management. *International Journal of Social, Behavioral, Educational, Economic, Business and Industrial Engineering, 8*, 3358–3363. https://waset.org/publications/9999996/the-use-of-scuba-diving-tourism-for-marine-protected-area-management

Munt, I. (1994). Ecotourism or ego-tourism. *Race and Class, 36*, 49–60. DOI: 10.1177/030639689403600104

Nelson, S., & Braun, B. (2017). Autonomía in the Anthropocene: New challenges to radical politics. *South Atlantic Quarterly, 2*, 223–235. https://doi.org/10.1215/00382876-3829368

Padilla, N.S. (2015). The environmental effects of tourism in Cancun. *International Journal of Environmental Sciences, 6*, 282–294. DOI: 10.6088/ijes.6032

Pérez Villegas, A. (2000). El Desarrollo Turístico de Cancun, Quintana Roo y las Consecuencias sobre la Cubierta Vegetal. *Investigaciones Geográficas, 43*, 145–166. http://www.redalyc.org/articulo.oa?id=56904309

Puls, S.L., de la Rosa, R.S., & Olivares Urbina, M.A. (2013). Analysis of the room supply in the hotel zone of Cancun, Mexico, *EMU 9, Journal of Tourism Research Hospitality, 2*, 2–8. DOI: 10.4172/2324-8807.1000114

Richard, A. (2008). Withered Milpas: Governmental disaster and the Mexican countryside. *Journal of Latin American and Caribbean Anthropology, 13*, 387–413. DOI: 10.1111/j.1935-4940.2008.00043.x

Sayre, N.F. (2012). The politics of the anthropogenic, *The Annual Review of Anthropology, 41*, 57–70. https://doi.org/10.1146/annurev-anthro-092611-145846

Secretaria de Desarrollo Social (SEDESOL). (2010). *Información urbana Ciudad Cancun.* Mexico D.F.: Secretaría de Desarrollo Social.

Secretaria de Turismo (SECTUR). (2015). *Boletin Cuatrimestral de Turismo. Análisis Integral del Turismo* Mexico: Secretaría de Turismo.

Sheller, M., & Urry, J. (2004). *Tourism mobilities: Places to play, places in play.* London: Routledge.

Spalding, M.D., Ruffo, S., & Lacambra, C. (2014). The role of ecosystems in coastal protection: Adapting to climate change and coastal hazards, *Ocean and Coastal Management, 90*, 50–57.

Steffen, W., Persson, Å., Deutsch, L., Zalasiewicz, J., Williams, M., Richardson, K., … Svedin, U. (2011). The Anthropocene: From Global Change to Planetary Stewardship. *Ambio, 40*, 739–761.

Swyngedouw, E. (2013). The non-political politics of climate change. *International E-Journal for Critical Geographies, 12*, 1–8.

Torres, R., & Momsen, J. (2005). Gringolandia: The construction of a new tourist space in Mexico. *Annals of the Association of the American Geographers, 95*, 314–335.

The Nature Conservancy, 2011 Mexico. The Mesoamerican Reef. At: https://www.nature.org/ourinitiatives/regions/latinamerica/mexico/placesweprotect/mesoamerican-reef.xml

Thomson, G., & Newman, P. (2016). Geoengineering in the Anthropocene through regenerative urbanism. *Geosciences*, *6*, 46–57.

United Nations, International Strategy for Disaster Reduction (UN). (2007). *Towards a culture of prevention: disaster risk reduction begins at school. Good practices and lessons learned*. Retrieved from http://unesdoc.unesco.org/images/0018/001898/189857e.pdf

United Nations Climate Change Conference (UNFCCC). (2011). *United Nations Framewrok Convention on Climate Change*. Accessible at: https://unfccc.int/sites/default/files/resource/docs/2011/cop17/eng/07.pdf

Vicuña Gonzalez, V. (2013). *Securing paradise. Tourism and militarism in Hawai'i and the Philippines*. Durham and London: Duke University Press.

Wissenburg, M. (2016). The Anthropocene and the body ecologic. In P. Pattberg & F. Zelli (Eds.), *Environmental politics and governance in the Anthropocene. Institutions and legitimacy in a complex world* (pp. 15–30). New York: Routledge.

Ecotourism after nature: Anthropocene tourism as a new capitalist "fix"

Robert Fletcher

ABSTRACT

How does ecotourism – conventionally characterized by its pursuit of a "natural" experience – confront assertions that "nature is over" attendant to growing promotion of the "Anthropocene"? One increasingly prominent strategy is to try to harness this "end of nature" itself as a novel tourism "product". If the Anthropocene is better understood as the Capitalocene, as some contend, then this strategy can be viewed as a paradigmatic example of disaster capitalism in which crises precipitated by capitalist processes are themselves exploited as new forms of accumulation. In this way, engagement with the Anthropocene becomes the latest in a series of spatio-temporal "fixes" that the tourism industry can be seen to provide to the capitalist system in general. Here I explore this dynamic by examining several ways in which the prospect of the loss of "natural" resources are promoted as the basis of tourism experience: disaster tourism; extinction tourism; voluntourism; development tourism; and, increasingly, self-consciously Anthropocene tourism as well. Via such strategies, Anthropocene tourism exemplifies capitalism's astonishing capacity for self-renewal through creative destruction, sustaining itself in a "post-nature" world by continuing to market social and environmental awareness and action even while shifting from pursuit of nonhuman "nature" previously grounding these aims.

Introduction

In the 2009 climate change docudrama *The Age of Stupid* (Armstrong, 2009), a British couple travels to France to take a tour of a shrinking glacier. Their guide, a local man who has led excursion to the same glacier for decades, relates how he has witnessed the glacier progressively recede over the time he has observed it. Deeply disturbed by their experience, the British couple return home to continue their work as environmental activists and resolve to stop traveling by airplane as a result of what they have witnessed.

In this way, as the glacier they visited shrinks, the commercial tourism industry designed around it shifts from selling an encounter with the glacier *per se* to selling an experience of its imminent disappearance. In other words, the tour becomes less about getting in touch with a spectacular "nature", as in the past, than of experiencing the *loss of* this nature in the face of human-induced change. A subtle difference with profound implications, this dynamic points to the subject I wish to address in this article: how ecotourism, as a form of tourism centered on selling an encounter with natural spaces, is responding to increasing assertions that we now live in a new age – the so-called "Anthropocene" – portending the "end of nature" itself. As with the French glacier tour described above, I contend, a common strategy seems to entail shifting from selling an encounter with nature to selling an experience of the *end of this nature* as a novel tourism "product". In this way, I argue, the rise of Anthropocene tourism can be understood as a significant form of "disaster capitalism" seeking to transform the ostensive threat posed by Anthropocenic changes to the future of (eco)tourism into new opportunities for further tourism expansion. Consequently, engagement with the Anthropocene becomes the latest in a series of spatio-temporal "fixes" that the tourism industry provides to the capitalist system in general.

While my analysis is grounded in two decades of empirical research concerning a variety of tourism dynamics in diverse locations (e.g., Fletcher 2011a, 2011b, 2014), the material presented here is drawn primarily from secondhand sources. I begin by reviewing growing discussion of the Anthropocene in both natural and social sciences and the introduction of the concept into tourism studies specifically. I then outline the conceptual framework, a synthetic Marxist-post-structuralist political ecology, informing my analysis. After this I describe how tourism develop-ment tends to function as a form of "fix" or "disaster capitalism" in relation to assertions that the Anthropocene should be more properly labeled the "Capitolocene" due to the epoch's imbrica-tion within industrial capitalism. I then outline several emerging forms of tourism that can be seen to market "the end of nature" ostensibly signaled by the Anthropocene as one of their cen-tral "products". Some of these relate to ecotourism specifically while others concern efforts to sell an experience of the end of nature more broadly. I end by exploring the implications of ana-lysis for future research and practice concerning tourism development in this bold new "post-nature" (Wapner, 2010) age.

Touring the Anthropocene

The arrival of the "Anthropocene" has become a substantial focus of critical social scientific dis-cussion in recent years (see e.g. Braun, 2015; Castree, 2014a, 2014b; Cook, Rickards, & Rutherfurd, 2015; Lorimer, 2015). First advanced by geologists, most notably Paul Crutzen, at the turn of the century (Crutzen, 2002; Crutzen & Stoermer, 2000), the Anthropocene thesis is essentially the assertion that human influence has so come to dominate all nonhuman processes that it can now be potentially identified as a distinct layer in the geological record and thus should desig-nate our movement from the Holocene into a new epoch characterized by this pervasive human signature. It builds on an earlier contention by journalist Bill McKibbon (1989) that expansion of human influence – particularly in terms of anthropogenic climate change – has precipitated the "end of nature" as a distinct self-willed force altogether. Contemporary discussion of the Anthropocene contains quite similar proclamations that "Nature is Over" (Walsh, 2012) or that "Nature no longer runs the earth. We do" (Lynas, 2011, p. 12; in Castree, 2014b, p. 13). The con-cept is also caught up in growing discussion of the "Great Acceleration" threatening "planetary boundaries" (Steffen et al., 2011) as well as fears about a worsening "sixth extinction crisis" (Kolbert, 2014) or even the prospect of "biological annihilation" (Ceballos, Ehrlich, & Dirzo, 2017) altogether. Despite early debate concerning its validity as a scientific descriptor, over the last decade the Anthropocene concept has become increasingly accepted in both popular and aca-demic fora wherein exploration of its implications for both the natural and social sciences has exploded (see e.g. Castree, 2014a; Lorimer, 2015; Moore, 2015a; Ogden et al., 2013; Wapner, 2014; Zalasiewicz et al., 2008).

Yet the concept has also been criticized on a variety of grounds. Malm and Hornborg, among others, call attention to how it may obfuscate the ways that different groups are disproportion-ately responsible for and impacted by Anthropocenic changes, asserting that a homogenous "humanity seems far too slender an abstraction to carry the burden of causality" (Malm &

Hornborg, 2014, p. 65). In particular, some Indigenous peoples claim that the Anthropocene frame overlooks not only the ways in which they bear a disproportionate burden of recent impacts but also how "in the nineteenth and twentieth centuries, we already suffered other kinds of anthropogenic environmental change at the hands of settlers, including changes associated with deforestation, forced removal and relocation, containment on reservations" (Whyte, forthcoming, p. 3). In a similar vein, some suggest that the label is too ahistorical in neglecting to point out that the vast majority of the changes it identifies have occurred within and been caused by the age of industrial capitalism (Moore, 2016). Others have critiqued the concept as a fiction of human hubris refuted by evidence demonstrating the very limited control humans actually exert over many ecosystemic processes. Still others accept the concept yet contend that it points to the need for humans to reign in their actions rather than extending them further to exercise the "planetary stewardship" many call for in response to Anthropocene concerns (see Wuerthner, Crist, & Butler, 2014, 2015).

Only recently has the Anthropocene become a subject of tourism studies specifically (see Gren & Huijbens, 2014, 2016; Moore, 2015b, 2015c). In a first foray into the discussion, Gren and Huijbens asserted that "tourism policy and practice in the Anthropocene … implies that tourism needs to be measured up in specific relation to the boundaries and limits vis-à-vis the Earth and humanity at the global scale" (2014, p. 12). Indeed, from this perspective tourism development can be understood as itself a significant driver of Anthropocenic changes: "In Anthropocene understanding, modern tourism is a geophysical force which has contributed to the reshaping of the Earth for human purposes and to climate change" (Gren & Huijbens, 2014, p. 4). A subsequent edited volume then expanded substantially on these themes in various ways (Gren & Huijbens, 2016).

One of the consequences of considering the Anthropocene from a tourism perspective, for many, is the demand to take seriously the industry's impact upon planetary systems and the consequently urgent need to develop a far more sustainable tourism. Thus Gren and Huibens assert that it is "necessary to deepen the debate on sustainability in and of tourism by addressing the existing problems from the perspective of the geophysical forces of humanity and the Earth in the Anthropocene" (2014, p. 13). Amelia Moore (2015b, p. 8), similarly, finds "the Anthropocene idea reflected in contemporary sustainable tourism via new products and processes within the industry as it attempts to expand into new areas and produce new clients". Among other dynamics, she points to emergence of what she styles "the new ecotourism" in "what is still commonly referred to as 'sustainable tourism' … characterized by tourist ventures that now consciously address some of the contradictions inherent in the industry and the growing preoccupation with global change through destination design or branding" (Moore, 2015b, p. 9). She asserts, "In this way, the advent of ecotourism … over the past few decades is but the beginning of a rearticulation of development design in the name of coming to terms with anthropogenic planetary change" (Moore, 2015c, p. 516).

In this article I seek to contribute to this burgeoning discussion by building particularly on Moore's work to further interrogate some of the "new products and processes" that she highlights as characterizing an emergent Anthropocenic tourism. Specifically, I explore how novel forms of Anthropocene tourism seek to sustain capital accumulation in the face of new challenges posed by Anthropocenic changes themselves. These challenges – in the form of climate change calling into question the viability of long-haul air transport upon which the global tourism industry depends, for instance, leading to calls for reduced visitation or even "stay-at-home" tourism (see e.g. Gren & Huibens, 2016b; Hall, 2016) – affect the tourism industry as a whole of course but have particular implications for ecotourism – tourism focused specifically on selling "nature-based" experiences to support biodiversity conservation and community development (see Honey, 2008) – that have yet to be fully interrogated.

One of the main implications of the Anthropocene concept, after all, has been to call into question the characteristically modern, Western conceptual divide between "nature" and "culture" understood as diametrically opposite realms (Castree, 2014b; Latour, 2014). And it is of course in this very nature-culture dualism that the central appeal of ecotourism, entailing a quest to cross the conceptual divide from "culture" into an ostensibly unpeopled "wilderness" at the heart of an autonomous "nature", has long been grounded (Fletcher, 2014; West & Carrier, 2004). As Hall explain, "The commodification of nature as spectacle by tourism is clearly integral to nature-based tourism, where representations of, and connection to, places, people and causes has long been mediated through commodified images" (2016, p. 56). If the advent of the Anthropocene means that we are now "after nature", as Purdy (2015) contends, what does this then imply for the future of an ecotourism that has always entailed pursuit of this very nature, however elusive or even illusory it may actually be? And how might this ostensive "end of nature" forms the basis for other novel Anthropocenic tourism "products" as well?

This question is particularly pertinent given that ecotourism is increasingly appealed to as a key means to address growing environmental and social problems in pursuit of sustainable development. It is also commonly considered one of the main forms of financial and institutional support for protected areas seeking to preserve ostensibly pristine natural spaces and the often endangered nonhuman species they harbor (Fletcher, 2014; Hall, 2016; Honey, 2008). The "end of nature" thus paradoxically threatens not only ecotourism itself but the practice's contribution to help stave off the environmental degradation seen to precipitate this nature's end. Even more problematically, ecotourism has been described as contributing to exacerbating this very degradation. Hall observes, "Often despite, and in some cases perhaps because of, the very good intentions of those who seek to use tourism as a conservation tool, tourism is deeply embedded within processes of human-driven species loss that look set to become Earth's sixth great extinction event" (2016, p. 54). given that "many of the factors linked to biodiversity loss such as land clearance, pollution and climate change are also related to tourism development" (2016, p. 56). In multiple ways, therefore, the advent of the Anthropocene can be seen to significantly threaten the future of the ecotourism industry.

All may not be lost, however. In this article, I argue that tourism promoters may be working to sustain the industry in the face of Anthropocenic threats by shifting focus from selling an encounter with the "wild nature" the Anthropocene threatens to render obsolete to selling a confrontation with the end of this nature itself. In this way, the ostensive threat posed by the Anthropocene to the ecotourism industry's future may be paradoxically transformed into an opportunity for further expansion. The spectre of the Anthropocene is increasingly used to sell other experiences beyond ecotourism specifically as well. Below, I explain how this occurs. First, however, I describe the conceptual framework in which my analysis is grounded.

Political ecology after nature

In line with this special issue's overarching theme, my analysis is grounded in a political ecology perspective. Political ecology is an interdisciplinary field of study that explores the interconnections among social, cultural, political, economic and environmental processes at different levels and scales (see Bryant, 2015; Perreault, Bridge, & McCarthy, 2015). While the field comprises many different streams of analysis, two of the most prominent draw on Marxist and poststructuralist (mostly Foucaultian) perspectives, respectively. In this first stream, researchers have long sought to describe how a capitalist political economy harnesses natural resources in pursuit of capital accumulation (Heynen, McCarthy, Robbins, & Prudham, 2007). In the second stream, meanwhile, one of the main efforts has been to problematize the modern nature-cultural dichotomy and the way it shapes both human–environment relations and the forms of natural resource management these inspire (Braun & Castree, 2001; Escobar, 1999). In a formative article not-coincidentally titled "After Nature", Arturo Escobar (1999) endorses an "antiessentialist political ecology" that does not take a stable external "nature" as its referent, as in much conventional ecological science, but instead explores how distinctions between different processes are defined and materialized via environmental policy and practice. While at times these two streams of analysis have proceeded in parallel, increasingly they have been brought together to describe strategies for environmental management as the expression of a particular discourse and mode of production simultaneously. Combined, the two perspectives thus afford a synthetic perspective on the dynamics via which different cultural perspectives and political economic programs come together to shape environmental practices in particular places and times.

Political ecology has a long pedigree of application within tourism studies (Stonich, 1998) but only recently has the perspective become prominent with the publication of several books employing it in sustained series of analyses (Fletcher, 2014; Mostafanezhad, Norum, Shelton, & Thompson Carr, 2016; Nepal & Saarinen, 2016). Within this growing political ecology of tourism one finds a similar overlap between Marxist and poststructuralist perspectives as in political ecology more generally. Bringing these two perspectives together, then, affords an understanding of tourism development as both the embodiment of a particular discursive perspective and approach to human–environment relations and a political–economic process entailing pursuit of capital accumulation. In describing the ways that growing discussion of the Anthropocene simultaneously complicates a nature-culture dichotomy and harnesses this complication as the basis for renewed accumulation via creation of novel tourism "products", this article's analysis can thus be seen to follow Escobar in exploring the potential of a "political ecology after nature" to productively elucidate contemporary socioecological processes.

Disaster capitalocene

As previously noted, one of the main criticisms of the Anthropocene concept is that it presents an apolitical frame obscuring the fact that the transformative effects it highlights are not due to human activity as a whole but more specifically to the relentlessly colonizing expansion of industrial capitalism. Donna Haraway, Jason Moore and others thus assert that this epoch should more properly be termed the "Capitalocene" (Moore, 2016).[1] From this perspective, the rise of Anthropocene ecotourism may be viewed as a form of "disaster capitalism" *par excellence*. Disaster capitalism is Naomi Klein's (2007) popular term for a particular form of capitalism that seeks to harness crises created by capitalist processes themselves as opportunities for further accumulation. She thus defines disaster capitalism as "orchestrated raids on the public sphere in the wake of catastrophic events, combined with the treatment of disasters as exciting marketing opportunities" (2007, p. 6). As this definition implies, Klein sees this dynamic as particularly characteristic of capitalism's current neoliberal phase, wherein in addition to being harnessed as sources of new accumulation, disasters are also often used to justify further neoliberalization (i.e. privatization, marketization, commodification – see Castree, 2008) in a vicious cycle of escalating crisis.

The global tourism industry can be seen as one of the world's most effective and creative forms of disaster capitalism, by means of which a variety of problems precipitated by capitalist development are transformed into new "products" for tourist marketing and consumption (see Büscher & Fletcher, 2017; Fletcher, 2011a). Paradigmatic examples include slum tourism, in which the poverty created by unequal geographic development (Harvey, 2006; Smith, 2006) becomes the basis of voyeuristic tourism focused on experiencing this same poverty (Büscher & Fletcher, 2017; Dürr & Jaffe, 2012), and war tourism, in which tourists are transported to the conflict zones fueled by a massively lucrative weapons industry (Fletcher, 2011b). In this way, tourism can be understood as providing a series of spatial, temporal and other "fixes" to (temporarily) transcend limitations to capital accumulation created by capitalist processes themselves (Duffy, 2015; Fletcher, 2011a; Fletcher & Neves, 2012; Hall, 2014).

Ecotourism development can be viewed as a particular form of disaster capitalism, pursuing among others an "environmental fix" (Castree, 2008) for the ecological damage wrought by conventional capitalist development (Fletcher, 2011a; Fletcher & Neves, 2012). As a quintessentially neoliberal practice emphasizing creation of private markets for commodification of natural resources for sale as tourism "products" (Duffy, 2015; Fletcher & Neves, 2012; West & Carrier, 2004), ecotourism promotion can indeed be understood to fulfill both dimensions of Klein's definition.

In the long term, of course, the transformations precipitated by the Anthropocene will likely produce significant constraints upon tourism development as well as (human) life more broadly. As Huijbens and Gren assert, "Under the terms of the Anthropocene it seem unlikely that tourism can carry on in its modern register, that is, as a section of the Anthropos's geo-force which potentially undermines its own safe operating space by today's carbon-fuelled travelling" (2016, p. 5). Urry thus darkly predicts

> the substantial breakdown of many mobility, energy and communication connections currently straddling the world. There would be a plummeting standard of living, a relocalization of mobility patterns, an increasing emphasis upon local warlords controlling recycled forms of mobility and weaponry, and relatively weak imperial or national forms of governance ... Systems of repair would dissolve, with localized recycling of bikes, cars, trucks, computers and phone systems. Only the super-rich would travel far, and they would do so in the air, within armed helicopters or light aircraft, with very occasional tourist-type space trips to escape the hell on Earth in space, the new place of excess. (Urry, 2010, p. 207, in Gren & Huibens, 2014, p. 4)

In the short term, however, these transformations may offer particular opportunities for disaster capitalism via tourism expansion. In this way, Anthropocenic tourism may be understood as an important aspect of the process that Tsing (2015) describes as actors striving to forge livelihoods within the "capitalist ruins" of the Anthropocene more generally.

In what follows I explore this potential by focusing on a number of practices that can be seen to exemplify this dynamic: disaster tourism; extinction tourism; voluntourism; scientific tourism; development tourism; and, finally, self-consciously Anthropocene tourism itself. As previously noted, not all of these practices necessarily qualify as ecotourism *per se*, but rather exemplify the way in which the Anthropocenic changes are harnessed as the basis of touristic experience more broadly. The first two categories are most paradigmatic of the way Anthropocenic tourism may function as a form of disaster capitalism. The following three categories are less so, containing some elements that conform to this characterization and others that do not.

Selling the end of nature

Disaster tourism

"Disaster" or "dark" tourism entails the touring of post-disaster sites (Sion, 2014). These can be obviously human-induced disasters, such as the site of the nuclear meltdown at Chernobyl (Chubb, 2016), or ostensibly "natural" disasters, like New Orleans post-Katrina (Klein, 2007). Of course, as numerous researchers have pointed out, such disasters can rarely be considered wholly "natural" when one takes into account the contribution of human processes and structures to precipitating them (Blaikie, Cannon, Davis, & Wisner, 2014). Yet in their framing as purely natural disasters, such events might be understood to form the basis of ecotours seeking to explore such spaces. A common rationale for such tours in the face of critique denouncing them as tasteless voyeurism is thus that their directly experiential character confers a lasting impact leading to positive change. As Chubb (2016) writes, "What's more likely to make a profound impact; fire you up to care about big things? Someone telling you how terrible the Cambodian genocide was, or seeing the bones of murdered victims peeking through the soil for yourself?"

Disaster ecotourism of this sort takes a variety of forms. First is the mere viewing of the awe-inspiring power of the natural forces at work in such cases. The second is to raise awareness of their impacts on humans as well as other forms of life. The third is to raise awareness concerning the ways in which human processes, such as anthropogenic climate change, have contributed to such situations. In this way, disaster tours are increasingly linked explicitly to Anthropocenic processes to which the events and processes they feature are attributed.

A prime example of this dynamic concerns Hurricane Katrina, in response to which disaster tourism was promoted as a self-conscious form of disaster capitalism. As Gould and Lewis describe, "Hurricane Katrina (and hurricane Rita which followed) brought New Orleans' tourism industry to a screeching halt… To fill the void and to reap economic opportunity, tour operators developed Katrina 'disaster tours'" (Gould & Lewis, 2007, p. 177). Yet unlike in many disaster tourism sites, "human intervention in the natural environment is clearly implicated in the Katrina disaster causation narrative" (2007, p. 182). As the authors describe of one disaster tour company's spiel,

> The Gray Line narrative unambiguously places blame for the ecological disaster on human manipulation of the natural environment. It attributes responsibility for the flooding to the role of the oil and gas industry in cutting straight canals through the wetlands to facilitate the quickest possible river-gulf access, thus creating a vector through which the storm surge was directed into the city. (2007, p. 190)

The great irony in this case is that "New Orleans was not primarily an ecotourism destination prior to the ecological disaster. That is, while New Orleans's economy was dependent on tourism, the ecotourism segment of that industry was minimal" (2007, p. 182). Paradoxically, therefore, the destruction wrought by a hurricane seen as the product of human action became the basis for a novel ecotourism industry exploring this "post-nature" landscape. In this way, via Anthropocenic disaster itself "a previously non-ecological tourism destination is transformed, in part, into an eco-disaster tourism destination" (2007, p. 183).

Extinction tourism

"Extinction" or "last chance" tourism entails visits to view or experience phenomena that are framed as in danger of imminent disappearance (Leahy, 2008). Common examples include visits to shrinking glaciers, as in the example from the introduction, or to witness endangered species such as polar bears. Visits to small island nations such as the Maldives in danger of submersion due to climate change-induced sea level rise fall into this category as well. In this way, as Leahy (2008) describes, "Tourism companies are now using climate change as a marketing tool".

As with disaster tourism, extinction tourism can focus on one or more of the three different themes described in the previous section. The glacier tour described in the introduction, for instance, combines all three of these, simultaneously evidencing: the "natural" power of both glaciers and the global warming diminishing them; how this shrinking impacts the local people who have become dependent upon their glacier for their livelihood via tourism work; and how all of this should serve as a wake-up call concerning the planet-altering gravity of climate change more generally, inspiring reform in other areas of one's life as well.

In extinction tourism, value is thus actually created by nature's ostensible diminution, given that "there is a strong, perhaps perverse, desire in many people to go and see rare things" (Leahy, 2008). This irony of this dynamic is increased when one considers the role of the tourism industry itself in helping to fuel the climate change causing the diminution subsequently harnessed as a source of enhanced value. A more perfect circle of disaster capitalism would be difficult to imagine.

Voluntourism

Volunteer tourism or "voluntourism" occurs when a visitor pays for a tour in the course of which they contribute unpaid labor to a related social or environmental cause. This has been a rapidly growing segment of the tourism industry in recent years and a significant focus of research in its own right (see esp. Mostafanezhad, 2016; Vrasti, 2013; Wearing, 2001). Voluntourism can occur in different ways. In its most immersive form, one's entire experience centers around the labor one contributes to the cause. Throughout Costa Rica, for instance, volunteers are recruited to assist in turtle conservation efforts, in the course of which they patrol beaches both to dissuade potential poachers and document turtle nesting sites (Gray & Campbell, 2007). In some sites volunteers work through the night for weeks on end in this way. In a less immersive form, volunteer work is balanced with purely pleasurable excursions to other places where no work takes place. Some Costa Rican turtle conservation organizations, for example, offer volunteers whitewater rafting, ziplining and other adventure tourism excursions on weekends. While in some project voluntourists' labor is actually put to productive use, in many it is actually their money that is most desired as a key source of funding, in return for which they are made to feel useful by offered tasks that at least do not damage the work in question (Brightsmith, Stronza, & Holle, 2008).

Increasingly, voluntourism in many places is framed explicitly as an effort to address Anthropocenic processes and their impacts. Clemmens (2010) writes, "If climate change wasn't something on your radar as a voluntourist or voluntourism operator in the first decade of the new millennium, then it most assuredly will be in Decade 2.0". He foretells that

> climate change will generate unprecedented interest in the environment – particularly projects that can assist the native vegetation in flourishing and offsetting carbon emissions. Carbon offset programs will start to "employ'" voluntourists in such projects as:

> 1. Minimizing the impacts of their travel via the removal of invasive species (which allows the local flora, oftentimes responsible for absorbing more CO_2, to thrive),
> 2. Planting and caring for native species that mitigate flooding, erosion, and similar threats,
> 3. Planting and caring for native species that are both edible and drought resistant, and
> 4. Planting species that allow local residents to realize alternative income streams and reduce the likelihood of impacts from slash and burn agriculture, for example

In this way, voluntourists can work to mitigate the impacts caused by their own implication in the travel industry – in the process providing a key source of value for the organizations they support to keep selling an experience of helping to *stave off* the end of nature.

A subset of voluntourism is so-called "scientific tourism". From a certain perspective, all scientific research involving travel can be seen as a form of tourism – even social science research concerning tourism itself (West, 2008). In its more paradigmatic form, however, scientific tourism involves travel to participate as a fee-paying, non-expert assistant in scientific research. This obviously overlaps somewhat with voluntourism, the key difference being that while some voluntourists may be scientific researchers many will not. Travel to participate in research concerning climate change and other Anthropogenic processes, such as documenting endangered species and threats posed to them, is a rapidly growing phenomenon, and hence a key means by which the end of nature is being incorporated into the Anthropocenic tourism industry.

Development tourism

"Development" tourism occurs when visitors tour sites of economic development projects, often those to which they have contributed funding (as in the ubiquitous donor visits arranged by development organizations) (Salazar, 2004). One increasingly popular form of such tourism entails university study abroad trips which seek to educate participants concerning development challenges and interventions intended to address them. When such tourism occurs in the context of projects intended to tackle climate change and its impacts, such as reforestation projects, drought resistant agricultural development, coastal adaptation or any other of a host of different foci, this can be included within the Anthropocene tourism complex.

Related to this is "developmentourism", a neologism coined by Baptista (2017) to describe a dynamic in which economic development and tourism are so tightly conjoined that they can no longer be distinguished. As distinct from development tourism, then, in Baptista's development-tourism tourism is in fact the main form of development, while the main focus of this tourism is the development impacts of the tourism itself. When this tourism is focused on communities using tourism revenue to develop climate change adaptation projects showcased during the tour, an Anthropocene disaster capitalism dynamic is clearly present. One project I have experienced in Costa Rica, for instance, offers a tour of a demonstration forest dedicated to carbon sequestration, wherein the financing for sequestration is not the offset credits normally sold to fund such a project but the fees charged to tourists coming to view it.

Anthropocene tourism

The latest development in the end-of-nature-tourism arsenal concerns the creation of tourism initiatives that self-consciously adopt the Anthropocene label to harness the sense of planetary emergency it embodies as the basis for the experience on offer. To date most such tourism takes place in museums and art galleries. A prime example is an exhibition called Gaia in the Anthropocene held at the Garage Rotterdam in the Netherlands, 2 February–31 March 2018. As the Rotterdam Tourism Information site describes:

> In the last decades we became aware of the unmistakable influence humans exert on the conditions of the earth. Paradoxically, we also face growing unpredictability and uncontrollability of nature. Earthquakes, hurricanes, floods and extreme droughts appear to occur more and more often. Gaia in the Anthropocene stems from the idea that this fearsome development raises the desire for an approach that recognizes emotions, intuition, faith and myths that envision an ensouled nature. The exhibition is in line with the growing attention to nature-related myths in contemporary art and is a place where the visitor can withdraw from reality.[2]

Similarly, the Durham Museum in Omaha, Nebraska, USA, offers an exhibit called "Omaha in the Anthropocene", explaining:

> The "anthropocene" is a proposed new geological era currently under consideration by the International Commission on Stratigraphy. It makes a bold claim that humans have become a geologically significant force in earth's history. Objects are also important material sources of these historical changes. The Durham Museum is partnering with Creighton University's History Department to produce an immersive, interdisciplinary experience for students in the fall semester of 2017. In conjunction with the curriculum of Dr. Adam Sundberg, Assistant Professor of History and Digital Humanities, museum staff will instruct and assist Creighton students with independent research related to The Durham Museum's collection, distillation of that research into a lecture to be presented near the end of the fall semester, and an exhibition to be presented at The Durham in Spring 2018.[3]

Robin et al. (2014) describe three further exhibitions invoking the Anthropocene. The first, titled *Welcome to the Anthropocene: The Earth in Our Hands*, [*Willkommen im Anthropozän: Unsere Verantwortung für die Zukunft der Erde*] was held at the Deutsches Museum in Munich from 2014 to 2016. It covered

> 1450 m² (ca. 15,600 square feet) and is structured in three parts. The first section provides a comprehensive introduction into the Anthropocene both as a geological hypothesis and new conceptual framework. The introduction includes a range of technological objects that highlight the eras of industrialization (from the late 1800s, building on Paul Crutzen's narrative of the origins of the Anthropocene) and the Great Acceleration from the 1950s. The second part of the exhibition consists of six thematic areas that present selected phenomena of the Anthropocene, looking particularly at systemic connections, global and local interdependencies, and temporal dimensions … The third and final part of the exhibition discusses the future in the Anthropocene. It looks at past visions of the future, emphasizing their transformative potential while simultaneously highlighting their fragility and ambivalence. It then discusses possible scenarios of the future for people to consider in a more relaxing space; the final installation invites people to listen to possible scenarios and to plant their own possible scenario in an evolving field of paper daisies. (Robin et al., 2014, p. 212–13)

The second exhibition, a collection of "Anthropocene Posters sponsored by the Art Museum, Haus der Kulturen der Welt (HKW), placed the Anthropocene in a 'museum without walls' in the streets of Berlin in 2013" (Robin et al., 2014, p. 207).

Their third example, however, is not a museum or galley at all but rather an entire community. As the authors describe,

> Pyramiden, a town established to mine coal well north of the Arctic Circle in the early 20th century, has been recently transformed as an attraction for climate change science and heritage tourism. Here the hybridized local landscape creates a snapshot of the Anthropocene, bringing together industrial coal-mining heritage buildings, polar tourism and science forged in the geopolitics of the changing Arctic environment. (2014, p. 207)

Newell, Robin, and Wehner (2016) go further to describe how a growing number of museums in a wide variety of places are developing exhibitions focusing on anthropogenic climate change and its effects and implications. In this way, selling the end of nature need not be limited to the fragments of "nature" seen to remain in rural spaces but may occur within the heart of industrial civilization itself.

Of course, not all museums and art galleries necessarily entail significant commodification, being publicly-funded or otherwise subsidized by sources beyond patrons' fees. Hence, the extent to which such places perpetuate disaster capitalism *per se* demands more context-specific investigation. They can, however, still be considered elements of the more general tendency I have highlighted to harness the Anthropocene as the basis of expanded touristic experience.

Another example of self-consciously Anthropocenic tourism may comprise academic conferences wherein the Anthropocene and its implications are a focus of discussion (Swanson, Bubandt, & Tsing, 2015). As Swanson et al. observe, "In the past few years, conferences with "Anthropocene" in the title have increased even faster than CO_2 levels. Nearly every major meeting within the social sciences and humanities has had multiple Anthropocene-related panels" (2015, p.1 50). Such conferences themselves can of course rarely be considered a form of tourism *in toto*, in that most (although not all – see e.g. Honey, 2008) are organized (and subsidized) by non-profit organizations and participants attend primarily for professional rather than reactional reasons. Yet when attending these conferences participants do tend to spend money on hotels, restaurants and other hospitality infrastructure while often engaging in more direct touristic activities (either organized by the conference or independently) as well. Hence, aspects of such conferences may be seen to contribute to the Anthropocene disaster tourism complex explored herein.

Conclusions

Via strategies such as those previously outlined, the rise of Anthropocene tourism exemplifies capitalism's astonishing capacity for self-renewal through creative destruction (Harvey, 2006). In this way, the practice of ecotourism can be sustained in a "post-nature" world by continuing to market social and environmental awareness and action even while shifting from the pursuit of nonhuman "nature" that previously grounded these activist aims. Such dynamics may thus allow the ostensive "limits to growth" (Meadows, Meadows, & Randers, 1972) posed by the environmental degradation wrought by industrial capitalism to be transformed into opportunities for further growth itself. Consequently, expanding Anthropocene tourism may provide a key "fix" for obstacles to accumulation via spatial-temporal displacement of accumulated capital into new avenues for investment and future return. Given that tourism is one of the largest capitalist industries in the world (UNWTO, 2018), this potential may not be insubstantial (Fletcher, 2011a). As I have previously described elsewhere, therefore, the tourism industry may continue to play a key role in sustaining not only itself but the capitalist system as a whole into the Anthropocene (Fletcher 2011a). While Amelia Moore predicts that "Anthropocene arguments about responsible travel practices and destination design are less likely to center on protecting pristine nature and more likely to revolve around devising 'innovative' means of managing socionatural relations in ways that are familiar and attractive to tourists and that confirm their understandings of global change" (2015c, p. 519), my analysis suggests that the remaining fragments of this ostensibly pristine nature may constitute – for a time at least – one of the principle "products" of Anthropocene tourism.

Of course there remain clear limits to this potential in the long run, which must eventually be reached and a new model developed for tourism management, as well as economic governance more generally, that does not depend on continual growth (Büscher and Fletcher, 2017; Hall, 2009, 2010; Higgins-Desbiolles, 2010, 2018). Hence, Higgins-Desbiolles asserts that achieving a truly "sustainable tourism necessitates a clear-eyed engagement with notions of limits that the current culture of consumerism and pro-growth ideology precludes" (2010, p. 125). Given that as an economic system capitalism as a whole is dependent on such growth (Fletcher, 2011a), particularly in its current neoliberal wherein growth constitutes the "one true and fundamental social policy" (Foucault, 2008, p. 144), this movement must thus of necessity be away from capitalism as a mode of production and form of exchange. Fortunately, as Robinson reminds us, tourism "need not be a *capitalist* activity" (Robinson, 2008, p. 133, emphasis in original). To realize its "post-capitalist" potential, tourism must, first and foremost, "move radically from a private and privatizing activity to one founded in and contributing to the common" (Büscher & Fletcher, 2017, p. 664). In this way, the practice may be harnessed as a force of progressive political, social and environmental justice, as Higgins-Desbiolles (2006, 2008, 2018) maintains. The success of tourism as such an instrument of post-capitalist politics, in sum, must be gauged by the extent to which it pursues: (1) forms of production not based on private appropriation of surplus value; and (2) forms of exchange not aimed at capital accumulation; that (3) fully internalize the environmental and social costs of production in a manner that does not promote commodification and (4) are grounded in common property regimes (Agrawal, 2003).

Some commenters remain hopeful that, notwithstanding the various dilemmas outlined in this article and elsewhere, embrace of the Anthropocene concept will assist in such a shift. Thus Gren and Huibens call for "an ethical Anthropocene a-tourism geo-gaze" that "ensures a subversive critical unmasking of the ideology of tourism" (2014, p. 12). Similarly, Hall contends that the Anthropocene must direct attention to issues "such as the structural imperatives of the capitalist economy that drive emissions, species exchange and biodiversity loss" (2016, p. 66). This is line with overarching aspirations that a "good" Anthropocene may inspire humans to take seriously the impacts of their actions upon earth systems and hence take responsibility for this reality by becoming conscientious planetary stewards (Lorimer, 2015; Ogden et al., 2013).

Meanwhile, however, the mainstream tourism industry appears to remain committed to a path of incessant growth, with the UN World Tourism Organization (2018) maintaining its consistent yearly prediction that international arrivals will expand continually to reach 1.8 billion by 2030. At the same time, however, concerns about tourism "overcrowding" in many areas as a result of such promotion has provoked increasingly confrontational protest over the last several years. Industry insiders have responded by labeling this backlash "tourism-phobia" and asserting that "growth is not the enemy; it is how we manage it".[4] Hence, they are likely to continue to embrace Anthropocene tourism and other forms of disaster capitalism as a "fix" to stave off economic and environmental crises for as long as they are able.

This strategy is understandable. To question growth as the basis for tourism development would be to question not only the industry's particular *raison d'etre* but also its function as a key pillar of the capitalist political economy more generally. This would be a tall order for organizations central to the industry's development itself. Short of this, as in many other realms (Amore, Hall, & Jenkins, 2017), redirection to "post-political" (Swyngedouw, 2010) discussion of the possibilities of preserving forms of "sustainable" or "green" growth via mere technical adjustment rather than fundamental transformation is a logical and pragmatic choice (see e.g. UNEP, 2011). Whether this can succeed, however, remains a fundamental question for critical scholarship concerning our prospects for developing a truly sustainable tourism.

Notes

1. While employing the same term, Haraway and Moore conceptualize it in quite different ways, the nuancess of which are beyond the scope of this article (see Haraway, 2016 and Moore, 2015 for extended discussions). In this analysis I follow Jason Moore's (2015) approach in understanding the Anthropocene as an expression of the consequences of capitalist production in particular.
2. https://en.rotterdam.info/agenda/gaia-in-the-anthropocene/; accessed 6/12/2017.
3. https://durhammuseum.org/event/omaha-in-the-anthropocene-a-learning-exploration-with-creighton-university-2/?instance_id =4179; accessed 6/12/2017.
4. http://london.wtm.com/en/events/wtm-ministerial-programme/Ministers-Summit/; accessed 25/9/2017.

Disclosure statement

No potential conflict of interest was reported by the author.

References

Agrawal, A. (2003). Sustainable governance of common-pool resources: Context, methods, and politics. *Annual Review of Anthropology*, *32*, 243–262.

Amore, A., Hall, M.C., & Jenkins, J. (2017). They never said 'Come here and let's talk about it': Exclusion and non-decision-making in the rebuild of Christchurch, New Zealand. *Local Economy*, *32*(7), 617–639.

Armstrong, F. (Director) (2009). *The age of stupid*. London: Spanner Films.

Baptista, J.A. (2017). *The good holiday: Development, tourism and the politics of benevolence in Mozambique*. New York: Berghahn Books.

Blaikie, P., Cannon, T., Davis, I., & Wisner, B. (2014). *At risk: Natural hazards, people's vulnerability and disasters*. New York: Routledge.

Braun, B. (2015). From critique to experiment: Rethinking political ecology for the Anthropocene. In T. Perrault, G. Bridge & J. McCarthy (Eds.), *The Routledge handbook of political ecology* (pp. 102–115). New York: Routledge.

Braun, B., & Castree, N. (Eds.) (2001). *Social nature: Theory, practice and politics*. Malden, MA: Blackwell.

Brightsmith, D.J., Stronza, A., & Holle, K. (2008). Ecotourism, conservation biology, and volunteer tourism: A mutually beneficial triumvirate." *Biological Conservation*, *141*, 2832–2842.

Bryant, R.L. (Ed.) (2015). *International handbook of political ecology*. Cheltenham, UK: Edward Elgar.

Büscher, B., & Fletcher, R. (2017). Destructive creation: Capital accumulation and the structural violence of tourism. *Journal of Sustainable Tourism*, *25*(5), 651–667.

Castree, N. (2008). Neoliberalising nature: The logics of deregulation and reregulation" *Environment and Planning A*, *40*, 131–152.

Castree, N. (2014a). Geography and the Anthropocene II: Current contributions. *Geography Compass*, *8*(7), 450–463.

Castree, N. (2014b). *Making sense of nature*. New York: Routledge.

Ceballos, G., Ehrlich, P., & Dirzo, R. (2017). Biological annihiliation via the ongoing sixth mass extinction signalled by vertebrate population losses and declines. *Proceedings of the National Academy of Sciences of the United States of America*, *114*(30), E6089–E6096.

Chubb, L. 2016. *Dark tourism: Why it's okay to visit disaster zones on holiday*. Independent, 11 October. Retrieved from   http://www.independent.co.uk/travel/news-and-advice/disaster-tourism-new-orleans-auschwitz-hiroshima-hurricane-katrina-fukushima-killing-fields-a7356156.html.

Clemmens, D. (2010). *How will global climate change impact voluntourism in decade 2.0?* VolunTourism Newsletter, *5*(4). Retrieved from http://www.voluntourism.org/news-feature254.htm.

Cook, B.R., Rickards, L.A., and Rutherfurd, I. (2015). Geographies of the Anthropocene. *Geographical Research*, *53*(3), 231–243.

Crutzen, P. (2002). Geology of mankind. *Nature*, *415*, 23.

Crutzen, P., & Stoermer, E. (2000). "The 'Anthropocene.'" *Global Change Newsletter*, *41*, 17–18.

Duffy, R. (2015). Nature-based tourism and neoliberalism: Concealing contradictions. *Tourism Geographies*, *17*(4), 529–543.

Dürr, E., & Jaffe, R. (2012). Theorizing slum tourism: Performing, negotiating and transforming inequality. *European Review of Latin American and Caribbean Studies*, *93*, 113–123.

Escobar, A. (1999). After nature: Steps to an antiessentialist political ecology. *Current Anthropology*, *40*(1), 1–30.

Fletcher, R. (2011a). Sustaining tourism, sustaining capitalism? The tourism industry's role in global capitalist expansion. *Tourism Geographies*, *13*(3), 443–461.

Fletcher, R. (2011b). 'The only risk is wanting to stay': Mediating risk in Colombian tourism development. *Recreation and Society in Africa, Asia and Latin America*, *1*(2), 7–30.

Fletcher, R. (2014). *Romancing the wild: Cultural dimensions of ecotourism*. Durham, NC: Duke University Press.

Fletcher, R., & Neves, K. (2012). Contradictions in tourism: The promise and pitfalls of ecotourism as a manifold capitalist fix. *Environment and Society: Advances in Research*, *3*(1): 60–77.

Foucault, M. (2008). *The birth of biopolitics*. New York: Palgrave MacMillan.

Gould, K., & Lewis, T.L. (2007). Viewing the wreckage: Eco-disaster tourism in the wake of Katrina. *Societies Without Borders*, *2*(2), 175–197.

Gray, N.J., & Campbell, L.M. (2007). "A decommodified experience? Exploring aesthetic, economic and ethical values for volunteer ecotourism in Costa Rica" *Journal of Sustainable Tourism*, *15*(5), 463–482.

Gren, M., & Huijbens, E.H. (2014). Tourism and the Anthropocene. *Scandinavian Journal of Hospitality and Tourism*, *14*(1), 6–22.

Gren, M., & Huijbens, E.H. (Eds.). (2016). *Tourism and the Anthropocene*. London: Routledge.

Hall, C.M. (2009). Degrowing tourism: Décroissance, sustainable consumption and steady-state tourism. *Anatolia*, *20*(1), 46–61.

Hall, C. M. (2010). Changing paradigms and global change: From sustainable to steady-state tourism. *Tourism Recreation Research*, *35*(2), 131–143.

Hall, C.M. (2014), You can check out any time you like but you can never leave: Can ethical consumption in tourism ever be sustainable? In C. Weeden & K. Boluk (Eds.). *Managing ethical consumption in tourism: Compromise and tension* (pp. 32–56). Routledge, Abingdon.

Hall, C.M. (2016). Loving nature to death: Tourism, consumption, biodiversity loss and the Anthropocene. In M. Gren & E.H. Huijbens (Eds.), *Tourism and the Anthropocene* (pp. 52–73). London: Routledge.

Haraway, D.J. (2016). *Staying with the trouble: Making kin in the Chthulucene*. Durham, NC: Duke University Press.

Harvey, D. (2006). *Spaces of global capitalism*. London: Verso.

Heynen, N., McCarthy, J., Robbins, P., & Prudham, S. (Eds.) (2007). *Neoliberal environments: False promises and unnatural consequences*. New York: Routledge.

Higgins-Desbiolles, F. (2006). More than an "industry": The forgotten power of tourism as a social force. *Tourism Management*, *27*, 1192–1208.

Higgins-Desbiolles, F. (2008). Justice tourism and alternative globalization. *Journal of Sustainable Tourism*, *16*, 345–364.

Higgins-Desbiolles, F. (2010). The elusiveness of sustainability in tourism: The cultureideology of consumerism and its implications. *Tourism and Hospitality Research*, *10*(2) 116–129.

Higgins-Desbiolles, F. (2018). Sustainable tourism: Sustaining tourism or something more? *Tourism management perspectives*, *25*, 157–160.

Honey, M. (2008). *Ecotourism and sustainable development: Who owns paradise?* (2nd ed.). New York: Island Press.

Huijbens, E.H., & Gren, M. (2016). Tourism and the Anthropocene: An urgent emerging encounter. In M. Gren & E.H. Huijbens (Eds.), *Tourism and the Anthropocene* (pp. 1–13). London: Routledge.

Klein, N. (2007). *The shock doctrine: The rise of disaster capitalism*. New York: Metropolitan Books.

Kolbert, E. (2014). *The sixth extinction: An unnatural history*. New York: A&C Black.

Latour, B. (2014). Agency in the times of the Anthropocene. *New Literary History*, *45*, 1–18.

Leahy, S. 2008. *Extinction tourism: See it now before it's gone*. Retrieved from http://stephenleahy.net/2008/01/18/extinction-tourism-see-it-now-before-its-gone/.

Lorimer, J. (2015). *Wildlife in the Anthropocene: Conservation after nature*. Minneapolis, MN: University of Minnesota Press.

Lynas, M. 2011. *The God species: How the planet can survive the age of humans*. London: Fourth Estate.

Malm, A., & Hornborg, A. (2014). The geology of mankind? A critique of the Anthropocene narrative. *The Anthropocene Review*, *1*(1), 62–69.

McKibbon, B. (1989). *The end of nature*. New York: Anchor.

Meadows, D.H., Meadows, D.L., & Randers, J. (1972). *The limits to growth*. New York: Universe Books.

Moore, A. (Ed.) (2015a). Environment and Society: *Advances in Research*, *6*(1), special issue on "The Anthropocene: A Critical Exploration."

Moore, A. (2015b). Tourism in the Anthropocene Park? New analytic possibilities. *International Journal of Tourism Anthropology*, *4*(2), 186–200.

Moore, A. (2015c). Islands of difference: design, urbanism, and sustainable tourism in the Anthropocene Caribbean. *The Journal of Latin American and Caribbean Anthropology, 20*(3), 513–532.

Moore, J. (2015). *Capitalism in the web of life: Ecology and the accumulation of capital.* London: Verso.

Moore, J. (Ed.). (2016). *Anthropocene or capitalocene?: Nature, history, and the crisis of capitalism.* New York: Pm Press.

Mostafanezhad, M. (2016). *Volunteer tourism: Popular humanitarianism in neoliberal times.* London: Routledge.

Mostafanezhad, M., Norum, R., Shelton, E.J., & Thompson-Carr, A. (Eds.). (2016). *Political ecology of tourism: Community, power and the environment.* New York: Routledge.

Nepal, S., & Saarinen, J. (Eds.). (2016). *Political ecology and tourism.* New York; Routledge.

Newell, J., Robin, L., & Wehner, K. (Eds.). (2016). *Curating the future: Museums, communities and climate change.* London: Taylor & Francis.

Ogden, L., Heynen, N., Oslender, U., West, P., Kassam, K-A., & Robbins, P. (2013). Global assemblages, resilience, and earth stewardship in the Anthropocene. *Frontiers in Ecology and the Environment, 11*(7): 341–347.

Perreault, T., Bridge, G., & McCarthy, J. (Eds.). (2015). *The Routledge handbook of political ecology.* New York: Routledge.

Purdy, J. (2015). *After nature: A politics for the Anthropocene.* Cambridge, MA: Harvard University Press.

Robin, L., Avango, D., Keogh, L., Möllers, N., Scherer, B., & Trischler, H. (2014). Three galleries of the Anthropocene. *The Anthropocene Review, 1*(3), 207–224.

Robinson, W. (2008). *Latin America and global capitalism: A critical globalization perspective.* Baltimore, MD: Johns Hopkins University Press.

Salazar, N.B. (2004). *Developmental tourists vs. development tourism: A case study.* New Delhi: Kanishka Publishers.

Sion, B. (Ed.). (2014). *Death tourism: Disaster sites as recreational landscape.* New York: Seagull Books.

Smith, N. (2006). *Uneven development: Nature, capital, and the production of space.* Athens, GE: University of Georgia Press.

Steffen, W., Persson, Å., Deutsch, L., Zalasiewicz, J., Williams, M., Richardson, K., … Molina, M. (2011). The Anthropocene: From global change to planetary stewardship. *AMBIO: A Journal of the Human Environment, 40*(7), 739–761.

Stonich, S.C. (1998). Political ecology of tourism. *Annals of Tourism Research, 25*(1), 25–54.

Swanson, H.A., Bubandt, N., & Tsing, A. (2015). Less than one but more than many: Anthropocene as science fiction and scholarship-in-the-making. *Environment & Society, 6,* 149–166.

Swyngedouw, E. 2010. Apocalypse forever? Post-political populism and the spectre of climate change. *Theory, Culture & Society, 27*(2–3): 213–232.

Tsing, A.L. (2015). *The mushroom at the end of the world: On the possibility of life in capitalist ruins.* Princeton, NJ: Princeton University Press.

United Nations Environment Programme (UNEP). 2011. *Towards a green economy: Pathways to sustainable development and poverty reduction.* Nairobi: UNEP.

United Nations World Tourism Organization (UNWTO). (2018). *World tourism highlights 2017.* Madrid: UNWTO.

Urry, J. (2010). Consuming the Planet to excess. *Theory, Culture & Society, 27*(2–3), 191–212.

Vrasti, W. (2013). *Volunteer tourism in the global south: Giving back in neoliberal times.* London: Routledge.

Walsh, B. (2012, March 12). Nature is over. *Time Magazine,* 83–85.

Wapner, P. (2010). *Living through the end of nature: The future of American environmentalism.* Cambridge, MA: MIT Press.

Wearing, S. (2001). *Volunteer tourism: Experiences that make a difference.* New York: Cabi.

West, P. (2008). Tourism as science and science as tourism: Environment, society, self, and other in Papua New Guinea. *Current Anthropology, 49*(4), 597–626.

West, P., & Carrier, J. (2004). Ecotourism and authenticity: Getting away from it all? *Current Anthropology, 45*(4), 483–498.

Whyte, K.P. Forthcoming. Our Ancestors' Dystopia Now: Indigenous conservation and the Anthropocene. In U. Heise, J. Christensen, & M. Niemann (Eds.), *Routledge companion to the environmental humanities.* New York: Routledge.

Wuerthner, G., Crist, E., & Butler, T. (Eds.). (2014). *Keeping the wild: Against the domestication of the Earth.* New York: Island Press.

Wuerthner, G., Crist, E., & Butler, T. (Eds.). 2015. *Protecting the wild: Parks and wilderness, the foundation of conservation.* New York: Island Press.

Zalasiewicz, M.C., Smith, A. Barry, T.L., Coe, A.L., Bown, P.R., Bentchley, P., … Stone, P. (2008). Are we now lining in the Anthropocene? *GSA Today 18*(2), 4–8.

Friction in the forest: a confluence of structural and discursive political ecologies of tourism in the Ecuadorian Amazon

Annie A. Marcinek and Carter A. Hunt

ABSTRACT

Tourism in the Anthropocene is a powerful driver of global connections that has direct consequences for social and environmental well-being across the planet. This political ecological analysis of tourism in the Ecuadorian Amazonian presents ethnographic vignettes to account for the ways that interwoven global discourses related to biodiversity conservation and community development are encountered, contested, and leveraged to advance particular approaches to tourism at the local level. We invoke Tsing's theory of friction to frame these discursive encounters in the context of tourism-related decision-making in the community of Misahuallí, including instances of discursive shifts being leveraged into improved well-being of local residents. This paper makes an important contribution to the scholarship on the political ecology of tourism by bringing the emic perspectives of local residents to the forefront and by demonstrating the value of Tsing's friction metaphor for analyzing the global connections inherent in tourism. Frictions between inequities and imbalances of power, perpetuated by both the structures and discourses associated with the use of tourism to address conservation and development objectives, remain at the vanguard of tourism research as we move through the Anthropocene.

Introduction

With over 1.2 billion travelers crossing international borders every year (UNWTO, 2015), the phenomenon of tourism is a powerful force for bringing different ideas, values, worldviews, and discourses into contact with one another. Tourism is thus a hallmark activity of the Anthropocene, largely due to the spike in global tourism in the mid-twentieth century, the period that scholars now refer to as the onset of the Great Acceleration (Steffen, Crutzen, & McNeill, 2007). With anthropogenic disturbances threatening the life-supporting systems of the biosphere, and with the identification of international tourism as one of the primary drivers of socioeconomic and environmental change in this Great Acceleration of the Anthropocene (Steffen, Broadgate, Deutsch, Gaffney, & Ludwig, 2015), it has never been more important to understand the influence that interconnected global activities such as tourism have on local social and environmental well-being.

There have long been debates over the abilities of tourism and ecotourism to achieve lofty environmental conservation goals (e.g. Boo, 1990; Buckley, 2010; Budowski, 1976; Honey, 2008) and sustainable community development objectives (e.g. deKadt, 1979; Mowforth & Munt, 2015). Yet as the "largest scale movement of goods, services, and people that humanity has ever per-haps ever seen" (Greenwood, 1989, p. 171), tourism by its very nature involves unequal power relations. For better and for worse, these unequal relations determine how global and national-level discourses about conservation and development manifest in local-level decision-making. With the United Nation's inclusion of tourism as a target in three of its Sustainable Development Goals, and the UN's declaration of 2017 as the International Year of Sustainable Tourism for Development, a global discourse persists about tourism being a solution to social and environ-mental challenges of the Anthropocene.

As one of the largest and most transformative industries on the planet, tourism "demands a much greater degree of theoretical and empirical interrogation that it is given at present" (Duffy, 2016, p. xvi). The purpose of this paper was to provide an ethnographic interrogation of the ways that international tourism is negotiated by local Indigenous peoples, and how both dis-course and structures about conservation and development manifest in their perspectives on tourism in the Amazonian rainforests of the Napo Province, Ecuador. Political ecology provides a useful theoretical framing to this analysis that accounts for where, when, and how conservation, development, and ecotourism discourses intersect in tourism in this region and lead to particular structural outcomes in and around the community of Misahuallí. While we endeavor to address both structural (e.g. Blaikie & Brookfield, 1987) and discursive (e.g. Escobar, 1996) aspects of the political ecology approach, our approach draws primarily upon the metaphor of "friction" (Tsing, 2005) to characterize the ways that conservation and development discourses are encountered and emically interpreted in the context of tourism development. Through varied ethnographic vignettes that each present a different manifestation of friction in rainforest tourism, we analyze the power dynamics of tourism-related decision-making and how such dynamics influence social and environmental realities in Misahuallí, Ecuador. Given the increased understanding of global and local connection that characterizes the Anthropocene, Tsing's "friction" metaphor is invalu-able as a means of conceptualizing these interfaces in the context of tourism. Introducing tour-ism scholars to the influential work of Tsing and the friction metaphor is thus a key contribution of this paper.

The sections that follow first briefly introduce the general structural and discursive approaches to political ecology theory. We then review literature on the political ecology of tourism, before giving a nod to Tsing's contribution on encounters of friction. This review leads to the research questions that guide the subsequent analysis. After describing the Ecuadorian study context and the ethnographic research from which this paper is derived, we introduce five varied vignettes that provide ethnographically and theoretically engaged accounts of several frictions occurring at the confluence of different ideas about conservation, development, and tourism in the Ecuadorian Amazon. These vignettes provide the context for addressing our research questions. We conclude this paper by outlining the implications of the friction concept for further research on the political ecology of tourism in the Anthropocene.

Dimensions of political ecology

Political ecology "involves a clarification of the impact of unequal power relations on the nature and direction of human-environment interactions in the Third World" (Bryant, 1997, p. 8). Brosius (1999) suggests two main approaches to political ecology. Integrating elements of human ecol-ogy and political economy, a "structuralist" approach to political ecology first emerged from the cultural ecology writings of Wolf (1972) and later came of age with the publication of Blaikie and Brookfield's (1987) book *Land Degradation and Society*. This early work critiqued the blame levied on developing nations by asserting that environmental problems in the "Third World" are not a simply result of policy failures in less-developed countries, "but rather are a manifestation of broader political and economic forces associated notably with the spread of capitalism" (Bryant, 1997, p. 8). This structuralist approach emphasizes the persistence of poverty resulting from exhaustion of natural resources (Blaikie & Brookfield, 1987; Painter & Durham, 1995). Writings on "structural political ecology situate environmental change and resource conflicts in political and economic contexts with multi-scalar dimensions, ranging from the local to the global, and emphasized the historical processes that influence environmental change" (Campbell, Gray, & Meletis, 2008, p. 202).

Complementing the structural approach, a post-structural (Foucault, 1980), discursive perspective emerged that pinpoints discourse as a means of legitimizing certain forms of development to the expense of others (Brosius, 1999; Campbell et al., 2008). Poststructuralist or discursive political ecologists thus focus on ways that the use of particular language and symbols privileges and empowers certain viewpoints, institutions, and forms of development over others. This perspective is perhaps best embodied in the writings of Escobar (1996, 1999) and the essays in *Liberation Ecologies* (Peet & Watts, 1996). Scholars working in this vein seek to "dethrone 'hegemonic' discourses – those stories that hold a lock on the imaginations of the public, decision-makers, planners, and scientists – so that other possibilities and realities are made possible" (Robbins, 2012, p. 70). Doing so requires better accounting of local, emic perspectives (Harris, 1976), such as those captured through ethnographic methods. In this political ecological analysis of tourism in the Ecuadorian Amazon, we emphasize the emic perspectives that local government officials, tour operators, lodge owners, tourism employees, and other local residents hold about tourism's contributions the region's conservation and development challenges.

Political ecology and tourism

Stonich (1998) first brought the political ecology perspective to bear on tourism, focusing on visitation to the Bay Islands of Honduras. Characterizing the structuralist perspective, Stonich (1998, 2000) demonstrated how locals have little influence on decisions related to tourism development in their own communities and how little improvement in quality of life results from their participation in tourism, except among previously wealthy elites. Campbell (2007) drew similar conclusions in her exploration of conservation and tourism interfaces in rural Costa Rica. Additional descriptions of local resident exclusion from tourism-related decision-making have been put forth through research in numerous contexts (e.g. Belsky, 1999; Gössling, 2003). Accounting for such global to local dimensions, or *glocal* dimensions (Salazar, 2010), as well as the historical background of a given context, is thus key to outlining the political ecology of tourism in a given context.

Some political ecologists have focused their critiques on a particular form of tourism – ecotourism – as a neoliberal, market-based strategy that provides privileged access to biodiverse environments for affluent tourists while doing little to address the structural inequalities that maintain poverty for those residing near tropical biodiversity (e.g. Büscher & Davidov, 2016; Duffy, 2008; Fletcher & Neves, 2012; Hunt, 2011). Whether addressing tourism or ecotourism, these studies make it clear that preexisting tensions over structural inequalities in access to resources are frequently exacerbated by international tourism visitation to rural biodiverse areas of the lesser developed world.

Focusing on the language, symbols, and imaginaries associated with tourism, other scholars describe an ecotourism "bubble" (Carrier & MacLeod, 2005) that places priority on Western cultural and environmental values, practices, and worldviews (Davidov, 2013; West & Carrier, 2004). These writings highlight the ways that discourse can enforce political power dynamics, promote particular conservation and development agendas, and prioritize particular scales at which tourism-related governance occurs. Such findings are presented in recent collections on the political ecology of tourism (e.g. Mostafanezhad, Norum, Shelton, & Thompson-Carr, 2016; Nepal & Saarinen, 2016), and they remain central to discussions of tourism in the Anthropocene.

In spite of these scholarly critiques, the global tourism industry is growing. Tourism continues to receive investment and policy support through entities such as the World Bank (World Bank Group, 2015), U.S. Agency for International Development (USAID, 2015), and the International Ecotourism Society (TIES, 2017), all of which promote tourism as a joint conservation and development tool. More broadly, in 2017 UNWTO is overseeing the International Year of Sustainable Tourism for Development, an agenda and associated discourse that unquestionably characterizes tourism as a policy tool for promoting social and environmental well-being across the globe ensures that tourism continues to have prominent role in discussions of the Anthropocene. Political ecology has clear value for analyzing the ways that tourism forces such abstract global discourses to manifest in local, often contested, realities

Friction

Tourism is an activity that by its very nature is based on encounters between vastly different cultures, classes, value systems, and discourses. One heretofore unexplored means of characterizing these global–local encounters in the context of tourism is through the metaphor of friction outlined by Tsing (2005). Responding to the idealized notion that in a globalized world, transnational flows of wealth, power, and governance should proceed unencumbered or without friction, Tsing (2005) reclaims the idea of friction to describe those places of encounter where such flows come into contact and are often contested. The concept of friction recognizes the ways that local histories and environments collide with the recent arrivals of globalized structures and discourses. Tsing's (2005) ethnographic analysis of East Kalimantan, Indonesia, outlines the ways that international capital and global interconnectivity manifest for local people and forests, how global connections alter traditional livelihood practices, and how external structures and discourses associated with rainforest destruction are both locally contested and occasionally leveraged into new ways of being by local populations.

The notion of friction is thus applicable to analyzing the intersections between newly emerging global phenomenon and long-standing local cultural and environmental practices. Despite clear relevance, Tsing's (2005) influential work and the concept of friction has yet to be cited among tourism scholars or applied to analyses of tourism. As an industry that accounts for 9% of global GDP, 6% of global trade, and one in every eleven jobs on the planet (UNWTO, 2015), tourism is the epitome of the type of globally interconnected activity that characterizes the Anthropocene. Thus, while tourism continues to merit attention from political ecologists (Mostafanezhad et al., 2016), the political ecology of tourism would be served by further attention to the empowering and disempowering frictions that result from tourism's global interconnectivity, to the challenges this presents for local communities, and to the ways that local emic understandings are leveraged into new opportunities provided by tourism-related friction (Tsing, 2005). Here, we analyze emic accounts of the frictions occurring at the intersection of structural and discursive elements of tourism in the Ecuadorian Amazon through a series of vignettes highlighting tourism's "makeshift links across distance and difference" that are increasingly responsible for shaping this region in the Anthropocene (ibid, p. 2).

Current research questions

To explore the globally and locally interwoven assertions and manifestations of ideas about conservation and development as they are understood and negotiated by local stakeholders in the context of international tourism in the rainforest of Ecuador, this analysis is guided by the following specific questions:

- How does ecotourism discourse rooted in global and national political institutions circulate in the Amazonian community of Misahualli, Ecuador, manifesting in transformations to local social and environmental conditions?
- Where have these discourses encountered resistance due to conflicting emic understandings (i.e. where does friction exist), and where are these discourses readily accepted, or even leveraged, into locally beneficial outcomes, if at all?
- What can the friction metaphor and the inclusion of the emic perspectives gathered via ethnographic research add to our understanding of the political ecology of tourism and its consequences in the Anthropocene?

Research context

The Napo Province is the only Ecuadorian province where over half of the 100,000 residents (56.8%) self-identify as *indígena* (INEC, 2017). Indigenous communities in Napo are comprised of several ethnic groups, including the Cofan (Lu & Bilsborrow, 2011), Shuar, Achuar, Kichwa (Davidov, 2013; Smith, 2014), and Waorani peoples (Lu & Bilsborrow, 2011). Common livelihood strategies across these ethnic groups include wage labor for petroleum companies, cash cropping, handicraft sales, cattle, timber, hunting, and fishing (Lu & Bilsborrow, 2011).

While tourism is currently a minor economic development activity for the region, the percentage of the population working in tourism is growing in recent years (Doughty, Lu, & Sorensen, 2010), partly because ecotourism is seen as having the potential for reconciliation between competing agendas of neoliberal development, ecological conservation, and cultural survival (Davidov, 2013). Yet Buscher and Davidov (2016) and Davidov (2013) critique tourism's influence on local culture and environments, drawing important attention to the ways that the very same local institutions and government entities discursively support both conservation-oriented tourism and environmentally destructive extractive activities.

Prior to the arrival of international tourism, the Napo region had long been a landscape where competing ideas about development and conservation played out. Early development efforts were characterized by the patronal Spanish *encomienda* system during the colonization period. Later rubber booms in the nineteenth and twentieth centuries led to intense environmental and social exploitation. In recent decades, the region has continued to encounter numerous poorly designed policies for land reform, unsuccessful agricultural reform, and unsound economic development decisions (Davidov, 2013). Also serving as a battleground for conservation discourse, the Napo region is often characterized as containing one of the world's last high-biodiversity wilderness areas, a large portion of which has been designated as the Yasuní National Park (Bass et al., 2010).

Today, no discussion of conservation or development in the region can occur without reference to the petroleum extraction industry (Sawyer, 2004). In addressing the history of this ecotourism–extraction nexus, Smith (2014) explains how oil firms including Dutch Shell, French Perenco, and Canadian Ivanhoe alienated local communities not on the basis of environmental degradation, but rather on the basis of the firms' ability to provide sustainable economic development. By touting petroleum revenues as the key means of developing the region, this industry jeopardizes not only the "peak" species of amphibians, birds, mammals, and trees that make "Yasuní ... probably unmatched by any other park in the world in total number of species" (Marx, 2010, p. 1171), but also the livelihoods and lifestyles of local Indigenous community members (Sawyer, 2004; Smith, 2014).

Ethnographic research

The data presented in this paper were derived from ethnographic research undertaken in and around Misahuallí, Ecuador. The second author first visited this region in 2003, and the first author spent two seasons of dedicated fieldwork in the Misahuallí region in 2013 and 2016. Both researchers were involved in the conceptualization and design of this research, and both authors collaborated closely on the preparation of this manuscript. Here, the overall scope of methods employed in this project is described.

Study site

Tena, Napo's provincial capital, is the region's principal urban zone. Here, an array of tour operators provides services to visitors looking for adventure activities (e.g. white water rafting, caving) and eco-cultural tourism (e.g. canoe sightseeing, Indigenous cultural demonstrations, local cuisine). Puerto Misahuallí (hereafter simply Misahuallí) is 22 kilometers, or a 25-minute taxi ride, from the center of Tena. This community sits at the intersection of the Misahuallí and Napo Rivers. Its population is a mix of Indigenous peoples, non-Indigenous Ecuadorian natives, and a smaller expatriate population of Americans and Europeans. Tours out of Misahuallí range in length from one to five days and generally include sightseeing throughout riverside communities, rainforest hikes, tubing and kayaking, and camping in rustic cabins. Several Indigenous community tourism projects dot the nearby riverbanks. These projects also offer cultural demonstrations, traditional dance, lance throwing, cooking, and interpretation of Indigenous knowledge of rainforest ecology. Nearly all tours require travel via motorized canoes that depart from Misahuallí multiple times daily.

Data collection

Data were collected via semi-structured interviews, participant observation, and archival research. The semi-structured interviews were all conducted in Spanish, aside from one. They explored themes including but were not limited to a) leadership and entrepreneurship in tourism; b) representation of Indigenous culture in tourism; c) day-to-day operations of lodges and tour agencies; d) general characteristics of typical forms of tourism in and around Misahuallí; and e) the degree of local vs. foreign ownership and management of tourism projects. They targeted culturally specialized informants (Bernard, 2011) with expertise in the tourism industry in and around Misahuallí. A diversity of resident perspectives was sought to capture a breadth of local, primarily Indigenous, community stakeholder perspectives on tourism in the region. Additionally, frequent travel occurred to Tena to meet with regional Ministry of Tourism officials and participate in guided excursions to the municipal park, *Parque Amazonico la Isla*.

Twelve key informants were self-identified Indigenous residents, and four were local non-Indigenous Ecuadorian natives. An additional three informants were expatriate lodge owners hailing from outside of Ecuador (i.e. France, Colombia, and Chile/USA, the latter accounting for the only interview conducted in English). Although they represent a small proportion of our informants and an even smaller portion of the population in Misahuallí, expatriate residents represent an important manifestation of global interconnectivity and their views were also sought out. Qualitative data from 19 semi-structured interviews provide the bulk of the data analyzed here. These informants elaborated extensively on themes of local and national tourism policy, forms of tourism advertisement, notable successes and failures of tourism in the region, funding of regional development projects, and future directions for tourism in the region.

Ethnographic data were also obtained through spontaneous informal interviews that occurred with residents in Misahuallí on a daily basis. These encounters yielded emic insight into views of Ecuadorian national and international tourists, the varying ways that tourism businesses are structured and operated, and the influence of four main business-owning families in the town. Participant observation then consisted of tourism activities including boat trips to tourism projects, tours of lodge premises, meals with lodge owners and employees, and informal interactions with visitors to tourism projects. These informal interview and participant observation data were documented in detailed fieldnotes (Musante & DeWalt, 2011). Lastly, archival resources were gathered from Web sites of local tourism agencies, reports from locally active NGOs, and government sources including the Ecuadorian Ministry of Tourism.

Analysis

Field notes and interview transcriptions were compiled into a corpus of ethnographic data within MAXQDA. An initial open coding process identified emergent themes based on repetitions in responses, local resident or Indigenous categorization of phenomenon, use of metaphors and analogies to convey local perspectives, and theory-related materials (Bernard, Wutich, & Ryan, 2016). Iterative coding leads to more coherent categorization of these themes and the recognition of tensions and friction between competing discourses. As opposed to providing an exhaustive, in-depth account of a particular political–ecological friction in Misahuallí, the approach here instead presents a breadth of vignettes that each provides an emic view of the friction resulting from encounters with discourses associated with conservation, development, and tourism in this region of the Ecuadorian Amazon.

Friction in the forest

The five vignettes that follow each meant to represent a place of encounter (Tsing, 2005). Through vignettes, "an analyst builds a description of an event, activity, or person drawing the description directly from field notes" (Musante & DeWalt, 2011, p. 197). These vignettes were chosen as they provide maximum variation accounts of global–local friction that tourism generates and, in some cases, alleviates. The first vignette centers on emic Indigenous residents' interpretation – and support – of Ecuadorian president Rafael Correa's change of policy from biodiversity conservation to petroleum extraction in the highly biodiverse Yasuní National Park. A second vignette moves toward the regional level of Tena to provide an account of the ways in which the juxtaposition of petroleum-driven development and ideas about environmental conservation has unfolded in the recent history of an interpretive eco-park. In the third vignette, the local mayor is profiled. His perspective on social and economic development needs of his district is analyzed to reveal how shifting attitudes toward tourism are influencing the development investments in Misahuallí. A fourth vignette then compares emic Indigenous and expatriate views on the appropriate approaches to tourism. The fifth and final vignette illustrates an event unfolding over time – the creation of a women's cooperative – that transformed traditional livelihoods and gender roles into new cultural patterns among an Indigenous community near Misahuallí.

Vignette #1: globally significant biodiversity meets a crude industry

Discourse serves as a means of legitimizing certain forms of development to the expense of others (Brosius, 1999). Nowhere is this truer than in the greater Napo River region, home to Yasuní National Park, regarded as one of the most biologically diverse places on the planet (Bass et al., 2010; Finer, Moncel, & Jenkins, 2010). The only two existing voluntarily isolated Indigenous groups in Ecuador inhabit the park, along with an additional 3,000 contacted Indigenous people belonging to Kichwa and Waorani groups. Yasuní is currently one of three "recommended destinations" in Amazonia promoted on the government's official travel Web site (Ecuador Travel, 2017). Seemingly at odds with the publicity of this Amazonian rainforest as an idyllic destination for adventurous nature travelers is the occurrence of petroleum extraction, which has been the major driver of social and environmental change in the Ecuadorian Amazon since the 1970s when oil speculation swept across the region (Sawyer, 2004; Büscher & Davidov, 2016). Driven by large multinational investments and generating massive revenues for the country, oil has subsequently risen to become Ecuador's primary export (Sawyer, 2004; Smith 2014).

In 2007, the year of his election, President Rafael Correa announced to the United Nations the presence of large deposits of crude oil in the Ishpingo-Tambococha-Tiputini (ITT) field of Yasuní National Park. Correa declared that these reserves would remain underground indefinitely under the condition that the international community supports the Yasuní-ITT initiative (Bass et al., 2010). Through this attempt to "put social and environmental efforts first" on an international stage, Correa offered to set up a trust to collect $3.6 billion, or half of the revenues the state would have otherwise collected if the oil were actually extracted. This money was to then be invested in renewable energy projects throughout the country. The Yasuní-ITT initiative called upon international governments, civil society organizations, socially and environmentally responsible private sector companies, and citizens worldwide to contribute to the trust and keep the oil underground (Larrea & Warnars, 2009).

The Yasuní-ITT initiative brought Ecuador and Correa's conservation and development discourse to the global stage. At home, the president leveraged this plan into strong support among some Indigenous residents of Misahuallí. In the early summer of 2013, several interviewees expressed respect for their country's elected leader. At that time, Correa had made two visits to the community since his election as president. One woman, a local community leader, believed Correa to be a savior of sorts. She recalled that during his visits to the community, Correa made sure to describe how his administration had procured an existing private multinational petroleum company in nearby Waorani territory. Motivation behind this purchase, according to the president, was to provide communities with more than just cash payouts. Correa promised school buildings, fútbol fields, and potable water systems. Correa's efforts to campaign for, provide, and verbally reinforce more sustainable, lasting compensation for petroleum activities were commendable for this informant, despite the fact that contamination from extractive activities was already well documented in the region.

Another informant in her late 20s felt empowered by Correa's discursive tactics to bring global environmental issues to the local stage. His ability to speak the Kichwa language also earned admiration, and she quoted his statement made during one of these visits, "I am Kichwa, even though I don't have Kichwa blood, because we are a multicultural country". Elaborating on the influence of Correa's speeches, this respondent felt her Indigenous culture mattered to the president, to the larger Ecuadorian community, and through the initial efforts to establish the Yasuní-ITT, to the world at large.

Yet in August of 2013, the Yasuní-ITT fund had raised just $13 million in donations. Abruptly, Correa and the Ecuadorian government canceled the plan and liquidated the trust. The administration's explanation for this shift in policy was that without international support, biodiversity conservation was untenable. Oil exploration and drilling would have to move forward in the national park as an "economic obligation", that is, a means of bringing development to the country's poor. In a nationally televised speech, Correa argued:

> The world has failed us … it was not charity that we sought from the international community, but co-responsibility in the face of climate change … the fundamental factor of this failure is that the world is a grand hypocrisy. And the logic that prevails is not that of justice, but of power.

The Yasuní-ITT initiative, and especially President Correa's interpretations of it on the global stage, positioned Ecuador as a social and environmental leader among developing countries, one that had been victimized by the failure of more powerful and wealthy nations to act on behalf of the globally significant cultural and biological diversity in the country's Amazonian rainforest. As in his visits to Misahuallí, Correa situated conservation and development discourses in opposition, though this time to establish a hegemonic consensus among the Ecuadorian people that other nations' governments, companies, and individuals were to blame for the continued social and environmental disruptions in the Amazon.

The Yasuní-ITT initiative and its unrealized outcomes embody numerous frictions that result when well-meaning global discourse related to an unquestioned need to protect biodiversity encounters a strong nationalist development discourse that prioritizes economic development as a means of confronting poverty (Tsing, 2005). Initially emboldened by rhetoric about how the preservation of Yasuní National Park would "combat global warming by avoiding the production of fossil fuels in areas which are highly biologically and culturally sensitive", Correa craftily evaded political responsibility for proceeding with extractive activities in the Amazon. In a widely viewed speech, he maintained that civil and governmental entities across the globe share responsibility for the conservation, or not, of the Amazonian region; this is a similar message of Ecuadorian solidarity to that which he conveyed directly to respondents in Misahuallí during his 2013 visit.

As premiums on crude oil strengthen connections between the global markets and this rural biodiverse rainforest along the Napo River, Correa's handling of the Yasuní-ITT initiative provides an exemplar of the intertwining of conservation and development structures and discourses at a national level, and how an elected leader leverages them for political purposes, and how the resulting policy descends metaphorically from the national government in Quito down the rivers of the Amazonian region to directly affect emic perspectives of the social and environmental well-being of local communities.

Vignette #2: a regional manifestation of development in the tourism–extraction nexus

Political ecology is effective for situating environmental change and resource conflicts in political and economic contexts across local to the global scales and in ways that emphasize historical processes influencing change (Campbell et al., 2008). Such change is underway in the provincial branch office of the Ministry of Tourism in Tena. This office sits across the Pano River from *Parque Amazonico la Isla*, a 25-hectare municipal park that initially served as a valuable environmental interpretation center for visitors and Ecuadorian schoolchildren upon opening in 1995. Currently, the park is in disrepair and closed to the public. According to a local environmental engineer in the regional Ministry of Tourism office, for the past 20 years, funding for the project has intermittently come from several entities. It was established by the Ecuadorian government using revenues from oil exploration and drilling. Initial construction included a now-faded billboard standing over the park's entrance, boasting *"El Petroleo impulsa el Buen Vivir!"* (Oil drives the Good Life!). This expression invokes the *Buen Vivir* nationalist discourse and movement underway in Ecuador (and across South America) at the time that called for alternatives to predominantly "Western" forms of development (e.g. Gudynas, 2011). Yet the billboard's message, and its location in a nature park, is an illustration of contradiction and irony. When a Ministry of Tourism official was interviewed, he was asked about the *Petroleo* billboard and the paradox it represents sitting in an environmental interpretation center. Although he acknowledged that oil extraction is clearly at odds with the environmental education encouraged through the park's interpretive efforts, support persists for the sign since it represents *Ecuadorian* support for local development. Clearly, there is a friction between competing conservation and development discourses.

Another such frictious example occurs in the park's facilities. In 2009, a $50,000 grant from the US Embassy provided funding for refurbishments, new educational equipment, chairs, desks, cabinets, and projectors for library and laboratory buildings. The grant came with the arbitrary condition that US Civil Rights history be worked into the curriculum for Ecuadorian schoolchildren who visit the park. Evidence of the grant manifests in the form of plaques inscribed with the US Embassy logo that embellish all signs in the municipal park and in a photo exhibit of US Civil Rights leaders on display in the library. The project has been at the mercy of whichever institutional entity is willing to invest at a given time and a shifting set of stipulations attached to such investments.

Eliminating such stipulations, Tena's mayor cut funds to the park completely between 2009 and 2015. Operations at *Parque Amazonico* ceased for several years. Resident animals were re-homed, and small interpretive "bio-stations" dotting the trails around the park were disassembled. In 2015, a newly elected mayor once again reassessed the park and worked it back into the provincial budget. Yet interviewed officials still express concerns about a continued, reliable source of funding, even as the Napo provincial branch of government works to invigorate the park and prepare it to once again receive guests. The park is valued as a community resource, and earmarked provincial budget funds provide local Ministry of Tourism officials with the flexibility to organize its educational programs based around native species, biodiversity, and the importance of conservation.

With the US Embassy's gift expended, officials no longer feel inclined to teach US Civil Rights history. The inscribed plaques remain in place. Relics of petroleum industry funding persist, including the billboard. Coupled with a criticism of Western forms of development that is inherent in reference to *Buen Vivir*, this sign serves as a symbol of nationalism, which is more emically desirable than discourse originating in the "West" about the US Civil Rights movement, embodied in the US Embassy plaques. The competing messages on these two signs represent an encounter of friction. Further friction comes into play when visiting tourists, both national and international, view this billboard and draw their own conclusions about the discourse it conveys regarding funding for conservation and development in the region. Such short touristic encounters are likely to overlook the nationalist explanation of the sign's persistence and instead arrive at the conclusion that oil continues to fund the park's existence when, ironically, it is their own presence via the Ministry of Tourism that supports the park's operations.

Vignette #3: a discursive shift in local development priorities

Tsing's (2005) concept of friction recognizes the ways that local histories and environments collide with the recent arrivals of intertwined global structures and discourses. Tourism development is driving such a collision in the community of Misahuallí. The Napo provincial government's budget supports municipal projects, including maintenance and renovation to Misahuallí's central plaza. This plaza is the epicenter of the town's tourism activity and, as noted below, serves as a focal point around which frictious discourses around desired forms of development in the community are articulated. Not coincidentally, visitors to this plaza regularly encounter a troop of 30 capuchin monkeys perched among the surrounding trees. While these wild animals are free to roam throughout the town, they know enough to congregate where the tourists do. In the plaza, they find opportunities to harass newly arrived groups of foreign and domestic visitors, often conspiring to obtain sweet snacks while visitors snap a photo. They have become Misahuallí's accidental mascots, just as the plaza has become the accidental hub of tourism activity.

In the summer of 2016, the plaza was completely encapsulated in a light green tarp as reconstruction and remodeling were underway. The mayor of Misahuallí at that time had first been elected in 2009 and was reelected for a second 5-year term in 2014. He also owned and operated the largest convenience store in the town center. His family moved to Misahuallí when he was four months old, and he attended the local school system alongside Ministry of Tourism officials mentioned earlier. On a daily basis, the mayor could be observed inside his storefront. As he leaned on the cement ledge facing the sidewalk, residents frequently stopped by to talk throughout the day. During two separate interviews, it was noted that these visits consisted of both friendly greetings and more pressing discussions regarding town matters.

When first encountered at this daily post, the mayor was eager to discuss the plaza renovation project underway across the road from his storefront. He proclaimed the revitalization of the plaza as a victory for him and his employees. He provided a narrative about the tireless efforts of him and his staff to convince the provincial government that plaza upgrades were needed. In response to an inquiry about the problem with the plaza prior to the new construction, he replied:

> No, no, no. It was just old. Remodeling. Now the park will be more open, so people can walk through (pulls out a large, rolled-up map) trees, walkways, (he lays his finger over three structures dotting the center) these are for the monkeys to climb. This is a center area here. It's very different. It's going to take two months; in August it's done. The price was very high, from the provincial government budget.

Although these changes do not seem especially crucial, his enthusiasm for the project was evident.

The mayor later listed his duties for the forty surrounding villages, all part of the Misahuallí township, stating,

> Roads, electricity, we have to help them out with all of these things. Now we are doing projects to make sure there's clean water. In one or two more years, each community will have it. It's difficult, some of [the communities] are very far away. It's much easier to help the communities that are more organized and closer.

In recounting typical mayoral duties, a palpable shift in the enthusiasm was evident between his unimpassioned descriptions of basic development services needed by the region's residents and a more animated discussion of the more recent renovation projects related to increasing the number of tourists visiting the region. This contrast in enthusiasm reveals a negotiation of friction between competing development discourses. When asked whether the plaza was the best place to direct provincial government funds, he emphatically prioritized the needs of tourism, stating, "yes, but more. For water fountains, things like that. It's better than nothing. Tourists that come here want to see a beautiful park. This is our picture; this is what we show the world". When prompted to reflect back on the budget for plaza refurbishment, the mayor indicated he was increasingly disinclined to direct investment toward community projects related to electricity, roads, clean water, and other common development concerns.

That shifts have occurred in priorities for development, and the governance of tourism is not surprising given that the mayor, like many other residents in Misahuallí, attributes 90% of the community's employment opportunities to tourism. Given that Lu and Bilsborrow (2011) place regional estimates at no more than to 45%, this is an obvious overstatement. Yet the mayor's rhetorical purpose here is to emphasize that the money available from the provincial level government for typical development projects in Misahuallí is highly limited, hinders his effectiveness, and holds the community back. He feels inclined to base Misahuallí's development priorities on the money his local government can access consistently. For the time being, what is consistently available is for improvements that benefit the tourism industry, such as beautification of the town plaza. For better or for worse, this politician's efforts to align his personal and financial priorities with the global tourism market in an attempt to attract additional visitors to his community have pushed enthusiasm, and potentially funding, for basic development services to the side. Thus, existing frictions between colliding economic and social development discourses are renegotiated so as to procure any provincial funding available.

Vignette #4: locally contested ways of knowing and the production of tourism

As Tsing (2005) notes, competing structures and discourses associated with conservation and/or development are often locally contested. Such frictious contests are underway among Indigenous, local non-Indigenous, and expatriate residents engaged in tourism around Misahuallí. Although it is difficult to explicitly separate out these groups and explore the manifestations of tourism knowledge they apply to their services, contrasting narratives emerged from informants' responses to similarly framed questions regarding views on tourism and community. Interactions among people with unequal or different ways of knowing, and who work to accommodate themselves to the global force of tourism, may lead to the emergence of new or unprecedented cultural forms and discourses (Tsing, 2005). As one such example in the tourism context, one expatriate lodge owner shares his opinion on benefits or drawbacks of foreigners owning local tourism operations, stating,

> There is no product like this one in the whole area, okay? The way we do tourism, cultural tourism, I haven't seen. There's around 27 places here, I've stayed in all of them. And the first thing that you notice is that the owners don't live there. We live here the whole year. This is my house, this is where I have breakfast (*gestures around the guest dining area*)."

This owner espouses a sub-discourse related to tourism in which his operation prioritizes the experience for visitors and thus provides the ideal service, a service without a comparable alternative among the local counterparts.

This expatriate also espouses a particular stance toward the lifestyle choices appropriate for lodge owners. He elaborates on his lifestyle as follows,

> You see I don't have a TV; here we don't have TV. I am un-contacted with the world … here you are un-contacted with everything. You don't know what is happening out there (*gestures upriver toward Misahuallí and Tena*). At least I don't give a s**t. If Trump gets elected it's not my problem. If Greece is in crisis, it's f*****g their problem. If Europe is in crisis, it's their problem.

This informant actively sought out a particular lifestyle that allows a separation from "out there", which seems to represent not just the USA, Greece, and Europe, but also more developed countries in general.

He later explains a personal motivation and reasoning for his marginal living, extending this personal discourse into a rationale for extensive solo trips into Waorani territory for months at a time. He feels that this authenticates the opportunities his lodge provides for guest interactions with other local Kichwa residents, their traditions, and their ecological knowledge. These web advertisements and personal reasoning seem to overlap, yet there is a notable disagreement between the owner's own isolationist rationale for inhabiting the lodge year-round and the cultural interactions encouraged for potential lodge visitors on the Web site.

Even more friction is evident in comparison with the approach taken by lodge owners and managers in Indigenous communities around him. In a region predominantly populated by Indigenous residents, expatriate sub-discourses related to lifestyle and isolation obtain little traction. When similar questions were posed to a Kichwa male who manages a small tour operator in the center of Misahuallí, his emic view openly contested foreign ownership in favor of the benefits that community-owned tourism yields. He states,

> When there is a foreign owner, they don't act … For example, we are guides here. Some of those other [owners] are in Quito, in Europe, and just the administration is here. If you need the management in community [tourism], you can go right to the people. You can communicate to their face, and converse. This is the difference.

While this local Kichwa manager exhibits little disagreement with the expat's concern for good products and services, he reveals friction with a discourse that favors foreign-owned projects, emphatically situating his own project in diametric opposition.

The reason for this friction is that community owned and operated tourism is emically viewed as contributing directly to the needs of communities that may otherwise be at risk under foreign ownership:

> And these people don't know what [communities] need, while we, if I bring my group, I already know what I can do for this location. This is the difference. For example, if you are an [outside] owner, you visit the communities on trips. Sometimes you neglect to give the help that is really needed. While I'm here, I am a

guide, I have more, how can I say ... I have the idea to help. So a community that we solicit, we give directly to the community.

The alternative narrative offered here indicates that tourism should prioritize the obligation to local Indigenous communities engaging in tourism. As a local guide and expert in Amazonian biological and cultural diversity, this informant claims an authority to providing services or deciding where to spend group money in the communities when he arrives at a particular location with tourists.

While by no means exhaustive examples or the only contested discourses between foreign owners and local Kichwa tour operators, the contrasting perspectives presented here provide evidence of ongoing friction between different, contested ways of approaching tourism. A clear friction exists between expatriate discourse that prioritizes the visitor experience and a competing emic discourse that prioritizes the development outcomes for local communities. This friction may exacerbate the differences between these business owners and the market segments they serve. Tourists visiting these projects encounter misconceptions and misinformation embodied in particular discourses about who has the most moral authority to operate tourism projects in the region.

Vignette #5: empowering encounters of friction

The interface between global discourses and local traditions does not always result in the exacerbation of conflict between people and environments involved in tourism. As Tsing (2005) notes, "the effects of encounters across difference can be compromising or empowering" (p. 6). As traditional subsistence lifestyles of Indigenous communities across the Amazon have come into contact with globalized markets, namely natural resource extraction, small-scale tourism has occasionally proven to be a mechanism through which communities can effectively conserve both natural and cultural resources (Stronza, 2010). As such resources are renegotiated and realigned, new cultural patterns can emerge (Tsing, 2005). This is the case near Misahuallí with the Shiripuno Lodge. This operation provides a local example of how Indigenous community institutions can shift over time to bring about new systems and discourses about gender empowerment, leading to greater autonomy and sustainable income for women.

Located a few minutes from Misahuallí via river transport, Shiripuno is a traditional Kichwa community. In 2005, with assistance from a small French NGO, community members established the Association of Kichwa Women of Shiripuno, Misahuallí, or *Amukishmi*. The genesis of this local women's association corresponded to the formation of a new ecolodge just outside the community. At that time, community members submitted a legal statute to the Ministry of Economic and Social Inclusion to formalize both the women's association and the ecotourism project. The statute's wording details leadership roles in the association, expectations of members, and sanctions for neglecting duties. Structural engagement horizontally with a foreign NGO and vertically with the Ecuadorian Ministry of Economic and Social Inclusion has facilitated Shiripuno's entrance into the tourism market.

A notable institutional outcome that emerged from these transformations is a consensus among *Amukishmi's* women about the need to confront machismo attitudes. The registered statute outlines specific steps to be taken when instances of domestic abuse are brought to the group. A member of Amukishmi notes that this new discourse related to gender empowerment is atypical among communities in the region and encounters friction in a region where men historically, "wanted the women to stay in their houses ... they didn't want the association ... there were fights, hits, mistreatment because of the project; [the men] didn't want it" (Marcinek & Hunt, 2015). As has been further elaborated elsewhere, a global connection to the NGO and the resulting increase in regional engagement for women involved in the administration of an ecotourism project have fostered an increase in social capital (Marcinek & Hunt, 2015), and thus a reduction in the friction for Indigenous women to participate in local and even regional deci-sion-making. This has in turn supported Shiripuno's efforts to maintain tradition, conserve natural and cultural resources, and transform existing conditions into newfound structured institutions.

As events have unfolded around the development of this women's cooperative and the asso-ciated ecolodge, newfound institutions brought about through entrance into tourism are Shiripuno's effort to reject an oil pipeline through their land in 2012. This decision required the community to weigh the offerings of the oil contract, including water wells, food, and other trin-kets. These considerations echo a discourse associated with petroleum leading to *"el buen vivir"*, mentioned in the first vignette. Yet the community president took Amukishmi's *"voz y voto"* (voice and vote) into consideration along with their verbalized encouragement to consider the benefits of refusal, notably "a different lifestyle for his children ... [and] rich cultivated lands". Through this alternative discourse, their new legal recognition, and their adherence to a Kichwa value of respect and coexistence with nature, Amukishmi overcame the discursive friction and influenced the community-level decision to eschew petroleum contracts in favor of an ecotour-ism project aimed at conserving local natural resources (Coria & Calfucura, 2012; Smith, 2014; Stronza, 2010). In this example, confronting existing gender structures with new forms of political empowerment provided transformative and empowering friction leading to the unmaking of the machismo hegemony and that now stands "in the way of the smooth operation of global power" of the oil industry (Tsing, 2005, p. 6).

Discussion

This paper has examined the encounters and frictions between ideas about conservation, devel-opment, and tourism in the Napo province of the Ecuadorian Amazon. To explore the globally and locally interwoven assertions and manifestations of conservation and development as they are emically understood and negotiated by local stakeholders, we asked three broad questions. First, how does ecotourism discourse rooted in global and national political institutions circulate in the Amazonian community of Misahuallí, Ecuador, manifesting in transformations to local social and environmental conditions? The vignettes presented here provide several sharp relief examples of global conservation and community development discourses intertwining both within and outside the context of tourism in and around Misahuallí. Rather than reinforcing the divergent structural and post-structural threads within political ecology scholarship, these exam-ples highlight not only the intersections of these two threads, but also how inseparable these threads are from one another in a highly interconnected Anthropocene. Structures emerge from discourses, and discourse flows back and forth between global and local scale via particular structural channels. In this analysis, the phenomenon of ecotourism produces structures and also provides a channel that introduces alternative and competing discourses about the proper way to conserve and develop the Ecuadorian Amazon. The political ecology of tourism in the Anthropocene should thus reconcile the contributions of both its structural and post-structural, discursive traditions.

Second, we asked where have these discourses encountered resistance due to conflicting emic understandings (i.e. where does friction exist), and where are these discourses readily accepted, or even leveraged, into locally beneficial outcomes, if at all? The vignettes provide sev-eral examples of where resistance and friction occur, with at least some evidence of discursive shifts being leveraged into locally beneficial outcomes in the last two vignettes. While we acknowledge that tourism is far from devoid of environmental consequences, as numerous envir-onmental anthropologists have noted (e.g. Büscher & Davidov, 2016; Duffy, 2008; Fletcher & Neves, 2012; Hunt, 2011), those consequences are likely to miniscule in relation to the wide-spread devastation to ecosystems created by extractive activities such as the petroleum industry, and thus a more sustainable option for local Indigenous populations struggling to maintain cultural integrity amid aspirations for improved quality of life. Exploring additional examples of the loosening or elimination of friction that has limited the capability of Indigenous peoples to achieve those aspirations will be a fruitful avenues for further research.

Finally, we asked what can the friction metaphor and the inclusion of the emic perspectives gathered via ethnographic research add to our understanding of the political ecology of tourism and its consequences in the Anthropocene? The political ecology of this biologically and cultur- ally diverse region accounts for numerous instances of friction between those espousing compet- ing discourses on the proper approaches to tourism, conservation, and development. The data presented here demonstrate how discourse related to the development of tourism privileges par- ticular uses of biodiversity and approaches to development, and how these discourses are reflected in local emic understandings of the transforming social and environmental conditions in Ecuadorian Amazon. Friction provides a powerful metaphor for assessing not only the global connections between the structures and discourses involved in the production of tourism, but also for the dynamics of Indigenous cultural change as new ways of knowing and being are assimilated into traditional understanding and practices. Given the loss of biological and cultural diversity underway at the global level in the Anthropocene, the ability to negotiate friction between old and new is likely to be directly related to the likelihood that particular Indigenous cultures persist in the face of such threats. Future research will be needed to test these hypo- thetical relations suggested by the analysis here.

Conclusion

In this era of the *Anthropocene*, understanding of the global interconnectedness of human activ- ity and it consequences for the planet is at an all-time high (Steffen et al. 2007). There is growing consensus that we are approaching ecological and climatic thresholds as represented by the nine Planetary Boundaries (Steffen et al., 2015), yet other "boundaries" are evaporating via increased global to global, global to local, and local to local connections. Political ecology remains a valuable theoretical framework for assessing such global–local connectivity and its consequences. As the biggest movement of people across international boundaries outside of war, tourism is a massive industry involving more than one billion international travelers each year (UNWTO, 2015). As such, tourism is a powerful force for both crossing and breaking down barriers and for creating global connections between the discourses and structures that influence social and environmental well-being. The strong structural and post-structural, discursive tradi- tions in political ecology ensure it has continuing value for analyses of tourism in the intercon- nected Anthropocene.

This political ecological analysis of tourism in the Ecuadorian Amazonian presents ethno- graphic vignettes to account for the ways that interwoven global discourses related to biodiver- sity conservation and community development are encountered, contested, and leveraged to advance particular approaches to tourism at the local level. Tsing's theory of friction manifests in each of these vignettes involving tourism-related decision-making in the community of Misahuallí. Of particular note are the instances of discursive shifts being leveraged by Indigenous peoples into improved social and environmental well-being. Additionally, the approach here makes an important contribution to the scholarship on the political ecology of tourism by plac- ing the spotlight directly on the emic perspectives of local residents. Inequities and imbalances of power, perpetuated by both the structures and discourses associated with use of tourism to address conservation and development objectives, have long been at the vanguard of political ecology research. Specifically for tourism scholars, frictions between old and new, local and glo- bal, and discourses and structures now also move into the vanguard of research on the political ecology of tourism in the Anthropocene.

Acknowledgments

This work was supported by a M.G. Whiting Indigenous Knowledge Research Grant from the Pennsylvania State University Interinstitutional Center for Indigenous Knowledge.

Disclosure statement

No potential conflict of interest was reported by the authors.

References

Bass, M. S., Finer, M., Jenkins, C. N., Kreft, H., Cisneros-Heredia, D. F., McCracken, S. F., ... Di Fiore, A. (2010). Global conservation significance of Ecuador's Yasuní National Park. *PloS One*, *5*(1), e8767. doi:10.1371/journal.pone.0008767

Belsky, J. M. (1999). Misrepresenting communities: The politics of community-based rural tourism in Gales Point Manatee, Belize. *Rural Sociology*, *64*(4),641–666. doi:10.1111/j.1549-0831.1999.tb00382.x

Bernard, H. R. (2011). *Research methods in anthropology: Qualitative and quantitative approaches*. Lanham, MD: Rowman Altamira.

Bernard, H. R., Wutich, A., & Ryan, G. W. (2016). *Analyzing qualitative data: systematic approaches*. Thousand Oaks, CA: SAGE Publications.

Blaikie, P., & Brookfield, H. (1987). *Land degradation and society*. London, UK: Methuen.

Boo, E. (1990). *Ecotourism: The potentials and pitfalls: Country case studies*. Washington, DC: WWF.

Brosius, J. P. (1999). Analyses and interventions: Anthropological engagements with environmentalism. *Current Anthropology*, *40*(3), 277–309. doi:10.1086/200019

Bryant, R. L. (1997). Beyond the impasse: The power of political ecology in third world environmental research. *Area*, *29*(1), 5–19. doi:10.1111/j.1475-4762.1997.tb00003.x

Buckley, R. (2010). *Conservation Tourism*. Wallingford, UK: CAB International.

Budowski, G. (1976). Tourism and environmental conservation: Conflict, coexistence, or symbiosis? *Environmental Conservation*, *3*(01), 27–31. doi:10.1017/S0376892900017707

Büscher, B., & Davidov, V. (2016). Environmentally induced displacements in the ecotourism–extraction nexus. *Area*, *48*(2), 161–167. doi:10.1111/area.12153

Campbell, L. M. (2007). Local conservation practice and global discourse: A political ecology of sea turtle conservation. *Annals of the Association of American Geographers*, *97*(2), 313–334. doi:10.1111/j.1467-8306.2007.00538.x

Campbell, L. M., Gray, N. J., & Meletis, Z. A. (2008). Political ecology perspectives on ecotourism to parks and protected areas. In K.S. Hanna, D.A. Clark, & D. Slocumbe (Eds.). *Transforming parks and protected areas: Policy and governance in a changing world* (pp. 200–221). New York, NY: Routledge.

Carrier, J. G., & Macleod, D. V. (2005). Bursting the bubble: The socio-cultural context of ecotourism. *Journal of the Royal Anthropological Institute*, *11*(2), 315–334. doi:10.1111/j.1467-9655.2005.00238.x

Coria, J., & Calfucura, E. (2012). Ecotourism and the development of indigenous communities: The good, the bad, and the ugly. *Ecological Economics*, *73*, 47–55. doi:10.1016/j.ecolecon.2011.10.024

Davidov, V. (2013). *Ecotourism and cultural production: An anthropology of indigenous spaces in Ecuador*. New York, NY: Palgrave McMillan.

deKadt, E. (1979). *Tourism: Passport to development?* Oxford, UK: Oxford University Press.

DeWalt, K. M., & DeWalt, B. R. (2011). *Participant observation: A guide for fieldworkers*. Plymouth, UK: Rowman Altamira.

Doughty, C., Lu, F., & Sorensen, M. (2010). Crude, cash and culture change: The Huaorani of Amazonian Ecuador. *Consilience: The Journal of Sustainable Development, 3*(1), 18–32.

Duffy, R. (2008). Neoliberalising nature: Global networks and ecotourism development in Madagasgar. *Journal of Sustainable Tourism, 16*(3), 327–344. doi:10.1080/09669580802154124

Duffy, R. (2016). Forward. In M. Mostafanezhad, R. Norum, E.J. Shelton, & A. Thompson-Carr (Eds.), *Political ecology of tourism: Community, power and the environment* (p. xvi). London, UK: Routledge.

Escobar, A. (1996). Construction nature: Elements for a post-structuralist political ecology. *Futures, 28*(4), 325–343. doi:10.1016/0016-3287(96)00011-0

Escobar, A. (1999). After nature: Steps to an antiessentialist political ecology. *Current Anthropology, 40*(1), 1–30. doi:10.1086/515799

Finer, M., Moncel, R., & Jenkins, C. N. (2010). Leaving the Oil Under the Amazon: Ecuador's Yasuní-ITT initiative. *Biotropica, 42*(1), 63–66. doi:10.1111/j.1744-7429.2009.00587.x

Ecuador Travel. (2017). Retrieved from www.ecuador.travel.

Fletcher, R., & Neves, K. (2012). Contradictions in tourism: The promise and pitfalls of ecotourism as a manifold capitalist fix. *Environment and Society: Advances in Research, 3*(1), 60–77.

Foucault, M. (1980). *Power/knowledge: Selected interviews and other writings, 1972-1977*. New York: Pantheon.

Gössling, S. (2003). *Tourism and development in tropical islands: Political ecology perspectives*. Northampton, MA: Edward Elgar.

Greenwood, D. J. (1989). Culture by the pound: An anthropological perspective on tourism as cultural commodification. In V.L. Smith (Ed.), *Hosts and guests: The anthropology of tourism* (pp. 171–186). Philadelphia, PA: University of Pennsylvania Press.

Gudynas, E. (2011). Buen Vivir: today's tomorrow. *Development, 54*(4), 441–447. doi:10.1057/dev.2011.86

Harris, M. (1976). History and significance of the emic/etic distinction. *Annual Review of Anthropology, 5*(1), 329–350. doi:10.1146/annurev.an.05.100176.001553

Honey, M. (2008). *Ecotourism and sustainable development: Who owns paradise?* (2nd ed.). Washington, DC: Island Press.

Hunt, C. (2011). Passport to development? Local perceptions of the outcomes of post-socialist tourism policy and growth in Nicaragua. *Tourism Planning & Development, 8*(3), 265–279.

INEC: Instituto Nacional de Estadística y Censos. (2017). Retrieved from www.ecuadorencifras.gob.ec/resultados/

Larrea, C., & Warnars, L. (2009). Ecuador's Yasuní-ITT Initiative: Avoiding emissions by keeping petroleum underground. *Energy for Sustainable Development, 13*(3), 219–223. doi:10.1016/j.esd.2009.08.003

Lu, F., & Bilsborrow, R. E. (2011). A cross-cultural analysis of human impacts on the rainforest environment in Ecuador. In *Human Population* (pp. 127–151). Berlin, Heidelberg: Springer.

Marx, E. (2010). Conservation biology. The fight for Yasuni. *Science (New York, N.Y.), 330*(6008), 1170–1171.

Marcinek, A. A., & Hunt, C. A. (2015). Social capital, ecotourism, and empowerment in Shiripuno, Ecuador. *International Journal of Tourism Anthropology, 4*(4), 327–342.

Mowforth, M., & Munt, I. (2015). *Tourism and sustainability: Development, globalisation and new tourism in the third world*. London, UK: Routledge.

Mostafanezhad, M., Norum, R., Shelton, E. J., & Thompson-Carr, A. (2016). *Political ecology of tourism: Community, power and the environment*. London, UK: Routledge.

Nepal, S., & Saarinen, J. (2016). *Political ecology and tourism*. London, UK: Routledge.

Painter, M., & Durham, W. (1995). *The social causes of environmental destruction in Latin America*. Ann Arbor, MI: University of Michigan Press.

Peet, R., & Watts, M. (1996). *Liberation ecologies: Environment, development, social movements*. London, UK: Routledge.

Robbins, P. (2012). *Political ecology: A critical introduction*. Oxford, UK: John Wiley & Sons.

Salazar, N. B. (2010). Studying local-to-global tourism dynamics through glocal ethnography. In Michael C. Hall (Ed.), (2011) *Fieldwork in Tourism: Methods, Issues and Reflections*. New York and London: Routledge.

Sawyer, S. (2004). *Crude chronicles: Indigenous politics, multinational oil, and neoliberalism in Ecuador*. Durham, NC: Duke University Press.

Smith, T. J. (2014). Crude desires and 'green' initiatives: indigenous development and oil extraction in Amazonian Ecuador'. In Bram Büscher and Veronica Davidov (Eds.), (2013) *The Ecotourism-Extraction Nexus: Political Economies and Rural Realities of (un)Comfortable Bedfellows*. New York and London: Routledge.

Steffen, W., Broadgate, W., Deutsch, L., Gaffney, O., & Ludwig, C. (2015). The trajectory of the Anthropocene: the great acceleration. *The Anthropocene Review, 2*(1), 81–98. doi:10.1177/2053019614564785

Steffen, W., Crutzen, P. J., & McNeill, J. R. (2007). The Anthropocene: are humans now overwhelming the great forces of nature. *AMBIO: A Journal of the Human Environment, 36*(8), 614–621.

Stonich, S. C. (1998). Political ecology of tourism. *Annals of Tourism Research, 25*(1), 25–54. doi:10.1016/S0160-7383(97)00037-6

Stonich, S. C. (2000). *The other side of paradise: tourism, conservation and development in the Bay Islands*. Elmsford, NY: Cognizant Communication Corporation.

Stronza, A. (2010). Commons management and ecotourism: Ethnographic evidence from the Amazon. *International Journal of the Commons, 4*(1),56–77. doi:10.18352/ijc.137

TIES. (2017). Retrieved from http://www.ecotourism.org/

Tsing, A. L. (2005). *Friction: An ethnography of global connection.* Princeton, NJ: Princeton University Press.

UNWTO. (2015). *Annual Report 2015.* UNWTO: Madrid. Retrieved from http://cf.cdn.unwto.org/sites/all/files/pdf/annual_report_2015_lr.pdf

USAID. (2015). *Biodiversity and development handbook.* Washington, DC: Agency for International Development.

West, P., & Carrier, J. G. (2004). Getting away from it all? Ecotourism and authenticity. *Current Anthropology, 45*(4), 483–498. doi:10.1086/422082

Wolf, E. (1972). Ownership and political ecology. *Anthropological Quarterly, 45*(3), 201–205. doi:10.2307/3316532

World Bank Group. (2015). *Tanzania's tourism futures: Harnessing natural assets.* Washington, DC: World Bank.

Tourism and community resilience in the Anthropocene: accentuating temporal overtourism

Joseph M. Cheer ⓘ, Claudio Milano and Marina Novelli

ABSTRACT

Global tourism growth is unprecedented. Consequently, this has elevated the sector as a key plank for economic development, and its utility is deeply embedded in political, economic and social-ecological discourse. Where the expansion of the sector leverages natural and cultural landscapes, this applies pressure to social and ecological underpinnings that if not reconciled, can become problematic. The way this plays out in Australia's Shipwreck Coast and the wider Great Ocean Road region, especially the implications for community resilience, is the focus. Emphasis is placed on the vulnerability of peripheral coastal areas to development that withdraws from destination endowments, yet fails to provide commensurate economic yield as a suitable trade-off. This is obvious where tourism intensification has led to concerns about the breach of normative carrying capacities. Temporal overtourism driven by seasonal overcrowding is countenanced as emblematic of tourism in the Anthropocene where focus tends to be largely growth-oriented, with much less attention given to bolstering social-ecological resilience, especially community resilience. At stake is the resilience of regional areas and their communities, who in the absence of garnering commensurate economic returns from tourism expansion find themselves in social and ecological deficit.

Introduction

The essence of the Anthropocene is captured by Walker and Salt's declaration that "humanity has been spectacularly successful in modifying the planet to meet the demands of a rapidly growing population" (2012, p. xi). With that in mind, that the Anthropocene has become embedded in the contemporary and critical tourism discourse is unsurprising; after all, the human-in-nature dimension central to the unfolding of the epoch is very much exemplified in global systems (Crutzen & Stoermer, 2000; Zalasiewicz & Waters, 2016), especially concerns that it represents a "threshold marking a sharp change in the relationship of humans to the natural world" (Hamilton, Gemenne & Bonneuil, 2015, p. 3). Unprecedented expansion underlines the global tourism status quo and coincides with a long period of unparalleled economic growth and affluence. The implications of this is greater global mobility (Brown & Wittbold, 2018) and the opening of new destinations, as well as improved access to more established ones.

With consideration to the scaffolding of this paper from a theoretical standpoint, "The Anthropocene has become a differential lens through which disciplines across the academy are reviewing, debating and reinventing their conceptions of humanity and nature" (Bauer & Ellis, 2018, p. 209). Tourism and the Anthropocene is framed by Gren and Huijbens (2014, p. 7) as "a geophysical force that is part of the relationship between humanity and the Earth". Apropos to that, the upshot of travel as a marker of the experience economy in the Anthropocene is manifold and includes implications for destination development, triple-bottom line impacts, policy and planning and natural resource management (Gren & Huijbens, 2016). The Anthropocene is accentuated by concerns regarding climate change, resource depletion, increased securitization and momentum shifts to the digital environment (Steffen, Crutzen & McNeill, 2007). This leads to questions around how practicable the pursuit of sustainable tourism in the Anthropocene might be and the extent to which it undermines the resilience of tourism dependent communities (Bec, McLennan & Moyle, 2016; Calgaro & Lloyd, 2008; Cheer & Lew, 2017; Lew, 2014; Lew & Wu, 2017).

Antarctica and the Arctic, once out of reach, are more accessible today and doubtless driven by so-called last chance tourism (Lemelin, Dawson, Stewart, Maher, & Lueck, 2010). This is a clarion call to reinforce that more than ever, tourism must align more closely with sustainability concerns (Bramwell, 2006; Saarinen, 2013). Moreover, this dovetails neatly into the Anthropocene that speaks of humans making hitherto unprecedented change to earth systems (Hamilton, Gemenne & Bonneuil, 2015), and as Gren and Huijbens implore, "For the first time in history, humanity is confronted with the task of having to carry the Earth on its shoulders" (2014, p. 15). Growth in tourism, as personified in visitation to Antarctica, is at the vanguard of the emergent contemporary mobility that emphasises the dilemma of the Anthropocene (Schillat, Jensen, Vereda, Sánchez, & Roura, 2016).

This prompts the question: what are the limits to tourism growth (O'Reilly, 1986)? Fundamentally, the link between tourism and the Anthropocene concerns the extent to which global travel undermines earth systems and raises the question: under what circumstances can this development be better positioned for more sustainable and resilient outcomes (Lew & Cheer, 2017; Hall, Prayag, & Amore, 2018)? Ushered in are broad considerations regarding how tourism growth elevates concerns about the provisioning of social and ecological systems for tourism (Mosedale, 2015; Mostafanezhad, Norum, Shelton, & Thompson-Carr, 2016; Nepal & Saarinen, 2016; O'Reilly, 1986).

Accordingly, of particular focus here is community resilience to tourism-induced transformations at the coastal periphery (we link community resilience with social resilience and assume the two to align). While we engage with the Anthropocene, reconceptualising the epoch and arguing its finer theoretical and ontological threads is beyond the scope of this undertaking. Instead, we make fundamental and precise connections between tourism and the emergent concept, overtourism and examine how this impacts the resilience of peripheral coastal communities. Stonich's (1998) stridence that "unbridled tourism development" represents a real risk for communities is acknowledged. The risk alluded to here is what Hall et al. (2018) refer to as change and disturbance in the tourism system. The principal question we pose asks: to what extent are tourism impact concerns shaped by community resilience as exemplified by tourism in peripheral coastal contexts in the Anthropocene?

In the main, we zero in on community resilience and leverage qualitative data that is community stakeholder focused and extracted via a longitudinal study between 2015 and 2017 in the Shipwreck Coast region of southern Australia. Fittingly, we employ social-ecological systems (SES) resilience as a broad theoretical framework from which we examine community resilience and argue that it is central to the Anthropocene and enmeshed in political, economic, social and ecological dimensions, which, in turn, impinge on and help shape nascent institutional structures (Gren & Huijbens, 2014; Hall et al., 2018). Importantly, we overlay this discussion with the master planning process, specifically the Shipwreck Coast Master Plan (Parks Victoria, 2015).

Overtourism is now part of the popular and scholarly lexicon; emblematic of tourism in the Anthropocene where the capacity of destinations to cope has reached tipping points (Milano, 2017; Sheivachman, 2017). In particular, we hone in on temporal overtourism which occurs in response to concentrated, occasional (e.g. special events), daily or seasonal visitation spikes (Gössling, Ring, Dwyer, Andersson, & Hall, 2016). Such situations are ubiquitous when management regimes fail (McKinsey & Company, 2017), and overtourism occurs when destinations breach tolerable thresholds that communities can absorb (Milano, 2018; Milano, Cheer, & Novelli, 2018). Also, overtourism raises objections against tourism that has outgrown its initial conceptualisations (Papathanassis, 2017; Seraphin, Sheeran, & Pilato, 2018). As Papathanassis (2017) argues, the problem is about governance and not tourism itself, and about planning and management and the extent to which communities remain amenable to tourism (Cheer, Coles, Reeves, & Kato, 2017; Rifai, 2017; Saarinen, 2013, 2018).

Case study

The 28-kilometre Shipwreck Coast study area, from Princetown to the Bay of Islands, is a magical place. The spectacular limestone stacks and coastal formations, including the Twelve Apostles and Loch Ard Gorge, are among Australia's best- known features, drawing millions of visitors each year. This narrow, fragile environment encompassing the Port Campbell National Park, the Twelve Apostles Marine National Park, The Arches Marine Sanctuary and the Bay of Islands Coastal Park is also home to a rich and diverse natural and cultural heritage, townships and their communities.

The Shipwreck Coast Master Plan (Parks Victoria, p. 4, 2015)

The Twelve Apostles Marine National Park and the Twelve Apostles drive visitation to the Shipwreck Coast region (Figure 1) (Cheer, 2018). As alluded to in the above quote from The Shipwreck Coast Master Plan, growing visitation sits awkwardly alongside pressures to maintain

Figure 1. Aerial view of the Twelve Apostles and Twelve Apostles Marine National Park boundary (Inset map – excerpt of south and south-eastern Australia). (Source: Google).

Figure 2. The Twelve Apostles Marine National Park with Twelve Apostles in the background – circa 2016 (Used with permission from Tourism Australia Image Library).

Figure 3. The Twelve Apostles Marine National Park with Twelve Apostles in the centre and top left – circa 1940s. This photograph suggests noticeable and gradual decline and erosion occurring over the course of the last 70 years (Used with permission from Matt O'Kane).

the region's natural values while also expanding tourism-driven economic development (Parks Victoria, 2015). Awareness of the region is centred on The Twelve Apostles, one of the most iconic images of the state of Victoria. The Twelve Apostles, a grouping of limestone stacks that lay adjacent to the coast at Princetown and 19 kilometres from the township of Port Campbell, is the main drawcard because of the high natural values *in situ*, including the Otway Ranges National Park, Port Campbell National Park and the Twelve Apostles Marine National Park (See Figures 2 and 3), and it is the most visited region in the state and third most in Australia (Parks Victoria, 2015).

The Great Ocean Road and, in particular, the Shipwreck Coast's early settlement dates back to colonial Australia and vessels that sank offshore. Stretching for a few hundred kilometres in the southwest of the state of Victoria, the Shipwreck Coast is a 3-to-4 hour drive from the capital, Melbourne (See Figure 3). Large-scale agriculture, sheep and dairy farming, cattle and wheat frame are the economy of the region. This harks back to its early establishment where the region relied on agriculture and logging, and fisheries once had a strong presence. Akin to many small, regional coastal towns, peripherality introduces constraints to social and economic development, including depopulation, corporatization of family farms, vulnerability to natural disasters including bush fires and droughts, and infrastructure shortcomings, especially public transport, roads, and communications (mobile telephone coverage is unavailable or compromised in some places) (Cheer, 2017; Green, 2004).

Tourism is the key economic impetus, and with close proximity to Melbourne, the Shipwreck Coast has become a day trip destination (Cheer, 2017; Han & Cheer, 2018). The failure to opti-mise tourism growth for greater local level benefit lies in reconciling competing priorities of eco-nomic expansion versus preserving sense of place, as well as adapting to the shifts away from agrarian livelihoods to a service-driven economy underlined by tourism (Ibid.). The constraints to optimising tourism are found in the bottlenecks that detract from the expansion of tourism infrastructure, especially related to funding of critical infrastructure (roads especially), plus ongoing perturbations that occur including bush fires, landslides and rockfalls along the Great Ocean Road (Pearson, 2017). At June 2017, the Great Ocean Road region of which the Shipwreck Coast is central, generated over 6 million unique visitations with more than $AU1.3 billion (Warrnambool City Council, 2018). Moreover, this unprecedented growth is expected to increase by over 50% to 2025.

The development of tourism infrastructure, including commercial accommodation (hotels and resorts), has lagged growth in visitation (Parks Victoria, 2015). Consequently, this has stymied efforts to increase overnight stays and curtailed tourist expenditure. The paradoxical circum-stance that arises is framed by high seasonal visitation in the Australian summer (November to February), with much comprised of groups and individuals passing through the region en route elsewhere. Unsurprisingly, this has raised local-level angst where expenditure spent on maintain-ing public amenities such as toilets and public areas exceed direct returns to local stakeholders. Visitation is presently characterised by growing international tourist presence, especially Chinese tourists, and has amplified infrastructure and tourism service deficiencies (Cheer, 2017; Han & Cheer, 2018; Pearson, 2017).

All of this raises questions about the resilience of the region's social and ecological backdrop to cope with increased visitation, especially during seasonal and daily peaks and whether the government agency charged with the protection of National Parks, Parks Victoria, can cope with the impact of growing visitation (Koob, 2017). Of note is the clash between efforts from the tour-ism sector to drive further expansion and local community angst over the low rates of economic return (Tyler, 2016; Cheer, 2018). This highlights the paucity of strategic governance where visit-ation is the key performance indicator, with less attention given to yield per visitor and length of stay (Koob, 2017; Zwagerman, 2016). The Shipwreck Coast Master Plan acknowledges the aforementioned tensions; however, its implementation has been stymied by politicking and stakeholder contention about ways forward.

Conceptual framework

Preceding the emergence of the Anthropocene in the tourism patois, the employment of nature to service growing leisure classes had already raised concerns. As Christaller opined in the 1960s, "tourism is drawn to the periphery of settlement districts as it searches for a position on the highest mountains, and in the most lonely woods, along the remotest beaches" (1964, p. 95). The human desire to be in and among nature remains intrinsic to the touristic endeavour, and in the 1960s, there was little to suggest that this was problematic. The 1960s ushered in the beginning of mass travel and the full-packaged holiday, and as Christaller (1964, p. 105) exclaimed, "Thanks to airplanes and thanks to our prosperity, destinations in Africa, in west and south Asia and in the Caribbean Sea are competitive to the countries in Europe". Christaller (1964) was prescient in advocating caution and offered a caveat suggesting that "helping induce the passage of such regions along the same path of former islands or forgotten places ... " might actually not be so laudable.

In the 1970s, as mass tourism hastened, the use of nature for tourism intensified, creating new utilities and exigencies for what were once mostly adaptable contexts. As Overton critiques in his 1973 depiction of the opening up of national parks for tourism, this led to the creation of a "new set of social relations which is [sic] imposed creates conflict and only marginal development" (Overton, 1973, p. 34). The social relations mentioned were centred on the political, economic and social ramifications of turning nature into commoditized touristic experiences. The idea to "save" natural areas led to what Overton (1973, p. 35) describes as "necessary to neutralize discontent and respond to protest from the many groups which make up the ecology movement".

Moore's exposé of tourism in the Anthropocene argues that "contemporary relations of nature and culture" are central to understanding the Anthropocene in tourism, and that tourism must transform and recalibrate to adapt to the evolving status quo (2015, p. 191). Moore's thesis is framed by adaptation and guided by the question: "To what extent do emergent ventures green wash their involvement in global assemblages of socioecological exploitation" (2015, p. 195)? Ultimately, as Stonich outlines, political ecology drivers demand "integrative policy approaches" that "ensure equitable and environmentally conservative development" (1998, p. 50). This conundrum is evident in the Shipwreck Coast where the competing and conflicting priorities of development and tourism sidle up against the desire to protect natural values and sense of place.

While the Anthropocene hastens the urgency for sustainable development, "there are significant limitations in the extent to which societal actors can respond to the challenges of environment resource management and sustainability" (Knight, 2015, p. 153). This is discernible in tourism, especially where disparities in stakeholder influence and agency over the scope and nature of development occur. For example, local-level capacities to deal with externally driven tourism interventions are very often curtailed (Seraphin et al., 2018). This calls for "a different kind of global social-politics" (Knight, 2015, p. 156) to underline sustainable tourism in the Anthropocene, characterised by an understanding that destinations have tipping points beyond which diminishing returns occur.

Globalisation and neoliberalism loom large in the Anthropocene and underlines global economic systems that enforce downstream impacts at the micro-level. As Soriano (2017, p. 5) argues, "environmental degradation is essentially a material problem", one that stems from consumptive practices tied to resource use and depletion. Related to this are global flows of capital and predicated on exploiting destination endowments. This relies on the provisioning of elements within the tourism system and whether a balance between profit optimisation and social and environmental integrity can be negotiated is central to sustainability concerns. Soriano sees this as "a sort of vicious cycle that grows as a snowball and shows the inherent unsustainability of this production mode" (2017, p. 9).

Whether enquiries into the Anthropocene–tourism nexus display "a panicked political imperative to intervene more vocally and aggressively in an earth transformation run amok" (Robbins & Moore, 2013, p. 9) bears consideration, for within tourism, amenity decline is evidence of system failure. This is obvious where overcrowding is evident, where the dominance of the built environment overwhelms the natural and where social-ecological transformations are characterised by amenity decline. For Autin, the Anthropocene introduces multiple dichotomies where processes in tourism are part of a "culture war about the recognition of environmental process" (2016, p. 222). This mirrors advocacy for more resilient and sustainable tourism and against tourism that diminishes adaptive capacities (Lew & Cheer, 2017).

Tourism as a production process is profoundly connected to transformative elements; for example, if the most obvious enabler of global tourism is the burning of fossil fuels, tourism must confront "decisions about production systems and investment priorities intermeshed with political maneuverings in an increasingly artificial, crowded and changing biosphere" (Dalby, 2016, p. 34). That tourism and social-ecological resilience is tied underlines that planning and management regimes for sustainable tourism must negotiate the dual spheres effectively (Saarinen, 2018). The notion of a good or bad Anthropocene is symbolic of concerns over the selection of modes of tourism that best enable sustainable tourism.

The term overtourism is an intrinsic hallmark of the Anthropocene and emerged because of what Milano refers to as "unsustainable mass tourism practices" (2017, p. 5). This describes the rapid and unprecedented growth of global tourism and particularly in marquee destinations where adverse impacts on local communities is evident (Goodwin, 2018; Seraphin et al., 2018). Novy and Colomb argue that this emerged because "tourism is fundamentally political" and that "the way tourism is accounted for and made sense of locally, for instance, has usually been shaped by the hotel industry and associated businesses" (2017, p. 6). The sense that local level agencies are very much diminished underlines the overtourism movement where the argument is not for a diminishment or removal of tourism, but more about governance that prioritises local well-being.

Fundamentally, overtourism references carrying capacity in a twofold manner (O'Reilly, 1986, p. 254): (1) "the capacity of the destination area to absorb tourism before negative impacts of tourism are felt" and (2) "levels beyond which tourist flows will decline because certain capacities as perceived by the tourists themselves have bene exceeded". The former relating to destination capacities prevails, whereas global tourist flows continues to surge despite overcrowding, hyper-inflation and the frenzied industrialization of tourist experiences. This calls for "the restructuring of processes of recent decades" (Novy & Colomb, 2017, p. 11) where the emphasis has been top-down and focused on growing visitation. Much of the nascent discourse on overtourism is focused on city contexts with cities such as Barcelona and Venice struggling to adjust rapid tourism growth (Milano, 2017; Misrahi, 2017; Sheivachman, 2017).

In defining temporal overtourism, this is most evident when seasonal variations arise as in the case of the Mediterranean summer and Golden Week in China, and during peak holiday periods and hallmark events such as the Olympics. This temporary state overwhelms local infrastructure and services, resulting in mass overcrowding and diminishment of sense of place. This is problematic because such temporary surges can have damaging social and ecological consequences (Cheer, 2018). Temporal overtourism, if driven externally and led by global tourism supply chains, is difficult to plan for and sudden visitation surges, while welcomed by local authorities, can lead to social-ecological stress. This is acute where ecological assets support visitation such as in National Parks, coastal and marine sanctuaries and small islands, and where human interactions with animals are integrated (Nepal & Saarinen, 2016; Saarinen, 2018). Hall (2016) refers to this as typical of the Anthropocene where "loving nature to death" has become more obvious in tourism systems. As Hall points out, "the extent to which tourism contributes to biodiversity loss through tourism urbanization, habitat loss and fragmentation and climate change is also dramatic" (2016, p. 66).

Temporal overtourism varies in its impacts with Goodwin (2018) arguing that it provides ideal support to struggling peripheral economies. Conversely, Seraphin et al. (2018) contend that overtourism manifests as overcrowding leading to the diminishment of sense of place. The emergence of overtourism and what to do about it is underlined by governance and management and effective policy formulation (Milano, 2017). Former UNWTO Secretary General Taleb Rifai argued that overtourism is not about tourism per se, but more about the way it is managed: "It's the failure of management not of the sector" (Rifai, 2017). Consequently, current policy responses include demarketing, redirection of tourist flows, new planning regimes, capping of tourism numbers and price hikes (Misrahi, 2017).

Community resilience concerns are intimately linked to SES that are described variously and meditates on striking a balance between social and ecological concerns, and related to the extent to which communities adapt and respond to shocks that test their capacity to adapt (Cheer & Lew, 2017; Hall et al., 2018). Resilience has its genesis in ecology, and more specifically, "ecological and social resilience may be linked through the dependence on ecosystems of communities and their economic activities" (Adger, 2000, p. 347). SES also references human-in-nature concerns, as exemplified by the Anthropocene, and locates the extent to which human induced change on SES undermines adaptive capacities of communities (Ruiz-Ballesteros, 2011).

In general terms, resilience is the ability of systems to bounce back after perturbations and/or disturbances that destabilise the steady state (Davoudi, 2012). Indeed, when characterising resilience, Holling (1973) demarcated between engineering and ecological resilience; the former related to returning to equilibrium after turbulence, while the latter concerns adaptation within critical thresholds arguing that there is capacity for systems to flip or morph into other stable states. Espiner and Becken's (2014) benchmarking for the examination of social-ecological resilience in protected areas surrounded by small tourism-centred communities argues that there are inconsistencies present in the way resilience is defined, conceived of and applied. That vulnerability and resilience are two sides of the same coin is not lost on Espiner and Becken (2014), who assert that returning to previous equilibria might not be possible.

In setting an agenda for building resilience in SES, the Stockholm Resilience Centre suggests moving "beyond viewing people as external drivers of ecosystem dynamics and rather looks at how we are part of and interact with the biosphere" (2016, p. 3). In particular, the call is for polycentric governance systems that leverage traditional and local knowledge that are linked to social and political processes. This emphasizes urgencies in the Anthropocene toward what Cleveland describes as "the unresilient epoch humans have created" (2014, p. 2).

Adger infers that "Social resilience is an important component of the circumstances under which individuals and social groups adapt to environmental change" (2000, p. 347). This is an important nod drawing on the changed conditions that underline the Anthropocene, and the capacities of communities to adapt and respond to changing social and environmental conditions, especially that generated externally. Social resilience is at the forefront of how policy-makers and broader stakeholder groups institute and construct frameworks for adaptation to climate and non-climate-related transformations. At a local level, the extent to which destination communities exercise agency and influence at developing resilience capabilities is questioned (Boonstra, 2016). This is evident where tourism expansion rides roughshod over local coping and control mechanisms.

That humans are testing the upper limits of the social-ecological ceiling in the Anthropocene is obvious, thus, calling for adaptive and resilience measures is pressing. As Adger argues, "Both sustainability and resilience recognise the need for precautionary action on resource use and on emerging risks, the avoidance of vulnerability, and the promotion of ecological integrity into the future." (2003, p. 1). This underscores provocations about what must be done to cope with unprecedented earth system changes. After all, the catch cry for greater social-ecological resilience centres on not only halting further change, but instead assessing how the human-in-nature might best adapt to and mitigate the rate and pace of change. This is vital for tourism communities, especially where tourism intensity is heightened (Lew, 2014)

If transformation underlines the Anthropocene, resilience as a framework to assess responses to perturbations must recognize that "people and their institutions are integral components of ecological systems" (Chapin et al., 2004, p. 344). This underlines how social-ecological resilience thinking is essential to transition processes and where sustainable transitioning to evolving steady states is fundamental (Lew & Cheer 2017). Hence, the question of resilience to what and for whom resounds for as Cretney argues, "it makes sense to tackle the root causes of social and environmental issues rather than perpetually react to disaster and crisis events" (2014, p. 636). Thus, linking overtourism in the Anthropocene to human-in-nature systems outlines that system harmony is at stake. Questions about whether the global tourism system is net contributor or net extractor from earth systems are heightened in the era of peak tourism. Where disjunctures arise, they contradict attempts at sustainable tourism and question governance regimes preoccupied with maximising visitor numbers and query how to make way for proportionate focus on social and ecological inheritances.

Methods

The empirical focus is the Shipwreck Coast region of southern Australia and the hugely popular Twelve Apostles Marine National Park. Longitudinal fieldwork from 2015 to 2017 frames the wider project underscored by mixed methods comprising of international visitor surveys ($n = 780$), in-depth interviews ($n = 72$), stakeholder focus groups ($n = 6$) and observational fieldwork as well as content analysis of the Shipwreck Coast Master Plan. Particular focus in this paper is given to industry, government and civil society views and aligns with that adopted by Espiner and Becken (2014) where the nuanced perspectives of tourism stakeholders were leveraged to understand the *in situ* dynamics of tourism expansion.

Overall, data collection and in-field participant observation employed is drawn from two initiatives: (1) longitudinal fieldwork related to a postgraduate field school conducted annually in September–October from 2015 to 2018 and (2) formal research conducted by final year graduate students (Tyler, 2016; Zwagerman, 2016). The base for this field school between 2015 and 2017 was the Shipwreck coast hamlet Port Campbell, with research activities based out of the Port Campbell Surf Life Saving Club. This experiential learning initiative comprised of in-field lectures over three days, stakeholder consultations and tourism sector and local government lectures.

The selection of tourism and community stakeholders was underpinned by the extent to which they were considered to be key informants (Espiner & Becken, 2014). This comprised of tour operators, accommodation providers, food and beverage businesses, local councils, community groups, park rangers, transport providers and allied tourism sector organisations. During the period 2015 to 2017, six phases of fieldwork were undertaken in October–November and March–April each year. These periods were chosen partly out of convenience to align with the teaching semester, and partly because it occurred outside of the school holiday periods when visitation is characterised by a seasonal influx of holiday or second home owners rather than tourists. Allied to this research but not included here was a large-scale survey of international tourist satisfaction in the region (Cheer, 2017; Han & Cheer, 2018).

For the broader project, frequency analyses utilising SPSS was used with quantitative data, while NVivo was employed to undertake deductive thematic analysis of qualitative data. In this paper, we place particular focus on qualitative data derived from in-depth interviews ($n = 72$) and focus groups ($n = 6$) with tourism stakeholders and contextualized this with the Shipwreck Coast Master Plan enacted in 2015. Interviews were not audio recorded, and instead, detailed transcriptions were noted at the time of interview. This was done to align with university research ethics requirements and to avoid potentially compromising participants given the smallness of the sector and that audio recordings would have been easily identifiable. This is why we draw only from in-depth interviews and not the focus groups as it would be difficult to provide anonymity from the latter.

The construction of emergent themes adhered to conventional coding approaches and follows Espiner & Becken's (2014) approach to contextualize stakeholder feedback in an environment where vested interests can potentially overwhelm the narratives embedded in data collected; the use of leading words such as "Anthropocene", "community resilience" and "overtourism" was not used at any point during the research. Instead, neutral and well-understood terms such as "sustainable tourism", "destination management", "sense of place" and "regional development" pervaded the interview discourse. Appleton's (1995) approach adopted from Miles and Huberman (1984) emphasising three key stages in regard to data analysis was applied: (1) data reduction, (2) data display and (3) conclusion drawing. Data reduction is framed around selecting, focusing, simplifying, abstracting and transforming emergent themes, and then repeating the process several times to outline the reliability and consistency of thematic catego-ries. This enabled the data display phase to advance where articulation of emergent themes was established and conclusion drawing was facilitated. Bazeley's (2009) point that analysing qualita-tive data is more than just highlighting themes is acknowledged alongside Strauss' (1987, p. 5) insistence that "the making of constant comparisons and the use of a coding paradigm, to ensure conceptual development and density' is vital in analysis of quality data" – this was central to the approach taken to data analysis throughout the wider project. Moreover, to ensure inter-coder reliability, the main author assumed responsibility for the development of the code book and allied category development.

Results

In juxtaposing temporal overtourism alongside community resilience, the inference is that within sustainable tourism development processes, both frameworks integrate the human-in-nature dimensions that characterize tourism in precarious contexts. Where the Shipwreck Coast is con-cerned, seasonal and daily overtourism has discernible long-term effects (Baum & Hagen, 1999). Drawing from the Master Plan process for the region, and empirical data collected for the broader longitudinal study, several key themes in the dual frameworks of social and ecological resilience emerged with implications for community resilience. In drawing from in-depth inter-views only ($n = 72$) in this instance, emergent themes articulate strong links between temporal overtourism and community resilience perturbations.

Temporal overtourism

Intense overtourism occurs in the Australian summer school holidays from the beginning of December to the end of January and coastal destination populations can swell to over 10 times the average population in the region (Cheer, 2017). Seasonal overcrowding while not unprece-dented has intensified in recent years as both domestic and international visitations have increased. While those who rely on tourism support heightened visitation, the so-called out-of-towners who own holiday houses and play a large role in decision-making reject expansion: "I bought a holiday house here because of the peace and quiet and reject any attempt to increase tourism here". (Holiday house owner)

Temporal overtourism also occurs on a daily basis where up to 90% of all visitations occur between 11AM and 3PM. This coincides with the drive duration from Melbourne to the Shipwreck Coast. This has the impact of peak time overcrowding, impacting visitor satisfac-tion, contributing to site hardening (walkways and roadways) and intense traffic volumes on a single-lane dual carriageway road. A common refrain underlined by a key stakeholder cited: "The way tour companies structure their itineraries is daft and impacts the quality of experi-ence, especially when 90% of all visitors come here between 11am - 3pm. This means long lines at the toilet, crowded car park and diminishing the desire to stay longer" (Local tour operator). Public amenities such as toilets, parking, walkways and viewing platforms are con-structed and maintained with little financial return because visitation is free of charge. Combined, these factors contribute to the low economic yield and limited length of stay: "The local councils and parks agencies spend millions of dollars providing public toilets, signage and waste management. What we get in return is a pittance and fails to cover the costs we incur". (Local tourism association member)

Occasional overtourism is especially notable during the Chinese Lunar New Year that occurs in the first quarter of the New Year. During this time, the number of tour buses and mostly independent Chinese travellers slow traffic to a trickle and coupled with poor cross-cultural road signage and variable driving conditions, risks to motorists and pedestrians are intensified. For local residents coming to grips with the changed traffic conditions, concerns are heightened: "It is getting increasingly dangerous on the Great Ocean Road – tourists are either parked on the road to take photographs or driving on the wrong side. Something has to be done before serious accidents start occurring" (Local resident). Similarly, sentiments of the commercial transport sector, vital to tourism and life in the region, suggest that: "Expanding the road network here is essential if we are to be able to cope with larger numbers". (Local bus driver)

Visitation data unreliable

Data used to estimate tourist visitation is extracted from the Australian Government's International Visitor Survey (IVS) and National Visitor Survey (NVS). This is derived from macro-level surveys of international and domestic travellers at national and state-wide levels and then extrapolated to the local level. An ongoing refrain is that the annual estimated visitation of 2.6 million tourists underestimates visitation to the region. This has serious implications for national and state government budgetary support and allocation of infrastructure maintenance and development funding. As reflected by representatives of the local tourism association: "The data at national and state level are so far removed from what is happening at a local level – it vastly underestimates visitation" (Local tourism association member). Another participant argued: "Unless we're collecting data at a local level, there is no way we can truly challenge the data that assumes to know what is going on - numbers visiting the region are far greater than we are led to believe". (Local tour operator)

Sense of place

In assessing sense of place, some of the key themes included maintaining seaside idyll of a quiet coastal town, limiting overcrowding in National Park and beachside areas, keeping the development of tourism modest with built structures restricted to no more than two levels, limiting traffic volumes and regulating the extent to which houses in the region are used for tourist letting in the peak summer holiday period. Sense of place accentuates the local and is what distinguishes one place from another (Kerstetter & Bricker, 2009). Here, sense of place is linked to maintaining the distinctive local character of towns – whenever new developments are proposed, especially at a large scale, this receives strong local opposition. Consequently, this stymies investment in tourism in the region. As one of the largest hospitality enterprises in the region lamented: "If we are to make the most from growing tourism demand, we need to ensure that our towns maintain their character – any development must be in sympathy with our identity and heritage and consider residents priorities first" (Local publican). Excess visitation is considered a detraction from the fundamental sense of place that underlines the costal idyll and with long-term enduring impacts. One of the most common grievances is the loss of access to lifestyle and leisure opportunities because of tourist generated overcrowding in the peak holiday periods – this is noticeable in National Parks and beachside locations. The extent to which the region's nature natural values are maintained is considered under threat: "The whole idea of a National Park is to enjoy nature in solitude – how can you do this when there are dozens of tourists milling around, taking photographs and talking loudly"? (Local Park Ranger)

Community resilience to natural hazards

The risk of natural hazards occurring is a constant threat in the wider region, and at a localised level in the Shipwreck Coast, this is more obvious (ABC, 2016; Pearson, 2017). Bush fires are a constant threat given that the intensity of land enclosed in National Parks is a feature of the region. Much effort to mitigate the impacts of natural hazards is undertaken at a community level through the volunteer firefighting services organisation, Country Fire Authority (CFA) and local-level community organisations such as the Committee for Lorne. However, the changing demographics at a local level tips the balance away from permanent residents to second home owners as a result of tourism expansion and the transience of seasonal workers is with-drawing from social cohesion and the resulting whittling down of community resilience: "As tourism increases, I see the social networks within the community declining as well. In the last bushfires, we had to rely on external assistance and that leaves us vulnerable". (Local store owner)

Humans social networks and social capital

The transformation of relational spaces is underscored in the Shipwreck Coast with the demarca-tion between tourist and local spaces blurring. This is more intense in smaller communities where the fracturing of human social networks has seen a steady decline with younger townsfolk moving to the capital city for education and employment, while older generations becoming deceased or transition to institutional care. The decline in social capital has also withdrawn from key community bodies including the local Surf Life Saving Club and volunteer organisations such as the Country Fire Authority (CFA). These organisations are the "glue that binds" and pivotal towards communities reorganizing in times of crisis (Cheer, 2017; Green, 2004). The severing of feedback loops undermines human social networks and social capital and the question remains: How can tourism led transformations and wider socio-demographic dynamics be shaped to enable strengthening community resilience? One participant articulated this succinctly: "In small towns, we rely heavily on volunteers to help support the vulnerable in our community and to deal with bush fires and other crises - the new settlers don't value these relationships in the same way". (Country Fire Authority volunteer)

Master plan inertia

The Shipwreck Coast Master Plan reveals discernible tensions between economic growth aspira-tions and the preservation of the social-ecological setting, including the safeguarding of social capital and sense of place, as well as optimisation of the region's national parks. As the State's Minister for Environment, Climate Change and Water (Ibid.) exclaimed: "The Master Plan will guide investment in facilities and infrastructure over the next 20 years to enhance the liveability of local communities, develop international quality visitor opportunities, and conserve and restore the region's biodiversity and landscape character". The region's extraordinary natural val-ues are a drawcard, although concurrently, this is also its Achilles heel preventing the expansion of tourism infrastructure and services to support growth. This is evident in the shortage, quality and range of accommodation, and adversely affects overnight stays. "The Shipwreck Coast suffers from low economic yield from the considerable number of visitors each year. People visiting the area place significant demands on infrastructure and the environment, but leave little in the way of a contribution to the local regional economy" (Parks Victoria, 2015, p. 4). Master planning is tied to political processes and it is acknowledged that the status quo is unworkable and unsus-tainable. The call is not solely about expanding the visitor economy; the appeal is for tourism that strikes a balance between economic expansion, economic yield and protection of social and ecological inheritances. This is evident in the Ministerial statement: "The environment and the economy go hand in hand in this Master Plan" (Ibid.). The Shipwreck Coast Master Plan demonstrates the embeddedness of community concerns, most of which acknowledges that optimisation of sectoral growth has failed to translate into enduring local gains (Cheer, 2018) and the capacity to cope with growing visitation (Han & Cheer, 2018).

Discussion and conclusions

Importantly, the Shipwreck Coast Master Plan calls for caution: "The future of this unique region for tourism, local communities and the environment is at a point of reinvention and necessary change" (Parks Victoria, 2015, p. 5). The interlinking of tourism, local well-being and the environment coalesces to highlight the unavoidable links and with vast implications for processes that shape policy and planning (Espiner & Becken, 2014; Holladay & Powell, 2013). This highlights the vulnerability of peripheral coastal areas to economic and political processes outside the region, reinforcing typical centre–periphery dynamics where central government prevarications come at a cost to peripheral area resilience against external shocks (Lapointe & Sarasin, 2017). Parks Victoria (2015, p. 6) describes the region as a "linear, fragile and vulnerable cultural landscape. A priority of the Master Plan is to protect and conserve this fragile coastal ecology in response to visitor usage and ongoing erosive natural processes" and it is stated: "The environment strategy takes a 'whole of landscape' approach to repair the parks, improve habitat, increase biodiversity and raise environmental awareness" (Ibid. p. 7). This suggests close alignment to the ideals of nature-based tourism and, if employed optimally, as an agent for the betterment of community resilience in the region.

Political processes that guide tourism development in fragile cultural and natural landscapes are highlighted, as well as the way government exercises overarching influence (Mosedale, 2015; Stonich, 1998). Recent Ministerial pronouncements reinforce urgency for political action: "The region attracts 2.6 million visitors a year but the average visitor stays less than 40 minutes and spends only 18 cents" (Pulford, 2017). A top-down, interventionist approach to planning is evident and excerbates the problematic process of regional cooperation (Cheer, 2018); regional actors compete for the same pool of tourists; consequently, individual self-interest takes precedence. The shift from growth-oriented priorities to non-economic impacts such as community well-being, environmental integrity and natural resource conservation is embedded in the Shipwreck Coast Master Plan. The inherent constraints that encumber tourism in peripheral areas and occur here are related to distance, duration of travel, tourism product or experience quality and density, and infrastructure capacity limitations. Where peripheral areas are hindered by economic stagnation, growing tourism inevitably becomes a mantra (Gössling et al., 2016; Lapointe & Sarasin, 2017). Yet, this is hindered by political processes that pit stakeholders against each other (Cheer, 2017).

Overtourism, and particularly temporal overtourism, is a fundamental theme that characterizes the interrelationship between community resilience and tourism growth in peripheral coastal locations where high natural values are normative (Lapointe & Sarasin, 2017). Temporal overtourism is increasingly evident as tourism becomes more entrenched in the region's economy. While the effects of temporal overtourism are palpable as seen in the intensification of crowding and escalation of traffic volumes, as well as impact on natural values, this has not carried over into commensurate economic gains (Cheer, 2017). This confirms stakeholder sentiments that the primary beneficiaries from tourism growth are not within the local area and instead in the travel supply chain beyond the region. Simultaneously, the costs of provisioning for tourism fall on local communities.

Temporal overtourism raises two key concerns: firstly, while human social networks and community resilience are impacted by temporal overtourism, no trade-off is attained leaving the local context in deficit (Cheer, 2017). Secondly, governance of tourism fails to negotiate more favourable approaches to tourism expansion through the political and planning processes (Bramwell, 2006; Gössling et al., 2016). This can be attributed to seemingly immovable structural constraints – how do you get the tourism industry to agree on reforms required? How do you influence the travel supply chain to see the region as something more than a day trip? This suggests that monumental reorganisation is required, but for this to occur, policy regimes must prioritise the discrepancies between economic return and costs of visitor servicing (Milano, 2017; Mosedale, 2015). This underlines wider concerns in the Anthropocene that speak of finding the "sweet spot" between developing and utilising social and ecological endowments for tourism-driven economic development with fundamental concerns about who foots the bill for the upkeep of social-ecological integrity (Stonich, 1998; Gren & Huijbens, 2016). Furthermore, how might tourism expansion privilege the building of social-ecological resilience within tourism communities to protect and conserve fragile ecological assets? Unless fundamental structural deficiencies are addressed, the rapidly growing tourist trade will underserve local stakeholders (Cheer, 2017; Green, 2004).

In returning to the central question framed at the outset: to what extent are tourism impact concerns shaped by community resilience as exemplified in tourism in peripheral coastal contexts in the Anthropocene? Moreover, the overarching line of enquiry essentially asks: What are the limits to tourism growth (Bramwell, 2006; McCool & Lime, 2001; Saarinen, 2013)? As global tourist numbers lurch towards the 1.5 billion visitor arrivals threshold by the end of 2018, evidence whereby testing of the upper limits of destination development is occurring has become all too common (Lemelin, Dawson, Stewart, Maher, & Lueck, 2010; Mosedale, 2015). In many ways, overtourism in the Anthropocene is a consequence of the growth orientation of tourism industry policy-makers and their industry counterparts (Novy & Colomb, 2017; Milano, 2017; Seraphin et al., 2018). As Béné et al. (2016, p. 166) are quick to point out, "resilience does not simply reflect the effects of tangible factors, but also has subjective dimensions". The subjective dimensions relate to how to come to terms with carrying capacity limitations and reconciling this with overall triple-bottom line considerations (Bramwell, 2006; Mostafanezhad et al., 2016; Saarinen, 2013).

The interplay between community resilience and the tourism system is of critical concern and is embedded in the question of limits to growth (Saarinen, 2013; Tobin, 1999). The bind is between optimizing economic returns and establishing the extent to which adopting growth strategies might diminish the cultural and natural values of the area, and by implication, community resilience (Kerstetter & Bricker, 2009; Ruiz-Ballesteros, 2011). The political inclination is conservative and shapes attitudes towards economic development exigencies. This aligns with Espiner and Becken's assertion that peripheral areas accentuated with high natural values hold "multiple ecological and sociocultural functions and are increasingly a focus for regional development" (2014, p. 646). If this is so, then the call is to ensure that a balance is struck between functions that serve economic urgencies and tourism expansion, and others that relate to strengthening of human social networks and community resilience (Cheer, 2017; Holladay & Powell, 2013).

This raises the question of whether community resilience is socially constructed and the roles that local stakeholders play in coastal periphery contexts where tourism expansion is more or less assured (Béné et al., 2016). This is a pressing enquiry given that so often tourism expansion impinges on extant community relations and networks vital to small peripheral communities. As Espiner and Becken (2014) point out, when tourism in small peripheral communities reaches a "non-return point", it is incumbent on all stakeholders to overcome such path dependencies and move toward adaptations that lead to new steady states. This has to be underpinned by tourism policy and planning that shapes tourism trajectories rather than be subject to the whims of the market. The Shipwreck Coast Master Plan seeks to do that but without political will and stakeholder cohesion, "non-return" points may become all too familiar.

The use of resilience thinking to assess the extent to which tourism is enhancing or detracting from more productive legacies is an ongoing refrain (Hall et al., 2018; Cheer & Lew, 2017). As Hooli (2017, p. 114) argues, "discovering the resilience of communities in the context of global tourism is a useful tool to analyse and reveal the challenging nature of the complex and relational transformation processes". Similarly, Herrschner and Honey (2017) point out that connectivity among tourism community members is vital in times of crisis yet very often this is compromised in tourism systems. The nature and scope by which the reproduction of tourism space takes place, and the implications this has for cultural and ecosystems embedded in relational spaces are related to reducing vulnerabilities within tourism systems (Espeso-Molinero, 2017; Lapointe & Sarasin, 2017).

Finally, we argue that discourse and praxis should place emphasis on the way political processes not only hoist ecological concerns as priority, but also consider how economic yields are commensurate with the costs associated with a given activity; in this case tourism (Stonich, 1998). As Hall et al. (2018, p. 42) affirm, "tourism as a socio-economic activity, is a major contributor to some of the markers of the Anthropocene", and while the sustainable tourism narrative predominates, the continuing global tourism juggernaut is at odds. As the Shipwreck Coast exemplifies, at stake is the resilience of regional areas (inclusive of communities and natural endowments), who in the absence of garnering adequate economic returns from the exploitation of their assets through tourism, will likely find themselves in social and ecological deficit. This signals implications for research that examines SES resilience, and by association community resilience within tourism systems.

Disclosure statement

No potential conflict of interest was reported by the authors.

ORCID

Joseph M. Cheer http://orcid.org/0000-0001-5927-2615

References

ABC. (2016). Great Ocean Road tourists welcomed back to fire-ravaged holiday spot 'with open arms'. ABC News, 6 January 2016. Retrieved from https://www.abc.net.au/news/2016-01-06/tourists-welcomed-back-to-the-great-ocean-road-after-fires/7071910

Adger, W. N. (2000). Social and ecological resilience: Are they related? Progress in Human Geography, 24(3), 347–364. doi:10.1191/030913200701540465

Adger, W. N. (2003). Building resilience to promote sustainability. IHDP Update, 2(2003), 1–3.

Appleton, J. V. (1995). Analysing qualitative interview data: Addressing issues of validity and reliability. Journal of Advanced Nursing, 22(5), 993–997. doi:10.1111/j.1365-2648.1995.tb02653.x

Bazeley, P. (2009). Analysing qualitative data: More than 'identifying themes'. Malaysian Journal of Qualitative Research, 2(2), 6–22.

Bauer, A. M., & Ellis, E. C. (2018). The Anthropocene divide: Obscuring understanding of social-environmental change. Current Anthropology, 59(2), 209–227. doi:10.1086/697198

Baum, T., & Hagen, L. (1999). Responses to seasonality: The experiences of peripheral destinations. International Journal of Tourism Research, 1(5), 299–312. doi:10.1002/(SICI)1522-1970(199909/10)1:5<299::AID-JTR198>3.0.CO;2-L

Bec, A., McLennan, C. L., & Moyle, B. D. (2016). Community resilience to long-term tourism decline and rejuvenation: A literature review and conceptual model. Current Issues in Tourism, 19(5), 431–457. doi:10.1080/13683500.2015.1083538

Béné, C., Al-Hassan, R. M., Amarasinghe, O., Fong, P., Ocran, J., Onumah, E., ... Mills, D. J. (2016). Is resilience socially constructed? Empirical evidence from Fiji, Ghana, Sri Lanka, and vietnam. Global Environmental Change, 38, 153–170. doi:10.1016/j.gloenvcha.2016.03.005

Boonstra, W. (2016). Conceptualizing power to study social-ecological interactions. Ecology and Society, 21(1).

Bramwell, B. (2006). Actors, power, and discourses of growth limits. Annals of Tourism Research, 33(4), 957–978. doi:10.1016/j.annals.2006.04.001

Brown, O., & Wittbold, B. (2018). Human mobility in the Anthropocene. In Robert McLeman, François Gemenne (Eds.), Routledge Handbook of Environmental Displacement and Migration (pp. 415–420). London: Routledge.

Calgaro, E., & Lloyd, K. (2008). Sun, sea, sand and tsunami: examining disaster vulnerability in the tourism community of Khao Lak, Thailand. Singapore Journal of Tropical Geography, 29(3), 288–306. doi:10.1111/j.1467-9493.2008.00335.x

Chapin, F. S., Peterson, G., Berkes, F., Callaghan, T. V., Angelstam, P., Apps, M., ... Whiteman, G. (2004). Resilience and vulnerability of northern regions to social and environmental change. AMBIO: A Journal of the Human Environment, 33(6), 344–349.

Cheer, J. (2017). Resilience in the visitor economy: Cultural economy, human social networks, and slow change in the regional periphery. In Joseph M. Cheer and Alan A. Lew (Eds.), Tourism, resilience and sustainability: Adapting to social, political and economic change (pp. 103–115). London: Routledge.

Cheer, J. M., & Lew, A. A. (Eds.). (2017). Tourism, resilience and sustainability: Adapting to social, political and economic change. London: Routledge.

Cheer, J., Cole, S., Reeves, K., & Kato, K. (2017). Tourism and Islandscapes-Cultural realignment, social-ecological resilience and change. Shima, 11(1), 40–54.

Cleveland, D. A. (2014). Resilience: Antidote for the Anthropocene. Resilience: A Journal of the Environmental Humanities, 1(1), 1–3.

Colomb, C., & Novy, J. (Eds.). (2016). Protest and resistance in the tourist city. London: Routledge.

Christaller, W. (1964). Some considerations of tourism location in Europe: The peripheral regions-underdeveloped countries-recreation areas. Papers in Regional Science, 12(1), 95–105. doi:10.1007/BF01941243

Cretney, R. (2014). Resilience for whom? Emerging critical geographies of socio-ecological resilience. Geography Compass, 8(9), 627–640. doi:10.1111/gec3.12154

Crutzen, P. J., & Stoermer, E. F. (2000). The Anthropocene. International Geosphere-Biosphere Programme Newsletter, 41, 17–18.

Dalby, S. (2016). Framing the Anthropocene: The good, the bad and the ugly. The Anthropocene Review 2016, 3(1), 33–51 doi:10.1177/2053019615618681

Davoudi, S. (2012). Resilience: A bridging concept or a dead end? Planning Theory & Practice, 13(2), 299–307. doi:10.1080/14649357.2012.677124

Espeso-Molinero, P. (2017). Collaborative capacity building as a resilience strategy for tourism development in indigenous Mexico. In Joseph M. Cheer and Alan A. Lew (Eds.), Tourism, resilience and sustainability: Adapting to social, political and economic change (pp. 184–201). London: Routledge.

Espiner, S., & Becken, S. (2014). Tourist towns on the edge: Conceptualising vulnerability and resilience in a protected area tourism system. Journal of Sustainable Tourism, 22(4), 646–665. doi:10.1080/09669582.2013.855222

Green, R. (2004). Sea change on the Great Ocean Road [The transformation of town character along the Victorian coastline.]. Landscape Australia, 26(4), 73.

Gren, M., & Huijbens, E. H. (2014). Tourism and the Anthropocene. *Scandinavian Journal of Hospitality and Tourism*, *14*(1), 6–22. doi:10.1080/15022250.2014.886100

Gren, M., & Huijbens, E. H. (Eds.). (2016). *Tourism and the Anthropocene*. London: Routledge.

Goodwin, H. (2018). Seasonality, Cruises and Overtourism: Coping with Success Retrieved from https://news.wtm.com/seasonality-cruises-and-overtourism-coping-with-success/

Gössling, S., Ring, A., Dwyer, L., Andersson, A. C., & Hall, C. M. (2016). Optimizing or maximizing growth? A challenge for sustainable tourism. *Journal of Sustainable Tourism*, *24*(4), 527–548. doi:10.1080/09669582.2015.1085869

Hall, C. M., Baird, T., James, M., & Ram, Y. (2016). Climate change and cultural heritage: conservation and heritage tourism in the Anthropocene. *Journal of Heritage Tourism*, *11*(1), 10–24. doi:10.1080/1743873X.2015.1082573

Hall, C. M., Prayag, G., & Amore, A. (2018). *Tourism and resilience. Individual, Organisational and Destination Perspectives*. Bristol: Channel View.

Hamilton, C., Gemenne, F., & Bonneuil, C. (Eds.). (2015). *The Anthropocene and the global environmental crisis: Rethinking modernity in a new epoch*. London: Routledge.

Han, X., & Cheer, J. M. (2018). Chinese tourist mobilities and destination resilience: Regional tourism perspectives. *Asian Journal of Tourism Research*, *3*(1), 159–187.

Herrschner, I., & Honey, P. (2017). Tourism and the psychologically resilient city. In Alan A. Lew and Joseph M. Cheer (Eds.), *Tourism resilience and adaptation to environmental change: Definitions and frameworks* (pp. 218–235). London: Routledge.

Holling, C. S. (1973). Resilience and stability of ecological systems. *Annual Review of Ecology and Systematics*, *4*(1), 1–23 doi:10.1146/annurev.es.04.110173.000245

Hooli, J. (2017). From warrior to beach boy: the resilience of the Maasai in Zanzibar's tourism business. In Joseph M. Cheer and Alan A. Lew (Eds.), *Tourism, resilience and sustainability: Adapting to social, political and economic change* (pp. 103–115). London: Routledge.

Holladay, P. J., & Powell, R. B. (2013). Resident perceptions of social–ecological resilience and the sustainability of community-based tourism development in the Commonwealth of Dominica. *Journal of Sustainable Tourism*, *21*(8), 1188–1211. doi:10.1080/09669582.2013.776059

Kerstetter, D., & Bricker, K. (2009). Exploring Fijian's sense of place after exposure to tourism development. *Journal of Sustainable Tourism*, *17*(6), 691–708. doi:10.1080/09669580902999196

Knight, J. (2015). Anthropocene futures: People, resources and sustainability. *The Anthropocene Review*, *2*(2), 152–158. doi:10.1177/2053019615569318

Koob, S. (2017). Tourism demands on the rise as 12 Apostles fall away. The Australian, 22 April. Retrieved from https://www.theaustralian.com.au/news/nation/tourism-demands-on-the-rise-as-12-apostles-fall-away/news-story/afe238a6a5589e433adcb9cbef995137

Lapointe, D., & Sarasin, B. (2017). (Re)production of resilient tourism space. In Alan A. Lew and Joseph M. Cheer (Eds.), *Tourism resilience and adaptation to environmental change: Definitions and frameworks* (pp. 141–156). London: Routledge.

Lemelin, H., Dawson, J., Stewart, E. J., Maher, P., & Lueck, M. (2010). Last-chance tourism: The boom, doom, and gloom of visiting vanishing destinations. *Current Issues in Tourism*, *13*(5), 477–493. doi:10.1080/13683500903406367

Lew, A. A., & Cheer, J. M. (Eds.). (2017). *Tourism resilience and adaptation to environmental change: Definitions and frameworks*. London: Routledge.

Lew, A. A., & Wu, T.-C. (2017). Cultural ecosystem services, tourism and community resilience in coastal wetland conservation in Taiwan. In Alan A. Lew and Joseph M. Cheer (Eds.), *Tourism resilience and adaptation to environmental change: Definitions and frameworks* (pp. 102–126). London: Routledge.

Lew, A. A. (2014). Scale, change and resilience in community tourism planning. *Tourism Geographies*, *16*(1), 14–22. doi:10.1080/14616688.2013.864325

McCool, S. F., & Lime, D. W. (2001). Tourism carrying capacity: Tempting fantasy or useful reality? *Journal of Sustainable Tourism*, *9*(5), 372–388. doi:10.1080/09669580108667409

McKinsey & Company. (2017). *Coping with Success: Managing overcrowding in tourist destinations*. McKinsey & Company and WTTC. Retrieved fromhttps://www.mckinsey.com/industries/travel-transport-and-logistics/our-insights/coping-with-success-managing-overcrowding-in-tourism-destinations

Milano, C. (2017). *Overtourism y Turismofobia. Tendencias globales y contextos locales*. Barcelona: Ostelea School of Tourism & Hospitality.

Milano, C., Cheer, J. M., & Novelli, M. (2018). Overtourism: A growing global problem. *The Conversation*, July 18, 2018. Retrieved from https://theconversation.com/overtourism-a-growing-global-problem-100029

Milano, C. (2018) Overtourism, malestar social y turismofobia. *Un debate controvertido*. PASOS: Revista de Turismo y Patrimonio Cultural, *16*(3), 551–564. doi:10.25145/j.pasos.2018.16.041

Miles, M. B., & Huberman, A. M. (1984). Qualitative data analysis: A sourcebook of new methods. In Miles and Huberman (Eds.) *Qualitative data analysis: a sourcebook of new methods*. London: Sage.

Misrahi, T. (2017). Wish you weren't here: What can we do about over-tourism? World Economic Forum Weekly Update, 19 September. Retrieved from https://www.weforum.org/agenda/2017/09/what-can-we-do-about-overtourism/

Moore, A. (2015). Tourism in the Anthropocene Park? New analytic possibilities, *International Journal of Tourism Anthropology, 4*(2), 186–200. doi:10.1504/IJTA.2015.070067

Mosedale, J. (2015). Critical engagements with nature: Tourism, political economy of nature and political ecology. *Tourism Geographies, 17*(4), 505–510. doi:10.1080/14616688.2015.1074270

Mostafanezhad, M., Norum, R., Shelton, E. J., & Thompson-Carr, A. (Eds.). (2016). *Political ecology of tourism: community, power and the environment.* Routledge.

Nepal, S., & Saarinen, J. (Eds.). (2016). *Political ecology and tourism.* London: Routledge.

Novy, J., & Colomb, C. (Eds.). (2017), *Protest and resistance in the tourist city.* London: Routledge

O'Reilly, A. M. (1986). Tourism carrying capacity: Concept and issues. *Tourism Management, 7*(4), 254–258. doi:10.1016/0261-5177(86)90035-X

Overton, J. (1973). A critical examination of the establishment of national parks and tourism in underdeveloped areas: Gros Morne National Park in Newfoundland. *Antipode, 11*(2), 34–47. doi:10.1111/j.1467-8330.1979.tb00128.x

Parks Victoria. (2015). *Shipwreck Coast Master Plan: Master Plan Report.* Melbourne: Parks Victoria. Retrieved from http://www.shipwreckcoast.vic.gov.au/shipwreck-coast-master-plan-an-overview

Papathanassis, A. (2017). Over-tourism and anti-tourist sentiment: An exploratory analysis and discussion. *Ovidius University Annals, Economic Sciences Series, 17*(2), 288–293.

Pearson, E. (2017). Landslide closes Great Ocean Road between Wye River and Kennett River. Geelong Advertiser, September 9, 2017. Retrieved from https://www.geelongadvertiser.com.au/news/geelong/landslide-closes-great-ocean-road-between-wye-river-and-kennett-river/news-story/a503cead0524c1c5ab59adfb6561dfdb

Pulford, J. (2017). EARLY WORK SET TO BEGIN ON SHIPWRECK COAST MASTER PLAN. Media Release, The Hon Jaala Pulford, Thursday, 12 October, 2017. Retrieved from https://www.premier.vic.gov.au/wp-content/uploads/2017/10/171012-Early-Work-Set-To-Begin-On-Shipwreck-Coast-Master-Plan-1.pdf

Rifai, T. (2017). Tourism: growth is not the enemy; it's how we manage it that counts. UNWTO Press Release 15 August 2017. Madrid: UNWTO. Retrieved from http://media.unwto.org/press-release/2017-08-15/tourism-growth-not-enemy-it-s-how-we-manage-it-counts

Robbins, P., & Moore, S. A. (2013). Ecological anxiety disorder: diagnosing the politics of the Anthropocene. *Cultural Geographies, 20*(1), 3–19. doi:10.1177/1474474012469887

Ruiz-Ballesteros, E. (2011). Social-ecological resilience and community-based tourism: An approach from Agua Blanca, Ecuador. *Tourism Management, 32*(3), 655–666. doi:10.1016/j.tourman.2010.05.021

Saarinen, J. (2018). What are wilderness areas for? Tourism and political ecologies of wilderness uses and management in the Anthropocene. *Journal of Sustainable Tourism,* 1–16. doi:10.1080/09669582.2018.1456543

Saarinen, J. (2013). Critical sustainability: Setting the limits to growth and responsibility in tourism. *Sustainability, 6*(1), 1–17. doi:10.3390/su6010001

Schillat, M., Jensen, M., Vereda, M., Sánchez, R. A., & Roura, R. (2016). *Tourism in Antarctica: A multidisciplinary view of new activities carried out on the White Continent.* Berlin: Springer.

Seraphin, H., Sheeran, P., & Pilato, M. (2018). Over-tourism and the fall of Venice as a destination. *Journal of Destination Marketing & Management, 9*, 374–376.

Sheivachman, A. (2017). Proposing Solutions to Overtourism in Popular Destinations: A Skift Framework. Retrieved from https://skift.com/2017/10/23/proposing-solutions-to-overtourism-in-popular-destinations-a-skift-framework/

Soriano, C. (2017). The Anthropocene and the production and reproduction of capital. *The Anthropocene Review, 1* 12.

Steffen, W., Crutzen, P. J., & McNeill, J. R. (2007). The Anthropocene: Are humans now overwhelming the great forces of nature. *AMBIO: A Journal of the Human Environment, 36*(8), 614–621.

Stockholm Resilience Center (SRC) (2016). *Applying resilience thinking: Seven principles for building resilience in social-ecological systems.* Stockholm: Stockholm Resilience Centre.

Stonich, S. C. (1998). Political ecology of tourism. *Annals of Tourism Research, 25*(1), 25–54. doi:10.1016/S0160-7383(97)00037-6

Strauss, A. L. (1987). *Qualitative analysis for social scientists.* Cambridge University press.

Tobin, G. A. (1999). Sustainability and community resilience: The holy grail of hazards planning?. *Global Environmental Change Part B: Environmental Hazards, 1*(1), 13–25. doi:10.1016/S1464-2867(99)00002-9

Tyler, B. (2016). Tourists happy to pay up to $10 to visit the Great Ocean Road's Twelve Apostles, study shows. Geelong Advertiser, June 11, 2016. Retrieved from https://www.geelongadvertiser.com.au/news/geelong/tourists-happy-to-pay-up-to-10-to-visit-the-great-ocean-roads-twelve-apostles-study-shows/news-story/63f7086cd895859c1fb46432d1be7b26

Walker, B., & Salt, D. (2012). *Resilience thinking: sustaining ecosystems and people in a changing world.* Island Press.

Warrnambool City Council. (2018). $58 million committed to implementing Shipwreck Coast Master Plan. Retrieved from https://www.warrnambool.vic.gov.au/news/58-million-committed-implementing-shipwreck-coast-master-plan

Zalasiewicz, J., & Waters, C. (2016). The Anthropocene. In *Oxford Research Encyclopedia of Environmental Science*. Oxford: Oxford University Press. Retrieved fromhttp://environmentalscience.oxfordre.com/view/10.1093/acrefore/9780199389414.001.0001/acrefore-9780199389414-e-7

Zwagerman, K. (2016). Putting Great Ocean Road hinterland on the map, The Standard, October 14, 2016. Retrieved from https://www.standard.net.au/story/4227740/putting-region-on-map/

Tourists and researcher identities: critical considerations of collisions, collaborations and confluences in Svalbard

Samantha M. Saville ⓘ

ABSTRACT

Svalbard is an "edge-of-the-world" hot spot for environmental change, political discourse, tourism and scientific research in the Anthropocene. Drawing on ethnographic and qualitative research, I use this context to critically explore the identity-categories of "researcher" and "tourist". Through the lens of political ecology, I draw out the uneven power relations of knowledge production that are attached to these labels and their consequences for ongoing efforts for managing sustainable tourism. By considering the experiences of tourists, researchers and "scientific tourists", both practically and from an embodied experiential perspective, I challenge the distinctions typically made between these roles. I bring to light several common aspects, goals and experiences these practices share. In doing so, I aim to disrupt the existing hierarchies of knowledge that champion an impersonal, rational scientific approach and call for a more varied array of knowledge and practices to be taken into account when considering the future ecologies of Svalbard and the broader Arctic region.

Introduction

> It is a not uninformative conceit to play with the scandalous suggestion that ethnographer and tourist are, if not the same creature then the same species and are part of the same continuum – that homo academicus might be uncomfortably closely related to that embarrassing relative turistas vulgaris. (Crang, 2011, p. 207)

In this article, I draw upon my potentially "embarrassing" experiences as a visitor to Svalbard, an increasingly popular tourist destination where problems and questions that dominate anthropocenic imaginaries are particularly prominent. Among these are climate change, energy security, transboundary pollution, invasive species, the very presence of humans in this extreme, remote, and yet accessible arctic archipelago. My central goal is to explore the power relations of knowledge-making and natural–social relations in this context. The political ecology of knowledge in Svalbard, as elsewhere, holds bearing on how society–environment relationships are managed, and according to which dominant discourses, interests and imaginaries. Sociocultural categories play important roles in the complex assemblage of environmental practice and policy (Mostafanezhad, Norum, Shelton, & Thompson-Carr, 2016). Moreover, our shifting, socially and spatially dependent identities are strongly linked to the ways in which we perceive the world, our imaginaries and our relationship with it (Natter & Jones III, 1997; Salazar, 2012). In this paper, I examine and challenge the categories of "tourist" and "researcher" as bounded and separate identities. These two "identities" and values associated with them are far more intertwined than we tend to acknowledge. This has consequences for what kinds of knowledge feed into imaginaries of the Arctic environment as a space to visit, reside in and manage. Indeed, what counts as knowledge and who is represented in decisions about how Svalbard's largely "wild" landmass is managed has been an area of contention and conflict in recent history (Nyseth & Viken, 2016). The article tracks examples of how the roles of tourist and researcher can collide and conflict, but more often meet and potentially collaborate. The categories of "researcher" and "tourist" are shown to be messy, fluid and indistinct, and yet simultaneously producing difference in terms of access to, and experience of, Svalbard.

In the first section, I introduce Svalbard as a site of conflicting anthropocenic imaginaries. The proceeding discussion is situated within the research context and methodological approach. I then explore how the roles of "tourist" and "researcher" relate to each other and produce socio-ecological relations through three broad themes: imagining and re-presenting Svalbard and the Arctic from afar; the practice and portrayal of tourist and researcher identities as "worlds apart", or otherwise; and individual, embodied experiences of encountering Svalbard. Throughout the article, I draw attention to the nature–culture relationships, discourses and knowledge regimes woven into the activities and identities described. Ultimately I contend that multiple perspectives of and from the Arctic are needed in contribution to a more effective discussion and action on global environmental change in the Anthropocene. Alone, science and awareness raising, as the current political climate indicates, is not the panacea to creating wide-scale action on climate change, the Arctic's "hot topic" (Brugger, Dunbar, Jurt, & Orlove, 2013; Hulme, 2014) or wider socio-ecological problems of the Anthropocene. Anderson and Bows (2012) argue relations between policy, politics and scientific communities need to radically change if meaningful climate change mitigation strategies are to be negotiated. Critically considering the boundaries between researcher and tourist identities, I argue, could help us to think differently, as Anderson and Bows suggest. In disrupting existing knowledge hierarchies, opportunities for new imaginaries of sustainable tourism in the Anthropocene open up.

Svalbard, emblem of the Anthropocene?

Svalbard is an archipelago within the Arctic Circle between 74 and 81 degrees north governed by Norway under a special international treaty since 1925 (see Figure 1). It is remote, the last stop before the North Pole, yet, especially since the arrival of a second commercial airline operating near daily flights from mainland Norway, surprisingly accessible. Unlike many Arctic areas of human settlement, Svalbard's population is non-indigenous, cosmopolitan and transient. The majority of the 2500 residents live in the Norwegian administrative centre of Svalbard, Longyearbyen. The active Russian mining town of Barentsburg houses 400 workers, and an abandoned Russian mining town (Pyramiden) hosts a handful of seasonal workers. There are also numerous research stations, including a small Polish base (Hornsund), and a collection of stations at Ny Ålesund, including bases for 13 different nations. Annual population turnover is 20%–25% and over 20% of the Longyearbyen population originate from more than 40 different nations other than Norway (Kristiansen, 2014).

Svalbard has a history of exploration dating back to the sixteenth century (Avango et al., 2011). Up until very recently, human activity and imaginaries of this region have been centred on the extraction and exploitation of natural resources, which also dominated the research agenda here (Roberts & Paglia, 2016). Tourism and other scientific research were very much on the margins. Since the 1990s, however, the Norwegian government has actively promoted both research and tourism as part of its efforts to diversify the economy away from mining and maintain a presence on the islands (Viken, 2011). This diversification was also a move to transition Longyearbyen from a mining company town to a more "normal", Norwegian family setting (Grydehøj, Grydehøj, & Ackren, 2012). To a large extent, this has been successful.[1] Longyearbyen now boasts many services and leisure facilities such as a sports centre, culture house, hotels and restaurants. Mining continues, but in both Norwegian and Russian operations, this is in decline. The new industries of tourism and research and education represent the "softer" version of extracting value from Svalbard's natural resources. Research time spent in Svalbard is increasing, nearly 700 students attended the University Centre (UNIS) in 2015 and there are plans for further growth. Tourism has expanded rapidly to attract around 70,000 visitors a year

Figure 1. Main islands and settlements of Svalbard.

(Statistics Norway, 2012). Late winter to spring is a busy period with activities such as skiing, snow-scooter excursions, ice-caving and husky-dog sled trips on offer. Whereas summer brings large cruise ships to shore and the appeal of wildlife tours, trekking, kayaking and boat trips.

Almost two-thirds of the nearly 62,000 km² land mass is glaciated with 65% protected as national parks or nature reserves. Environmental management and tourist marketing regimes alike resist the Anthropocene imaginary of a world without pristine nature through their ambitions to promote and protect the "last wilderness areas in Europe". The tourist imaginary of Svalbard enrols stereotypical notions of an arctic wilderness, where physical landforms, temperature and other species feature heavily. This is visible in promotional material from tourist companies as well as evident in the motivations for travel to Svalbard in visitor surveys. In my own survey of visitors to Svalbard, the "wilderness experience" or "Arctic nature" in general was a key pull factor in the decision to visit Svalbard, with 49% of the 55 respondents directly referencing features and terms related to this in an open question. These results correlate more broadly with previous visitor surveys and tourism research in the area (Enger, 2011; Enger & Jervan, 2010; Viken, 2006).

Within geopolitical discussions surrounding the future of the Arctic region, issues of the Anthropocene frequently become opportunities for Svalbard, which is often portrayed as a "hub" for future

Figure 2. Students contemplating the view on arrival to Petunia Bay (author's image, June 2013).

resource extraction and continued scientific research, especially about climate change (Brigham, 2007; Norwegian Ministry of Justice and the Police, 2010). The siting of the Svalbard Global Seed Vault, a backup storage facility for seeds of crop varieties from around the world, also known as the "Doomsday Vault" or "Noah's Ark", and the more recent development of The Arctic World Archive just outside Longyearbyen feeds into this "end of the world" discourse. These facilities also respond to global challenges of the Anthropocene, taking advantage of Svalbard's location and climate (both physical and political). Historical and political analysis of these activities argues that the expansion of tourism and research is really a continuation of struggles over sovereignty cloaked in discourses of energy security, climate change and local economic development (Avango et al., 2011; Grydehøj et al., 2012; Norum, 2016; Timothy, 2010). Alongside these interests, international NGOs work, sometimes in partnership with state-funded scientific institutes, to protect Arctic ecosystems (especially flagship species). For example, the Worldwide Wildlife Fund for Nature and the Norwegian Polar Institute operate a polar bear research and monitoring programme.

Svalbard represents something of a contradiction; Norway simultaneously promotes its territory as the last European wilderness where environmental protection is of utmost importance, a last stand against the Anthropocene, an age that embraces "novel ecologies" and acknowledges the impossibility of a pristine nature (J. Lorimer, 2015; Robbins & Moore, 2013). Meanwhile environmental exploitation continues. Although Svalbard's coal industry is now in decline, the logistical need for using coal to fuel the communities there has not diminished, nor has the appetite for potential development of further hydrocarbon and other mineral extraction that could be opened up through sea-ice retreat.

From a tourism perspective, Svalbard has been labelled a "disappearing destination" (Font & Hindley, 2017) on account of its association with climate change. Several tourist firms there are engaged in reducing the environmental impacts of their operations, a form of what Moore refers to as "the new ecotourism" (Moore, 2015, p. 190). Longyearbyen is now an accredited Norwegian "Sustainable Destination". Some firms operate carbon offsetting schemes and many cruise operators follow volunteer codes for responsible tourism (Bets, Lamers, & van Tatenhove, 2017). Tourist guides and expedition cruises often seek to educate tourists about not only the wildlife and landscapes they are visiting, but also environmental change; some employ scientists to assist with this. Scientific endeavours in Svalbard are strongly connected to tourism activities, whether directly through expert lectures onboard cruises, or indirectly through feeding into risk assessments of activities through research on snow and avalanche conditions or the impacts of tourist activities on Svalbard ecosystems, for example. Below, I briefly set out how I approached researching this field, before moving on to look in more detail at the roles of researchers and tourists in Svalbard and power relations of knowledge making at work.

Researcher tourism and "studying sideways"

I draw on extensive qualitative research undertaken between 2012 and 2016, including three field visits to Svalbard, which explored systems of value and nature–culture relationships in Svalbard. Whilst conducting this research I immersed myself in academic papers, fiction books, travel blogs and conversations with those who have visited or lived in Svalbard. In addition to documentary research and onsite ethnography, I held two focus groups with past and present students and staff of my home research institution, Aberystwyth University, which has several academic links to Svalbard, to discuss memories of Svalbard. During fieldtrips in 2013, 2014 and 2015, qualitative semi-structured interviews were conducted with over 70 Svalbard residents and stakeholders and a small-scale survey captured some motivations and impressions of visitors. The resulting data-set consisted of photographs, field notes, policy documents, promotional materials from tourist companies and other institutions, audio files from interviews and focus groups, which were transcribed and thematically coded along with materials generated from participatory exercises such as the "memory bag", described below. These materials were analysed on a rolling, iterative basis, shuttling back and forth from different sources over the course of the research (James, 2013). I present my findings in this article through an openly self-reflexive and value-laden account that is congruent with my argument that a broader range of communicative approaches is needed to initiating change in socio-natural relationships of the Anthropocene (Bramwell, Higham, Lane, & Miller, 2016).

In the summer of 2013, I made my first journey to the main island of Spitsbergen, spending two weeks getting to know Longyearbyen, and a week wild camping in Petunia Bay, a few kilometres away from the semi-abandoned Russian mining settlement of Pyramiden on an interdisciplinary, international field tour(see Figure 2). In Petunia Bay, I joined a multidisciplinary group of 40 students and staff from Sweden (KTH Royal Institute of Technology, Stockholm) and the USA (Illinois University) who were teaching and studying "Environment and Society in a Changing Arctic".[2] The group leaders were a team of Svalbard and Arctic experts with decades of field and scholarly experience of the region. Within this group, and over the course of my other fieldwork experiences in Svalbard, the role of my own identity as a researcher/tourist/visitor to the knowledge I was producing became important. During this trip, in particular, I was very much "studying sideways" (Hannerz, 2004), the other students of the group were having a parallel experience to my own. Like myself, the students had been reading about and received lectures on Arctic history, politics, literary traditions and geophysical processes, but this was their first trip to the Arctic and to Svalbard. I sought to approach them as allies in our investigation into this new landscape, which we were attempting to make sense of personally and academically. Including their reflections on the experiences we were sharing, as Hannerz (2004) points out, also aided my own understandings. As well as taking field notes during the camp and engaging in the group activities and conversations, I invited the students to leave short anonymous written comments detailing particularly memorable moments or strong feelings about their days in Petunia Bay. These mini-diary entries were collected in a "memory bag" and added to my data-set.

At the end of the field trip, the students commented that they had mixed ideas about their identity on Svalbard, another aspect I shared with them. Despite the dubious character of the colonial masculine hero-explorer being particularly associated with the discipline of (British) geography (see

Baigent, 2010; Phillips, 1997), as a "human geographer", I fell through the gaps of the existing identity templates on the island (and the fieldtrip group). The many contacts and research projects connecting my university and geography department to Svalbard focus exclusively on the "natural" or "physical" sciences. As a non-Norwegian social scientist, I was outside the conventional knowledge ecology and remit of these professional connections. This presented limitations, but also the freedom to negotiate my own liminal Svalbard-identity – mixing researcher and tourist strategies and practices as need be throughout all three fieldtrips – and it is this liminal identity that provided some of the motivation for this article.

Beyond an academic identity, personally, the excitement I had cultivated about visiting Svalbard (as described in the subsequent section) was coupled with guilt for betraying a part of my identity as an environmentalist. That my research would involve *flying* to Svalbard was a serious consideration as to whether or not to take on the project. I had previously been working for an environmental education charity that was serious about addressing climate change and improving human–nature relationships, ideally reducing the large impact of humanity that the Anthropocene heralds. For my own part, I had boycotted flying for the past seven years to reduce my carbon footprint. Climate change formed a key part of my intellectual journey that filtered my relationship with my environment through everyday practices. A field trip to the Arctic, where climate change is happening most conspicuously (Hassol, 2004) meant both engaging with academic and public debates linked to this region and its changes, but also compromising on some of my principles in order to do so. My hope was that bringing new perspectives to problems such as climate change could help us think differently and start to act differently to address the challenges of living in the Anthropocene. This is perhaps easier said than done – Woodyer and Geoghegan (2013) note the pressure to do so as academics, and as geographers in particular is high, and can easily lead to feelings of helplessness and despair in and outside of this role (Norgaard, 2006). They prescribe a course of enchantment, a re-engagement with the world and modes of thinking that recognise the value of descriptive work (Woodyer & Geoghegan, 2013). As literary critic James Bradley recently asserted, in order to face our environmentally insecure future, we may need "to step outside what we know and imagine ourselves in new ways" (Bradley, 2017).

By considering the role of identity and power relations in knowledge production, the following sections contribute to a potential new imaginary, one where categories of "tourist" and "researcher" are blurred and knowledge about environmental change is valued for its potential to helpfully contribute to open and progressive management decisions. As a starting point for drawing out the common experiences of tourists and researchers, the role of representations to both of these groups is, I argue, significant, as discussed in the following section.

The anticipatory arctic

> I sit at my desk, wearing my extra jumper, listening to the wind blowing around and through the building, to the dark rain, and try to convince myself I am not cold. It will get much colder in Svalbard. Trawling through images of ice, snow, glaciers and polar bears and a now-familiar chill descends. A chill that comes from imagining those icy landscapes, a chill induced by all the doubts these landscapes spark: will I be able to cope with the cold? Will I get eaten by a polar bear? Will I be able to function as a good researcher in these conditions? I turn to re-reading the safety guidelines and equipment lists, consider buying some more thermals and bolster my nerves with thoughts of the women who proceed me. (Research journal, Aberystwyth, January 2013)

Reflecting on this anxiety, I wonder why I repeatedly found myself going through the above process. To some extent, I was enacting my perceived role as a dutiful researcher. However, there was definitely more to it. I was excited. Scared, but also thrilled at the prospect of the "adventure" that lay ahead. The films, photos, novels, academic descriptions, enthusiastic colleagues and analysis I was engaging with all fed this growing enchantment.

There is a wide acknowledgement (Bloom, 1993, for example) that past representations of Arctic landscapes as a blank canvas work to legitimate a colonial impulse that takes many forms: claims of sovereignty, scientific exploration and touristic expeditions. The Arctic landscape can also be seen as one which is represented as a highly gendered space, the realm of "masculinist fantasy" (Dittmer, Moisio, Ingram, & Dodds, 2011, p. 203) and gentlemen heroes (Baigent, 2010) who explore the (feminine) pristine white landscapes. These representations and imaginaries can still be traced in current geopolitical actions such as resource and scientific exploration (Dittmer et al., 2011) as well as practical advice for the visitor to maintain such a visual landscape.

You are welcome to the Arctic – as long as you leave no signs of having been here! (Governor of Svalbard 2010, p. 8)

The "reading" of Arctic landscape representations as wilderness and invariably devoid of human presence has been highly criticised for masking a complex picture of social and natural relations (Wylie, 2007). Indeed, the idealised form of nature and responsible tourism represented in advice literature to tourists such as that above can be equally problematic. A broader conception of the visual as an embodied, emotionally engaged and conscious practice (Crang, 1997), that is "more- than-representational" (H. Lorimer, 2005; for example, Thrift, 2008) furthers these discussions. As Crouch (2005) and Edensor (2006) remind us, there is more to tourist (and other) engagements with place than "gazing" at the destination and representations of it. Practices of encountering representations of a would-be tourist destination or research site can evoke strong emotional affects and effects at the level of the individual body (Saville, 2008). My journal reflections are an example of how representations of Svalbard go beyond screens, brochures or journals, making them even more powerful. Whilst I am aware of the exclusions and elisions within the images and anecdotes, it does not render me immune to their affects.

Shared ground between tourist and researcher begins with anticipating Svalbard. Both will engage with promotional, historical, informational and/or scientific representations of Svalbard before embarking on a journey there, forming an eco-imaginary before arrival. Moreover, in their possible roles as travel writers, bloggers, sharers of photos, memories and research papers, both tourists and researchers have opportunities to add to this suite of representations. In other words, we are now in an age of prosumers, where we not only consume, but produce representations and traditional knowledge production hierarchies are not as clearly delineated as they once were (Norum, 2017).

Tourist and researcher: worlds apart?

Anderson (2004) tells us that identities such as "tourist" and "researcher", are not fixed but shifting continually along with and in response to place. Early polar voyages reported as much about the biographies of the men involved as their scientific activities. The categories of "science", "exploration" and "tourism" were far less distinct. At the turn of the nineteenth century, however, the professionalisation of academic disciplines, such as geography, led to the establishment of clear epistemological boundaries (Baigent, 2010). These social labels still differentiate practices and people in specific ways in Svalbard, which I discuss in the following section. I then turn to the collaborations and confluences between these roles.

Collisions and inequalities

Academics and other tourists routinely disparage the attractions of regulated tourist enclaves in contradistinction to the experiences gained through apparently more adventurous pursuits. (Edensor, 2006, p. 44)

Over the last few decades, there has been increasing criticism of mass tourism especially with the industry accused of exploitation, cultural contamination and environmental destruction. Whilst tourism brings income and infrastructure, it also requires management to address accompanying negative impacts. Although ethical alternatives to mass tourism that are "sustainable", "ethical", "eco" are now increasing, they represent a small segment of the market and attract ongoing critique (Buckley, 2004; Butcher, 2003; Coria & Calfucura, 2012). Scientific research has also been subjected to criticism at an academic and philosophical level (Collins & Pinch, 1998; Latour, 1999, 2004) and scientists appear to be fighting an increasingly difficult battle to be heard, the example of climate change being pertinent (Beck, 1992; Head & Gibson, 2012). However, the level of popular vilification is not commensurate and the extent to which individual responsibility for impacts on wilderness is attributed is uneven, with efforts to address and influence tourist behaviour becoming more and more evident through visitor education schemes (see Stanford, 2008, for a discussion).

There is a clear hierarchy of these scientific and touristic roles in Svalbard. Natural scientists operate with high budgets, enjoy a relatively prestigious position in the public consciousness and have social power. Conversely, tourists are generally looked down upon as nuisances by residents at least, yet are an important source of revenue all the same. The social scientist, ethnographer, anthropologist or human geographer sits (awkwardly) somewhere in the middle.

Examples of the separation and inequalities between the practices of science and tourism abound in this context. Although there have been some studies into the environmental impacts of scientific research activities in Svalbard (Krzyszowska, 1985; M. H. West & Maxted, 2000), these are in the minority compared to the consistent attention the impacts of tourism yields (Hagen, Vistad, Eide, Flyen, & Fangel, 2012; Kaltenborn, 2000; Roura, 2011; Viken & Jørgensen, 1998). These differences manifest politically and geographically with much discussion over the management zones, regulations and their exceptions, as long-term resident and tourist operator Andreas Umbreit describes:

> It is currently forbidden to pick flowers or camp unnecessarily on vegetation all over the archipelago, yet the bull-dozing of acres of land for road construction, mining and research facilities continues. (Umbreit, 2013, p. 47)

Access to Svalbard's "wilderness" through management zones is officially dependent on residency status, with residents needing to notify the Governor of visits to National parks, whilst visitors (scientists and tourists) need permission for trips outside of Area 10.[3] All visitors are obliged to "take care of Svalbard" (VisitSvalbard.com) and conform to the visitor guidelines and regulations. Visit Svalbard, the umbrella tourist organisation in Longyearbyen, summarises the tone of these guidelines thus, "It is impossible being an invisible tourist—but we do appreciate your trying:)." The emphasis shifts from "visitors" to "tourists", and in doing so, more explicitly differentiates the tourist and the research visitor. Scientists, especially physical scientists have reportedly enjoyed an easy passage through the Governor's permissions system in the past (Viken, 2011), though there is evidence of this changing with more recent regulations tightening access for all actors.

Local tourist operators have expressed frustration at the lack of evidence that tourist activities have had significant impacts in restricted areas and question the need and terms of access for scientists (Jørgensen, 2012; Umbreit, 2013). It has also been pointed out that evidence collected by tourist operators (such as species monitoring data) could effectively contribute to policy decisions, as they have previously, if there was a more open approach to what counts as "real" knowledge, rather than maintaining a strict divide between scientific and other "communities of practice" (Nyseth & Viken, 2016).[4] Moore describes these situations as a matter of "socioecologics", which "set the terms and narratives that count as relevant knowledge used for decision making" (Moore, 2015, p. 193). The contested nature of socioecologics in Svalbard has been a key factor in disputes over wilderness management in the past. Although scientists have also been unsatisfied with the use of scientific evidence in policy decisions, evidence from the tourist industry has not been considered on equal standing and access for tourism activities remains more restricted than that for science.

As participants in my focus groups described, tourists are, generally more limited in terms of movement outside of Longyearbyen. This news was delivered with some sympathy for my pending situation as a "tourist", which would likely be less rewarding than their time as students at UNIS. The majority of tourists will (sensibly) rely on organised tours for polar bear protection and essential local knowledge of the terrain:

> I think if you're a tourist you're quite restricted. You can't get out of the town by yourself, you have to go with someone with a gun, you have to pay someone for this that and the other. If you live here, you've got your own rifle and you know how to use it, it's a lot of freedom, you can do whatever you like. (Focus Group, 2013)

Tourist visits are, therefore, largely confined to the sites and schedules that the tour companies offer. Some portray Svalbard as a place where length of stay is the ultimate expression of social capital, closely followed perhaps by the extent of your adventures (Eliassen, 2009). From this perspective, the tourist on an organised tour is at the bottom of the social hierarchy. One research participant described this quite clearly in the case of my own emerging identity:

> The worst we know about is tourists. The best is people living here. In between you have visitors. And you are becoming a visitor – because you have been here several times and you stayed for a while, and you get some connections and get some friends and someone invites you for something... (Interview with long-term resident, 4 June 2014)

A smaller number of tourists who opt to rent rifles and go on independent tours in Svalbard, might occupy a higher social standing as "travellers" or "adventure tourists", but only if their skills, equipment and planning prove sufficient and the trip is safely concluded. Until more commercial tourist developments arrived in the late 1990s, this was the dominant tourist activity and is where there are more similarities between tourists and researcher practices and the respect held for them (Gyimóthy & Mykletun, 2004; Viken & Jørgensen, 1998).

Collaborations and confluences

Despite the political, practical, historical and socio-economic differences outlined above, there are also a great deal of shared experiences between the tourist and research visitor to Svalbard. Arvid Viken, a long-term researcher and consultant on tourism in Svalbard outlines how these groups share similar transport, logistics and services in Svalbard (Viken, 2011). In getting to and operating our camp in Petunia Bay, our large educational group made use of tourist transport infrastructure and local research institution equipment – demonstrating the kind of symbiotic relationships between sectors Viken describes. In fact, when exploring the extent of "last chance tourism" in Svalbard, Johnston, Viken, and Dawson (2012) note the surge in visitors (journalists, ministers, environment officers as well as scientific specialists) from 2004 onwards. This they connect to the Norwegian Polar Institute's work contributing to the International Panel on Climate Change reports and in doing so describe these visitors as tourists and disrupt the distinction.

There are also similarities in tourist and researcher *motivations* in coming to Svalbard. All share, to some degree, a romantic affinity with the character of the modern explorer, seeking to satisfy curiosity, escape the routines of everyday life for risky adventures and in the process test their character (Driver, 2010). The touristic and scientific "gazes" overlap and fall on similar features and objects: glaciers, cultural heritage, the Aurora and midnight sun, the (changing) Arctic climate, ecosystems and wildlife: Canada geese, polar bears, walrus, reindeer. The attractions of Svalbard can appeal equally to tourists and researchers. Not everyone is interested in the same things – polygon shapes in the permafrost surrounding Petunia Bay were exciting and educational for the glaciologists and geologist team members, whereas they saw the cultural artefacts that our polar history expert was looking for as "rusty junk", and as a social scientist, I found the differences in perspectives of interest. Many of the students in the group were also surprised and disappointed at the amount of industrial debris we saw in the Petunia Bay area, which provided powerful challenges to the "extreme wilderness" expected and promoted. There is a tension here between whether to focus on elements of the environment that did fit expectations of a pristine wilderness, or to hone in on those which did not. This tension raises questions as to how well we can separate our "quest for the sublime" and our "capture and dissect" agenda, as well as how our interests and specialisms affect our overall practice.

> Science and the quest for the sublime share the desire to move beyond everyday experience ... science, however, operates a different framework ... [it] sets out to capture and dissect a reality that the aesthetic of the sublime insists we cannot comprehend. (Spedding, 2004, p. 75)

There are different ways of valuing these phenomena, as West sets out:

> The value in the things seen and experienced [for the tourist] lies in the individual experiences, while for natural scientists, when they are in 'science mode', the value is in the careful cataloguing of one experience within a field of similar experiences in order to produce a sort of knowledge about the entity being seen. (P. West, 2008, p. 610)

Yet, tourists and researchers can both be curious to know more about Svalbard's environment, providing motivation to visit. As others observe (See Campbell, Gray and Hathaway in P. West, 2008), we cannot necessarily separate out being in "science mode" and being ourselves: the scientist is no less able or liable to enjoy an individual experience, nor is there any reason a tourist might not make connections to other experiences. Gyimóthy and Mykletun (2004) highlight the complex array of motivations involved in undertaking a trip to Svalbard, including identity forming, surviving in a challenging environment, testing one's skills and providing space away from modern stresses. As with the narratives and representations of the Arctic discussed in the previous section, it is unrealistic to assume a researcher would not be enticed by these same motivations to some extent. Narratives told about previous scientific trips are often couched in terms of adventure, adversity and excitement rather than distanced, inert data collection; in stark contrast to what is published in journals and textbooks.

Thinking about the kind of things tourists and researchers *do*, there are also similarities to observe in the practices of these roles, especially in the case of the social scientist or ethnographer. Whilst you do not generally see tourists taking precise snow samples, these scientific endeavours are a (high tech) method of collecting and documenting, which as Noy illustrates, is after all what tourists do too:

> Unlike tourists and visitors I (tell myself that I) did not travel to the site for pleasure and sightseeing ... Instead (so I continue) I went there to research, which is to say to collect data that would be relevant for my study and that is available only there (researchers collect data) ... But tourists, too, are great at collecting, as practices of both collecting and documenting (accessing, obtaining, photographing, transporting, etc.) are constitutive to the role of the tourist. (Noy, 2011, p. 923)

Although tourists may appear more enthusiastic for their versions of data collection such as photography, it is hard to imagine scientists being completely disinterested and unexcited about their own data collection activities. For the students on our Petunia Bay field trip, the (perhaps once in a lifetime) opportunity of going to the Arctic combined with pushing their boundaries of comfort and the further excitement and educational opportunities studying new subjects and ideas abroad. They saw themselves as part tourist, part researcher, part student, with the activities and practices they engaged with fitting all of those "categories", often simultaneously.

In this section, I have highlighted ways in which, not just the ethnographer and tourist are similar, at a general level, they are seeking to explore, document and collect various elements of Svalbard to feed into their eco-imaginaries of its spaces. The researcher in Svalbard has more in common with tourists than is generally assumed. These commonalities are extended in the following examination of the experiences of being in Svalbard, tourist, scientist, student or otherwise.

Being in the ineffable

> They've been here, they've seen it, they've felt it they've smelt it, they've heard it, yeah they know what it's about. To explain this place to someone who hasn't been here, it's not so easy, because it's something in the atmosphere I think. I guess you've noticed that as well. (Interview, research and education sector, 5 June 2014)

> It's such a different place, I think it leaves something with you, 'cos it's so different to anywhere you've ever been. (Focus group, 30 April 2013)

In this section, I draw on phenomenological, non-representational and humanistic approaches within geography[5] that have pushed us to consider the "full richness of subjective experiences of places and spaces" (Bondi, 2005, p. 436). Visiting Svalbard is something fundamental to both the tourist and researcher working there, neither could fulfil their role from a distance and that visit will necessarily affect the visitor.[6] This *being in* necessarily asks for a consideration beyond representation that extends to the bodies' senses, emotions, memories and imaginings through which Svalbard is experienced.

Following the "emotional" and "affective" turns in geography (B. Anderson & Harrison, 2006; Bondi, 2005, 2014; Davidson, Smith, & Bondi, 2007; Pile, 2010; Thrift, 2004) and the social sciences more generally, the experience of expressive failure is not new: non-representational, embodied nature of touristic experiences in many destinations has been well-theorised (Crouch & Desforges, 2003; Pons, 2003). Nevertheless, I encountered a range of views and experiences in Svalbard that are worth discussing, as they are relevant to the researcher/tourist hierarchy I wish to challenge. My research engagements within scientific and policy-making arena in Svalbard found that emotions were largely deemed inappropriate forms of knowledge and pitted against "rational" decision-making. However, many interviewees, including those presented above, made clear that Svalbard as a field site, home or tourist destination has an affective and emotionally embodied significance that undoubtedly filters into the eco-imaginary landscape of Svalbard. When talking to those who have been to Svalbard, finding it hard to relate these experiences to others was common. Having now been in this position, words like "amazing", "surreal", "awe-inspiring", "desolate", "interesting", "weird"; do not do justice to my encounters.

Delving deeper into this affective atmosphere of Svalbard revealed a general consensus that it was quite possible to catch what some call the "Svalbard Bug" (Heiene, 2009) – something akin to the "magic" tourists in Antarctica describe (Picard, 2015), yet distinct and different in Svalbard:

Mainly it's nature. It's a combination of people and nature. It's a beautiful combination actually. There is something called this polar erferi, I also call [it] Svalbard ergen – sun, the spots on your body that makes you hot.

SS: Like a fever?

Yeah, but it's connected with this nice [feeling] when you are in love. It's nature and it's being out. (Interview with long-term resident, 7 July 2014)

The conceptualisation of atmosphere as a reflection of the in-betweenness and meeting of personal and social identities with the material, external "nature" of Svalbard, goes some way to describing the embodied experience of the "Svalbard Bug". It is not something that is entirely due to oneself, others or Svalbard, but a combination of these. The landscape, temperatures, wildlife, small social community and sense of freedom are all elements that can feature in conceiving Svalbard in this way, though it is unlikely any two versions would encompass the same ideas. The Svalbard Bug does not "infect" everyone, and those who catch the "bug" may get different mutations of varying intensities.

Yet, this "being in" place as an engagement of all the senses does not occur in a vacuum; however, remote the location. Not only can representations be affectual and felt in the body as discussed earlier, but so too can those representations have a presence in place, within the body subject. We travel with our cultural baggage, "socially inscribed values and meanings [are] layered onto the landscape" (Crang, 2011, p. 9).

For me, going to Svalbard amplified the necessity of being there, in "the field", doing research/tourism/exploration. This enlarged the quandary over the environmental impact of travelling there, and created an uncomfortable expectation to feel moved and affected by the Arctic in pre-approved ways. Life in Longyearbyen could be the essence of everyday, filled with banal actions: going to the café, supermarket, library and dealing with a broken washing machine – these are not activities normally associated with a trip to the Arctic. By contrast, in Petunia Bay we spent spare time staring out at the snow-peaked mountains, fjord and glacier trying to drink in the atmosphere, to find a meaningful way to relate to where we were. We took thousands of photographs, pages of notes, perhaps in an attempt to authenticate our experience: we were *in* the Arctic (as hard as that was to believe sometimes), these were *our* notes and photographs, not someone else's:

The polar bear mum and two cubs we had been warned about were coming along the beach towards us! No one freaked out but all the 'gun people' got to the front with the dogs and flares ready. It was hard to believe this was happening. The parallels between this, and watching The Polar Bear Family and Me[7] were uncanny. As we watched them skirting the beach, just as I had done on the computer screen at home, was this any different? The pounding in my chest and hushed excitement of the group answers back, YES. We are IN the scene, we watch them, but they watch us back, sniffing the air, an unspoken conversation with potentially fatal conclusions on either side." (Field Diary, 3 July)

As Anderson, Adey, and Bevan (2010) argue, we should not forget the potential effects, affects and agency of place on research, whether in terms of the material and practical affordances of the research site or the affective nature of place on researcher or the researched. Indeed, recent work on emotions and in science studies has begun to explore more fully the entangled relationships between field science, enthusiasm and emotional engagements with place, non-human species and features such as glaciers and their retreat (Brugger et al., 2013; see also Cole, 2016; Geoghegan, 2013, for geographies of enthusiasm). Both Jamie Lorimer (2008) and Kristoff Whitney (2013) describe scientific practices of bird tracking, tagging and conservation efforts in terms of the affectual and emotional relationships to place and the species in question. They also observe how the adaptive skills and practice of performing objective, rational science can be fun, motivational in itself and a positive emotion, yet note how voicing emotions are seen as risking conservation efforts, silenced by institutional norms and written out of reports.

Many scientists working in Svalbard seem to have contracted some strain of the Svalbard bug. In some of the more open conversations I had with researchers, they described how they got "hooked" on Svalbard from their first visit. Many of these tales stem from a combination of experiencing extreme, sublime landscapes and conditions, and the opportunity and support for studying their phenomena of interest. Whilst at odds with the conventional construct of the model, objective, rational and neutral enlightenment scientist, researchers – as sensing, emotional beings – will react in some way to Svalbard's affective atmospheres. Their interaction with these atmospheres will have some bearing on the knowledge they produce (Livingstone, 2003); for example, finding new research questions to enable a continuation of visiting the same research sites.

Comparing tourist and research literatures in Svalbard, such affects and emotions are confined behind the lines of tourism and art, research cannot be tainted with such ideas; the research site is reduced to numbers and maps. Yet, as experiences of my colleagues confirm, affects and emotions of place and practice in Svalbard do not observe such disciplinary boundaries.

Conclusions

I have argued throughout, that whilst scientists and tourists both play an important economic and political role in contemporary Svalbard, they are treated differently, in terms of access to areas of Svalbard, and the treatment of the knowledge they produce from their visits. Scientists' knowledge and accounts have often been prioritised and distinguished, both practically and discursively, from those connected to tourism, whose identity and experiences are held to be separate, less culturally valuable, less authentic, less legitimate. This has bearing on the data and knowledge that then feeds into to environmental policy-making and management approaches. Here, my findings build on those of Nyseth and Viken (2016) and Bets et al. (2017), who argue for the inclusion of data gathered from tourist expeditions in environmental decision-making. My wider research critically assesses the processes of policy-making and the ways scientific and other knowledge are used when defining environmental management strategies.

Although there are clearly differences in the knowledge produced and approaches to creating that knowledge about Svalbard (Crang, 2011), I have sought to highlight the similarities and interdependencies between these two groups so often held apart. The visitor to Svalbard will engage with representations of Svalbard. In travelling to and carrying out their explorations of Svalbard, they utilise comparable resources and infrastructure. They are also likely to be attracted to Svalbard by similar motivations and values: to know more about the features of Svalbard that interest them; to experience an Arctic adventure, to test their abilities, to escape to the "wilderness". The visitor will, whether or not we get to hear about it, have an emotional, embodied encounter with Svalbard and carry this away with them, alongside the "data'" and knowledge they produce during their trip. One challenge in the Anthropocene as Jamie Lorimer (2015) suggests, is to be able to assess multiple forms of environmental knowledge, multiple truths as to how robust they are and who the politics and power relations behind their representation best serves.

Visitor numbers for leisure and research purposes in the Arctic continue to grow (Hall & Saarinen, 2010) alongside the environmental impacts of such visits, something which needs to be more fully and equally acknowledged. Yet, the experiences visitors to these changing landscapes have are affective, powerful and have potential to re-enchant us within our world (Woodyer & Geoghegan, 2013), initiate positive change, inspire different imaginaries and nature–culture relations, at the individual level or more broadly through an increased scientific understanding of the socio-natural processes at work in these regions.

As cognitive beings, we rely on categories to adapt to our environment (Lamont & Molnár, 2002) and construct our conceptual systems (Lakoff & Johnson, 1999, cited by Jones, 2009). However, further work needs to pay attention to how categories such as "scientist", "tourist" and "researcher'" are applied, by whom, to what end and how decisions based on the differences associated with them are made. There are multiple perspectives of the Arctic that should be heard: I suggest we need a more sympathetic approach to experiences and encounters with place. Letting the "life" into scientific research, through a more transparent acknowledgment of the emotions, affects and embodied relationship with the Arctic and valuing the experiences of others in landscapes of changing climates has potential to influence future imaginaries and raise new questions (Brace & Geoghegan, 2010). A "sustainable tourism" of the Anthropocene could incorporate such elements by building on the shared and varied experiences of researchers and tourists to places "at the edge" to develop approaches that encourage place advocacy in new, multidisciplinary ways. If the highly valued and prioritised accounts of trips to the Arctic remain in the realms of rational, objective and "placeless" academic papers, we risk losing the very experiences, affects and emotional engagements that can make such places meaningful.

Notes

1. Longyearbyen society remains distinct from that of mainland Norway: it is not envisaged as a "life-span community": there is no social support; births and retirement there are not encouraged.
2. The optional summer course is taught annually at different Arctic locations, see https://www.kth.se/social/course/AK1214/.
3. A zone including the main settlements and mining operations as well as two national parks in the Isfjorden area, central Spitsbergen and the area surrounding Ny-Ålesund in North-West Spitsbergen.
4. However, given data provided by tourist cruises has contributed to increased access restrictions for tourist operators in the past (Bets, Lamers, & Tatenhove, 2017), including other "communities of practice" in these kinds of decisions through, for example, species monitoring, would need to be carefully negotiated and a two-way level of trust developed.
5. For example, Yi-Fu Tuan, Merleau Ponty, David Seamon, Edward Relph, Martin Heidegger.
6. I follow recent discussions in treating affect as before emotion, "a quality of life that is beyond cognition and always interpersonal. It is, moreover, inexpressible: unable to be brought into representation…within and between bodies" (Pile, 2010, p. 8).
7. A TV documentary series which followed a polar bear and her two cubs, filmed in Svalbard. Wilkinson (2013), 'The Polar Bear Family & Me'. Available at http://www.bbc.co.uk/programmes/b01py74c#programme-broadcasts.

Acknowledgments

Thanks to the special issue editors and anonymous reviewers for their helpful feedback in improving this work. I am grateful to Gareth Hoskins, Kimberley Peters, Stephen Saville and colleagues at Aberystwyth University for comments on earlier drafts. I am also indebted to Dag Avango (KTH) and the students and staff who shared their field trip with me, and to research participants in Aberystwyth and Longyearbyen.

Disclosure statement

No potential conflict of interest was reported by the authors.

Funding

This work was supported by the Economic and Social Research Council; Department of Geography and Earth Science, Aberystwyth University.

ORCID

Samantha M Saville 🆔 http://orcid.org/0000-0001-6307-0573

References

Anderson, J. (2004). The ties that bind? Self- and place-identity in environmental direct action. *Ethics, Place and Environ- ment, 7*(1–2), 45–57.

Anderson, J., Adey, P., & Bevan, P. (2010). Positioning place: Polylogic approaches to research methodology. *Qualitative Research, 10*(5), 589–604.

Anderson, K., & Bows, A. (2012). A new paradigm for climate change. *Nature Climate Change, 2*(9), 639–640.

Anderson, B., & Harrison, P. (2006). Questioning affect and emotion. *Area, 38*(3), 333–335.

Avango, D., Hacquebord, L., Aalders, Y., De Haas, H., Gustafsson, U., & Kruse, F. (2011). Between markets and geo-politics: Natural resource exploitation on Spitsbergen from 1600 to the present day. *Polar Record, 47*(01), 29–39.

Baigent, E. (2010). "Deeds not words"? Life writing and early twentieth-century British polar exploration. In S. Naylor & J. R. Ryan (Eds.), *New spaces of exploration* (pp. 23–51). London: I.B. Tauris.

Beck, U. (1992). *Risk society towards a new modernity.* (M. Ritter, Trans.). London: Sage Publications.

Bets, L. K. J. V., Lamers, M. A. J., & van Tatenhove, J. P. M.. (2017). Collective self-governance in a marine community: Expe- dition cruise tourism at Svalbard. *Journal of Sustainable Tourism, 25*(11), 1583–1599.

Bloom, L. E. (1993). *Gender on ice: American ideologies of polar expeditions.* Minneapolis: University of Minnesota Press.

Bondi, L. (2005). Making connections and thinking through emotions: Between geography and psychotherapy. *Transac- tions of the Institute of British Geographers, 30*(4), 433–448.

Bondi, L. (2014). Understanding feelings: Engaging with unconscious communication and embodied knowledge. *Emo- tion, Space and Society, 10*, 44–54.

Brace, C., & Geoghegan, H. (2010). Human geographies of climate change: Landscape, temporality, and lay knowledges. *Progress in Human Geography, 35*(3), 284–302.

Bradley, J. (2017, February 21). *Writing on the precipice.* Retrieved 25 April 2017 from http://sydneyreviewofbooks.com/ writing-on-the-precipice-climate-change/

Bramwell, B., Higham, J., Lane, B., & Miller, G. (2016). Advocacy or neutrality? Disseminating research findings and driving change toward sustainable tourism in a fast changing world. *Journal of Sustainable Tourism, 24*(1), 1–7.

Brigham, L. W. (2007). Thinking about the Arctic's future: Scenarios for 2040. *Futurist, 41*(5), 27–34.

Brugger, J., Dunbar, K. W., Jurt, C., & Orlove, B. (2013). Climates of anxiety: Comparing experience of glacier retreat across three mountain regions. *Emotion, Space and Society, 6*, 4–13.

Buckley, R. (Ed.). (2004). *Environmental impacts of ecotourism.* Wallingford: CABI.

Butcher, J. (2003). *The moralisation of tourism: Sun, sand…and saving the world?* London: Routledge.

Cole, E. (2016). Blown out: The science and enthusiasm of egg collecting in the Oologists' record, 1921-1969. *Journal of Historical Geography, 51*, 18–28.

Collins, H., & Pinch, T. J. (1998). *The golem: What you should know about science.* (2nd ed.). Cambridge: Cambridge University Press.

Coria, J., & Calfucura, E. (2012). Ecotourism and the development of indigenous communities: The good, the bad, and the ugly. *Ecological Economics, 73*, 47–55.

Crang, M. (1997). Picturing practices: Research through the tourist gaze. *Progress in Human Geography, 21*(3), 359–373.

Crang, M. (2011). Tourist: Moving places, becoming tourist, becoming ethnographer. In T. Cresswell & P. Merriman (Eds.), *Geographies of mobilities* (pp. 205–224). Andover: Ashgate.

Crouch, D. (2005). Flirting with space: Tourism geographies as sensous/expressive practice. In C. Cartier & A. A. Lew (Eds.), *Seductions of place: Geographical perspectives on globalization and touristed landscapes* (pp. 23–35). London: Routledge.

Crouch, D., & Desforges, L. (2003). The sensuous in the tourist encounter introduction: The power of the body in tourist studies. *Tourist Studies, 3*(1), 5–22.

Davidson, D. J., Smith, M. M., & Bondi, P. L. (Eds.), (2007). *Emotional geographies* (New edition). Ashgate.

Dittmer, J., Moisio, S., Ingram, A., & Dodds, K. (2011). Have you heard the one about the disappearing ice? Recasting arctic geopolitics. *Political Geography, 30*(4), 202–214.

Driver, F. (2010). Modern explorers. In Simon Naylor & James R. Ryan (Ed.), *New spaces of exploration* (pp. 241–249). London: I.B . Tauris.

Edensor, T. (2006). Sensing tourist spaces. In C. Minca & T. Oakes (Eds.), *Travels in paradox* (pp. 23–45). Oxford: Rowman and Littlefield.

Eliassen, T. (2009). *An arctic aboriginal or an adventure tou r ist? studying the S valbardian in a perspective of mobility* (Master's thesis). University College Finnmark, Alta. Retrieved from Longyearbyen Town Library.

Enger, A. (2011). *Gjesteundersøkelse svalbard polarnattperioden 2010* [Svalbard guest survey polar night 2010]. Longyearbyen: Menon Business Economics.

Enger, A., & Jervan, M. (2010). *Rapport gjesteundersøkelse svalbard sommer 2010* [Report: Svalbard guest survey, summer 2010] (p. 49). Longyearbyen: Svalbard Reiselev.

Font, X., & Hindley, A. (2017). Understanding tourists' reactance to the threat of a loss of freedom to travel due to climate change: A new alternative approach to encouraging nuanced behavioural change. *Journal of Sustainable Tourism, 25*(1), 26–42.

Geoghegan, H. (2013). Emotional geographies of enthusiasm: Belonging to the telecommunications heritage group. *Area, 45*(1), 40–46.

Governor of Svalbard. (2010). *Svalbard: Experience Svalbard on nature's own terms* (5th ed.). Longyearbyen: Governor of Svalbard.

Grydehøj, A., Grydehøj, A., & Ackren, M. (2012). The globalization of the Arctic: Negotiating sovereignty and building communities in Svalbard, Norway. *Island Studies Journal, 7*(1), 99–118.

Gyimóthy, S., & Mykletun, R. J. (2004). Play in adventure tourism: The case of arctic trekking. *Annals of Tourism Research, 31*(4), 855–878.

Hagen, D., et al. (2012). Managing visitor sites in Svalbard: From a precautionary approach towards knowledge-based management. *Polar Research, 31*, 18432.

Hall, C. M., & Saarinen, J. (2010). *Tourism and change in polar regions: Climate, environments and experiences.* Oxon: Routledge.

Hannerz, U. (2004). *Foreign news: Exploring the world of foreign correspondents.* Chicago, IL: University of Chicago Press. Retrieved from http://www.press.uchicago.edu/ucp/books/book/chicago/F/bo3640452.html

Hassol, S. (2004). *Impacts of a warming arctic: Arctic climate impact assessment.* Cambridge: Cambridge University Press.

Head, L., & Gibson, C. (2012). Becoming differently modern: Geographic contributions to a generative climate politics. *Progress in Human Geography, 36*(6), 699–714.

Heiene, T. M. (2009). *Bitt av svalbardbasillen ? : Konstruksjon av lokalidentitet i longyearbyen* (Master's thesis). Universitetet i Oslo, Oslo. Retrieved from https://www.duo.uio.no/handle/10852/16828

Hulme, M. (2014, February 4). Science can't settle what should be done about climate change. Retrieved 24 March 2014, from https://theconversation.com/science-cant-settle-what-should-be-done-about-climate-change-22727

James, A. (2013). Seeking the analytic imagination: Reflections on the process of interpreting qualitative data. *Qualitative Research, 13*(5), 562–577. https://doi.org/10.1177/1468794112446108

Johnston, M., Viken, A., & Dawson, J. (2012). Firsts and lasts in arctic tourism: Last chance tourism and the dialectic of change. In H. Lemelin, J. Dawson, & E. J. Stewart (Eds.), *Last chance tourism* (pp. 10–24). London: Routledge.

Jones, R. (2009). Categories, borders and boundaries. *Progress in Human Geography, 33*(2), 174–189.

Jørgensen, M. (2012 January). *Suggestion for the latest management plan for East Svalbard.* Co-authored letter published in Svalbardposten. Retrieved from www.spitzbergen.de/wp…/East-Svalbard_Jorgensen-et-al_Jan2012.pdf

Kaltenborn, B. P. (2000). Arctic–alpine environments and tourism: Can sustainability be planned ? *Mountain Research and Development, 20*(1), 28–31.

Kristiansen, J. E. (2014). *This is Svalbard 2014. Statistics Norway.* Retrieved from http://www.ssb.no/en/befolkning/artikler-og-publikasjoner/_attachment/213945?_ts=14aca3dea58

Krzyszowska, A. J. (1985). Tundra degradation in the vicinity of the Polish polar station, Hornsund, Svalbard. *Polar Research, 3*(2), 247–252.

Lakoff, G., & Johnson, M. (1999). *Philosophy in the Flesh: The embodied mind and its challenge to western thought*. New York, NY: Basic Books.

Lamont, M., & Molnár, V. (2002). The study of boundaries in the social sciences. *Annual Review of Sociology, 28*(1), 167–195.

Latour, B. (1999). *Pandora's hope. Essays on the reality of science studies*. Cambridge, MA: Harvard University Press.

Latour, B. (2004). *Politics of nature: How to bring the sciences into democracy*. (C. Porter, Trans.). London: Harvard University Press.

Livingstone, D. (2003). *Putting science in its place: Geographies of scientific knowledge (Vol. Ebrary online)*. Chicago, IL: University of Chicago Press. Retrieved from http://site.ebrary.com/lib/aber/docDetail.action?docID=10381185

Lorimer, H. (2005). Cultural geography: The busyness of being 'more-than-representational'. *Progress in Human Geography, 29*(1), 83–94.

Lorimer, J. (2008). Counting corncrakes: The affective science of the UK corncrake census. *Social Studies of Science, 38*(3), 377–405.

Lorimer, J. (2015). *Wildlife in the Anthropocene: Conservation after nature*. Minneapolis: University of Minnesota Press.

Moore, A. (2015). Tourism in the Anthropocene Park? New analytic possibilities. *International Journal of Tourism Anthropology, 4*(2), 186–200.

Mostafanezhad, M., Norum, R., Shelton, E., & Thompson-Carr, A. (Eds.), (2016). *Political ecology of tourism: Community, power and the environment*. London: Routledge.

Natter, W., & Jones III J.-P. (1997). Identity, space and other uncertainties. In G. Benko & U. Strohmayer (Eds.), *Space and social theory: Interpreting modernity and postmodernity* (pp. 141–161). Oxford: Blackwell.

Norgaard, K. M. (2006). 'People want to protect themselves a little bit': Emotions, denial, and social movement nonparticipation. *Sociological Inquiry, 76*(3), 372–396.

Norum, R. (2016). Barentsburg and beyond: Coal, science, tourism, and the geopolitical imaginaries of Svalbard's 'New North'. In G. Huggan & L. Jensen (Eds.), *Postcolonial perspectives on the European High North: Unscrambling the Arctic* (pp. 31–65). London: Palgrave Macmillan.

Norum, R. (2017). Poach, scramble, then reheat: The mediated prosumption of Arctic geopolitical imaginaries. Paper presented at The Future of Polar Governance: Knowledge, Laws, Regimes and Resources, British Antarctic Survey, Cambridge.

Norwegian Ministry of Justice and the Police. (2010). *Report No. 22 to the storting: Svalbard (report to the storting No. No. 22)*. Norwegian Ministry of Justice and the Police. Retrieved from http://www.regjeringen.no/en/dep/jd/documents-and-publications/reports-to-the-storting-white-papers/reports-to-the-storting/2008-2009/Report-No-22-2008-2009-to-the-Storting.html?id=599814

Noy, C. (2011). The aesthetics of qualitative (re)search: Performing ethnography at a Heritage Museum. *Qualitative Inquiry, 17*(10), 917–929.

Nyseth, T., & Viken, A. (2016). Communities of practice in the management of an arctic environment: Monitoring knowledge as complementary to scientific knowledge and the precautionary principle ? *Polar Record, 52*(1), 66–75.

Phillips, R. (1997). *Mapping men and empire: A geography of adventure*. London: Routledge.

Picard, D. (2015). White magic: An anthropological perspective on value in Antarctic tourism. *Tourist Studies, 15*(3), 300–315.

Pile, S. (2010). Emotions and affect in recent human geography. *Transactions of the Institute of British Geographers, 35*(1), 5–20.

Pons, P. O. (2003). Being-on-holiday tourist dwelling, bodies and place. *Tourist Studies, 3*(1), 47–66.

Robbins, P., & Moore, S. A. (2013). Ecological anxiety disorder: Diagnosing the politics of the Anthropocene. *Cultural Geographies, 20*(1), 3–19.

Roberts, P., & Paglia, E. (2016). Science as national belonging: The construction of Svalbard as a Norwegian space. *Social Studies of Science, 46*(6), 894–911. https://doi.org/10.1177/0306312716639153

Roura, R. M. (2011). *The footprint of polar tourism: Tourist behaviour at cultural heritage sites in Antarctica and Svalbard*. Eelde, The Netherlands: Barkhuis.

Salazar, N. B. (2012). Tourism imaginaries: A conceptual approach. *Annals of Tourism Research, 39*(2), 863–882.

Saville, S. J. (2008). Playing with fear: Parkour and the mobility of emotion. *Social & Cultural Geography, 9*(8), 891–914.

Spedding, N. (2004). Glaciers. In S. Pile & N. Thrift (Eds.), *Patterned ground* (pp. 75–76). London: Reaktion Books.

Stanford, D. (2008). 'Exceptional visitors': Dimensions of tourist responsibility in the context of New Zealand. *Journal of Sustainable Tourism, 16*(3), 258–275.

Statistics Norway. (2012). *This is Svalbard: What the figures say. Statistics Norway*. Retrieved from http://uit.no/ansatte/organisasjon/artikkel?p_menu=28714&p_document_id=242164&p_lang=2&p_dimension_id=88183

Thrift, N. (2004). Summoning life. In P. Cloke, M. Goodwin, & P. Crang (Eds.), *Envisioning human geographies* (pp. 81–103). London: E. Arnold.

Thrift, N. (2008). *Non-representational theory space, politics, affect*. London: Routledge.

Timothy, D. J. (2010). Contested place and the legitimization of Sovereignty through tourism in Polar regions. In M. C. Hall & J. Saarinen (Eds.). *Tourism and change in polar regions* (pp. 288–300). Oxon: Routledge.

Umbreit, A. (2013). *Svalbard (Sptisbergen) with Franz Josef Land and Jan Mayen: The Bradt travel guide* (5th ed.). Chalford St Peter: Bradt Travel Guides.

Viken, A. (2006). Svalbard, Norway. In G. Baldacchino (Ed.), *Extreme tourism: Lessons from the World's Cold Water Islands* (pp. 129–142). Oxford: Elsevier.

Viken, A. (2011). Tourism, research, and governance on Svalbard: A symbiotic relationship. *Polar Record, 47*(243), 335–347.

Viken, A., & Jørgensen, F. (1998). Tourism on Svalbard. *Polar Record, 34*(189), 123–128.

West, P. (2008). Tourism as science and science as tourism: Environment, society, self, and other in Papua New Guinea. *Current Anthropology, 49*(4), 597–626.

West, M. H., & Maxted, A. P. (2000). An assessment of tundra degradation resulting from the presence of a field camp in Kongsfjorden, Svalbard. *Polar Record, 36*(198), 203–210.

Whitney, K. (2013). Tangled up in knots: An emotional ecology of field science. *Emotion, Space and Society, 6*, 100–107. https://doi.org/10.1016/j.emospa.2011.10.003

Woodyer, T., & Geoghegan, H. (2013). (Re)enchanting geography? The nature of being critical and the character of critique in human geography. *Progress in Human Geography, 37*(2), 195–214.

Wylie, J. (2007). *Landscape.* London: Routledge.

Entanglements in multispecies voluntourism: conservation and Utila's affect economy

Keri Vacanti Brondo

ABSTRACT

Through a case study of conservation voluntourism this article brings together critical political ecology, multispecies ethnography, and studies of humanitarian tourism to advance a political ecology of multispecies conservation voluntourism. The article presents multispecies conservation voluntourism as a field that produces and is produced by an "affect economy", or an economy based on the exchange or trade in the relational. Since the mid-1990s, life on Utila, Honduras, a popular discount backpacker scuba destination located along the Mesoamerican Barrier Reef, has been deeply transformed by the growth of dive tourism, the ecological destruction it has produced, and now the conservation voluntourism industry emerging in its wake. Seventy percent of Utila is comprised of mangroves and associated wetlands, home to several endangered and endemic species. Using examples of whale shark tourism, lionfish hunts, and iguana tracking, this article shows how conservation organizations operate as affect generators, enabling the privilege of engaging in multispecies encounters. Engaging in multispecies conservation voluntourism produces value in the form of cultural capital which is then exchanged for material outcomes by volunteers in the global economy; at the same time, this form of voluntourism obscures local relationships to nature and alters multispecies assemblages from past configurations.

Introduction

The articles in this special issue explore the political ecology of tourism in an era of anthropogenic change. While many political ecologists are focusing on the relationship between human livelihoods, conservation, and development in tourism contexts (e.g. Hoffman, 2014; Stonich, 2000; West, Igoe, & Brockington, 2006), new materialists in political ecology are working to comprehend the more-than-human aspects of the Anthropocene (e.g. Barad, 2007; Haraway, 2016; Kohn, 2013; Tsing, 2015). Assuming what I call "a political ecology of multispecies voluntourism", this article aims to theorize the emerging green economy of multispecies voluntourism, an industry predicated on both the exploitation and conservation of specific place-based species. Through a case study of "conservation voluntourism" on the Honduran Bay Island of Utila, this article connects two, often separate, literatures in conservation ethnography: neoliberal conservation and multispecies ethnography. These two bodies of work are brought together to confront and theorize a contemporary conundrum in the twenty-first century green economy: that the continuously expanding global tourism industry be positioned to mediate the capitalist contradictions to which it is also a central contributor (Igoe, 2017; Mostafanezhad, Norum, Shelton, & Thompson-Carr, 2017). Conservation voluntourism falls within Wearing's (2002, p. 240) early definition of volunteer tourism, which he described as travel focused on "aiding or alleviating the material poverty of some groups in society, the restoration of certain environments or research into aspects of society or environment". Conservation voluntourists pay host organizations to work on their own or others' research projects, to learn ecological monitoring methodologies, to feed and care for species as part of captive breeding programs, to capture and tag endangered species, or to manage invasive species through culling.[1] In Utila, this includes encounters with a range of threatened and invasive species and habitats, including iguanas, sea turtles, whale sharks, mangroves, and lionfish.

Conservation voluntourism is a field that is both generated by and produces what I call an "affect economy;" that is, an economy based on the exchange or trade in the relational or "becoming with" (Haraway, 2008, 2016), and the production and consumption of emotions and feelings that emerge through close encounters with another species. Through examples of whale shark, lionfish, and iguana encounters, I show how conservation organizations operate as affect generators, enabling the privilege of engaging in multispecies encounters, to be affected and to produce affect. Conservation voluntourists purchase these experiences, which rest on a sanitized and depoliticized suffering subject that circulates through spectacle, and exchange the cultural capital they accumulate through affective encounters with other species for a new social status.

Connecting neoliberal conservation and multispecies ethnography

Emerging in the mid-2000s, the critical literature theorizing the relationship between neoliberal market approaches and the proliferation of protected areas for biodiversity conservation is now quite vast, and includes many studies that explore the negative consequences of commercialized conservation (e.g. Arsel & Büscher, 2012; Berlanga & Faust, 2007; Brockington & Duffy, 2011; Brockington, Duffy, & Igoe, 2009; Brondo, 2013; Brondo & Bown, 2011; Büscher & Dressler, 2007; Büscher, Dressler, & Fletcher, 2014; Devine, 2014; Fletcher, 2010; Grandia, 2007; Holmes & Cavanagh, 2016; Igoe & Brockington, 2007; Sullivan, 2013; Sullivan, Igoe, & Büscher, 2013). Termed "neoliberal conservation" by Igoe and Brockington (2007), this now dominant approach to conservation includes the decentralization of environmental governance, and the introduction of new forms of commodification and commercialization of nature that emerge in order to fund conservation efforts. Or, as Fletcher, Dressler, and Büscher (2014) state, it is the "paradoxical idea that capitalist markets are the answer to their own ecological contradiction" (p. 29). A central revenue strategy of the mainstream conservation movement is the emerging conservation voluntourism industry.

Conservation voluntourism, like ecotourism, payments for ecological services (PES), carbon trade, species banking, and other market approaches, have all emerged out of the material transformation of nature within the context of capitalism. That is, under capitalism, nature has come to be seen as "a bountiful pool of resources that exist either in the form of material resources, or more recently, in the form of services that are meant to satisfy human needs" (Neves, 2010, p. 726). Conservation voluntourism, nature tourism, and ecotourism are service industries that create opportunities to satisfy human emotional and psychological needs.

The process through which nature becomes valued as a commodity, appearing in the market, as if by magic, detached from the social and material labor that produced the service, is described by Marx's concept of fetishization (Carrier & Macleod, 2005; Igoe, Neves, & Brockington, 2010; Marx, 1867; Neves, 2010). Under multispecies conservation voluntourism, when flagship, endangered, and vulnerable species are reduced to commodities to serve non-material human emotional needs or experiences desired by (mostly) western millennials to be competitive for employment in the global economy, the social and ecological costs of voluntourism are obscured from view (they are fetishized). For example, when tracking and monitoring whale shark movements became valued as a tourist attraction that contributes to the preservation of the species, and enabled voluntourists to encounter whale sharks up close and personal, their value to the island fishing community as locators for tuna diminishes, and the ecological costs of increased travel and development pressure on Utila are hidden from view. Katja Neves found similar fetishization in her work on cetourism.

In her work, Neves (2010) demonstrates that whale-watching, like many other forms of multispecies ecotourism or multispecies conservation voluntourism, presents nature as a service provider that is to be consumed in situ by tourists. The promotion of whale watching by international environmental NGOs and tourist companies alike obscures the contradictions of nature and capitalism. That is, Neves (2010) argues that by equating whale watching with ecologically sound cetacean conservation, tourists and even conservation NGOs themselves reduce their ability to recognize instances where whale watching actually ends up causing more harm, for instance, through underwater noise pollution, the stress of excessive numbers of boats around, and humans swimming with cetaceans.² Whale-watching, like the forms of multispecies voluntourism discussed in this article, present opportunities for humans to connect with nature, to experience and "become with" other creatures, and, presumably, through this capitalist endeavor of consuming such experiences that they will solve the problems created by capitalist extraction in the first place.

Thus far we have seen how Marx's work on commodity fetishism can be applied to conservation voluntourism to understand how the industry creates value in the trade of experiences in or with "nature" while detracting from the labor and value produced through grounded local interactions with nature. This new value is what underscores contemporary multispecies voluntourism and is central to Honduras "affect economy". I introduce the term "affect economy" to describe an emerging economic system based on the production and exchange of emotional services attached to "becoming with" another species (Haraway, 2008), and the subsequent cultural capital that emerges for the voluntourists.

What exactly might it mean to "become with" within a multispecies conservation voluntourism context? The multispecies literature (e.g. Barad, 2007; Haraway, 2008, 2016; Kirksey, 2012, 2014, 2015; Kohn, 2007, 2013; Rose et al., 2012; Schulz, 2017; Tsing, 2012, 2015) conceptualizes relationships between humans and nonhuman animals relationally, theorizing the body is if in a dynamic "dance of encounters" (Haraway, 2008, p. 4). Beings, therefore, "become with", rather than simply statically *be* in the world. This form of inquiry pushes beyond a narrow and anthropocentric understanding of the human condition, recasting the dualistic subject/object, nature/culture, mind/body, mental/material divisions into a more-than-human ontology that does not prioritize the agency of any entities over others. It extends the ongoing deconstruction of the Cartesian human/nature divide with a shift towards thinking of materialities not as passive "things" but as active, entangled, self-organizing, and vital (Schulz, 2017). The world, through this frame, consists of entanglements of many entities, organic and inorganic, unstable assemblages that are constantly in a state of emerging. While multispecies conservation voluntourism may present possibilities for emergent ecologies, in what follows, I speculate that most multispecies conservation voluntourists actually struggle to "become with", and this is because they have been conditioned by what Debord (1995/1967) terms "spectacle".

Spectacle refers to the rising commercialism that is supported by the widespread dissemination and mediation of commercial images that lack content and conceal inequities and conflicts, through which masses of people have become conditioned. Through the presentation of such spectacular media productions, individual subjectivity disappears and is replaced by a singular market consciousness (Debord, 1995/1967). In the realm of twenty-first century conservation, nature as spectacle is visually articulated and circulated through media presentations of celebrities, corporate leaders, and high-profile conservationists sharing a message that capitalism is the key to future ecological sustainability (Brockington, 2009, 2014; Igoe, 2016; Igoe et al., 2010, p. 487; Mostafanezhad, 2013). This is a worldview that goes largely unquestioned because its cause is also the solution: that is, society as spectacle, like nature as spectacle, is produced and reproduced by capitalist relations, as well as offered as a solution to the negative impacts associated with advanced capitalism and towards financial and ecological sustainability (Igoe, 2016; Igoe et al., 2010).

A large body of work in anthropology reveals how the humanitarianism industry, and associated volunteerism, rests on images of a depoliticized and dehistoricized suffering subject or category of person (e.g. hurricane victim, orphan, AIDs patient, etc.) that are meant to generate compassionate responses (e.g. Fassin, 2011; Feldman & Ticktin, 2010; Freidus, 2010; Malkki, 1996, 2010; Mostafanezhad, 2013; Ticktin, 2011). Along the same vein, the multispecies conservation voluntourism industry presents other forms of suffering subjects, that of the threatened whale shark, iguana, or sea turtle, for instance. Just as in humanitarian travel where volunteers are compelled to travel to the global South to participate in and then publicly perform their carework through social media posts and blogs, volunteers in conservation are called to address the immediate danger to biodiversity through images of beautiful, threatened species, which legitimize volunteer intervention. Both traffic in images of suffering innocent others; the photos of orphaned children in need raise similar moral imperatives to intervene as do photos of the nonhuman animal victims of human predatory behaviors.

In what follows, I explore how the value that is produced through the emerging multispecies voluntourism industry alters the material and affective economies that existed on the island prior to the growth of dive, nature, and conservation tourism. The multispecies becoming in the conservation voluntourism industry consist of new material and affective assemblages as conservation voluntourists purchase these experiences and exchange this accumulated cultural capital for a new social status.

Utila's social history and contemporary relations

Utila, Honduras, is a small but materially and socially complex place. It is the smallest of the major islands in the Bay Islands archipelago, which consist of three major islands (Roatan, Guanaja, and Utila), five small islands, and 53 cays. Utila was first inhabited by Indigenous Paya, who relied on natural resources in hunting, fishing, and small scale cultivation. With Spanish contact in the 1500s, the Indigenous population was killed or captured and enslaved. Population records are scarce, but Davidson's (1974) historical review found three sources that show a dramatic decline in population for all of the Bay Islands. In 1544, there were 150 houses found in Utila. In 1582, a census reported 40 married Indians on Utila. By 1639, the census reported 22 *tributarios* (tribute-paying) Indians. While the unit of analysis shifted between the accounts that Davidson could locate, what is clear is a consistent reduction of the Indian population overtime, much like had been found across Middle America (Davidson, 1974, pp. 34–35). As the Indigenous population slowly disappeared, England rose as a persistent disrupter of Spanish control of the Caribbean, with several attempts to settle permanently in what had been the Spanish Bay Islands occurring between 1638 and 1782. On top of organized English colonization attempts, buccaneering activity prevented permanent settlement by the Spanish throughout the seventeenth and eighteenth centuries. In fact, beyond the Garifuna, who were forcibly removed by the British from the island of St. Vincent and exiled to Roatan in 1797, the islands were largely unsettled.

Post-abolition brought new arrivals to the Bay Islands. In the 1830s, whites on the Cayman Islands were outnumbered by the slave population 5–1, and fearing that abolition was going to result in the loss of political and economic power, many white Cayman Islanders left for Belize and the Bay Islands. They were already familiar with the area since turtles from the Cayman

Islands often harvested in this region (Davidson, 1974, p.74). It was at this time that Utila's first Cayman families settled on the keys and south side of the island, joined by two other families from the United States. Slavery officially ended in the Cayman Islands in 1834, and after the 4 year "apprenticeship" that enabled former slave-owners to continue to hold former slaves ended, freed slaves immigrated in a second wave to the Bay Islands. These two waves of immigration – white and black – were inscribed geographically, with residential segregation continuing in Utilian neighborhoods until recent decades.

In 1852, despite the infringement of the Monroe Doctrine and Clayton-Bulwer Treaty[3], the British declared the islands a colony and the Bay Islands Colony lasted until 1858. The establishment of the British colony was brought to the attention of the United States, and eventually under a great deal of pressure, England agreed to surrender the islands to Honduras. At the time of its surrender to Honduras, the Utila's population was approximately 100 (Rose, 1904, p. 11). By the turn of the twentieth century, the population had grown to almost 800 (ibid.).

Utila's economy at the turn of the twentieth century was centered on the cultivation of coconuts, bananas, plantains, mangoes, and other agricultural products for export to the United States. The agricultural economy declined in the 1940s, and was replaced by men departing to work as merchant marines for large scale shipping services. Fishing for subsistence and sale has always been, and continues to be, an important source of income, especially for households in the Utila keys.

In the 1980s, with a population of around 1500, the island entered the global tourism industry, leading to profound changes to its population and environment. The Bay Islands were declared a priority tourism zone in 1982 through *Acuerdo Numero 87*. This act was followed by a series of other laws to promote tourism as a national economic development strategy. A number of specified "Tourism Zones" were established and tax and import incentives were provided to attract foreign investment. In 1992, tourism investors were given the same benefits as private export processing zones (e.g. tax exemptions, tax free imports for material supporting the industry, and 100% foreign ownership of property). The Bay Islands became one of Honduras' first Tourism Zones (Stonich, 2000, p. 10).

Once the government began to promote tourism development through new legislation, the industry grew exponentially. Between 1985 and 1996, the number of hotel rooms on Utila increased from 34 to 199. Several dive shops with attached lodging opened in the 1990s, and by 2001, the island had 11 dive shops and nearly 30 hotels (Currin, 2002). By the turn of the twenty-first century, Utila had established itself as a backpacker's paradise with a local community that was generally happy to host them. The construction of a new airport and associated highway further expanded visitations to the island. Whereas when I took my first trip there in 2000, transportation was almost exclusively by foot, bicycle, and only a handful of four wheelers, and airplanes still landed on a small dirt airstrip next to the beach, today tuk-tuks, motorbikes, golf carts, and ATVs travel the roads all day and night. With this increase in motorized traffic and infrastructure development along the bay, nagging motor beeps have replaced the scampering of blue crabs that one used to encounter on their walks along the main street. Despite its growth, Utila continues its tradition as a "budget" backpacker tourist destination. A central draw is that it is one of the least expensive locations in the world to obtain SCUBA certification.

The ethnic composition of the island is changing at a quickening pace, and with these changes come new material and affective relationships to nonhuman species. Descendants of the settler families still hold the majority of land, but are rapidly selling, with many leaving for the United States. Replacing them are foreign-born individuals, many of who were heavily involved in launching the dive industry in the 1980s or are recent arrivals living long-term on the island (who leave the country for 3 d every 3 months in order to keep continual residence on the island under a tourist visa). There is also an increasingly large number of Spanish-speaking Hondurans from the mainland who now call Utila home. Migration from the mainland grew significantly in the years since the 2007 coup d'état, which has brought growing levels of violence, political instability, and heightened poverty to the mainland (Phillips, 2015). Many come to Utila in hopes of finding employment in the tourism service industry – in construction, restaurants, and cleaning services – and for a safer environment to raise a family. In 2015, two ferry services got into a price war, continuously lowering their prices until one of them eventually went out of business. The "ferry war", as it has come to be known, led to a major spike in the number of Spanish-speaking Hondurans coming over from the mainland with their families, seeking refuge from the violence and poverty associated with mainland. In the past, the 715 lempira (US$35) was prohibitive. By 2016, the cost had dropped to 100 lempiras (US $4.25), and in spring 2016, one company was driven out of business. The current population is approximately 9000, although there is no reliable recent census (Mayor's Office, personal communication, 2018). The increase in population puts pressure on the island's biota.

The island's colonial legacy has much to do with contemporary race relations on the island, with some divisiveness between those who trace their roots back to the British Crown and Cayman Islands, the foreign-born community who launched the scuba dive industry in the 1980s, and those who have migrated over from mainland Honduras to work within tourism. Still to this day, descendants of the earliest settled families continue to enjoy the fact that their cultural makeup is distinctive from mainland Honduras, and many still identify with the British Crown and the United States. Now, with white and black islanders intermarrying more and more, and the spike in mainlander ladino settlement, the black-white racial tensions that previously marked island relationships are shifting towards "Utilian" (referring to both black and white descendants of early settlers) and "Spaniards" (a derogatory term referring to more recent settlers from the mainland). The island's wealth, in terms of industry and land, is held primarily in the hands of descendants of early settlers and foreign migrants. Spanish-speaking mainlanders fall at the bottom of the social and economic hierarchy. And those who are most often blamed as having the greatest negative impact on island resources or to be engaged in poaching endangered and vulnerable species are typically recent immigrants from the mainland.

Utila's biodiversity and conservation organizations

Utila has significant biodiversity for such a small place. Located approximately 30 km from the Honduran mainland and on the edge of the Mesoamerican Barrier Reef, the second largest reef system in the world, the archipelago consists of the main island of Utila and 11 small offshore cays. The reef system is home to more than 500 fish species, 350 mollusks, and 65 species of stony coral. The 17.37 square mile island is comprised largely of mangrove forests and associated wetlands, with estimates of up to 70% mangrove coverage (Canty, 2007), combined with lowland tropical forest. Utila has four species of mangrove trees: white mangrove (*Laguncularia race-mosa*), red mangrove (*Rhizophora mangle*), black mangrove (*Avicennia germinaus*), and button-wood mangrove (*Conocarpus erectus*). The red and black mangroves are most common and are habitat for an endemic species of spiny-tailed iguana (*Ctenosaura bakeri*), referred to locally as "swamper" or "wishiwilly" (Canty, 2007). *C. bakeri* is listed on the International Union for the Conservation of Nature's (IUCN) Red List as critically endangered for threats to population due to "habitat loss and fragmentation on associated with development for tourism and decreasing quality of habitat from introduced invasive vegetation", as well as local consumption and sale of eggs and meat (IUCN, 2017).[4] Utila's mangroves are also home to other iguana species, includ-ing the nonendangered *Iguana iguana* and the *Ctensosuara similis,* with the latter easily confused with the swamper by nonspecialists. In addition to the swamper, conservationists, biologists, and ecologists are also keenly interested in the endangered Green (*Chelonia mydas*) and Loggerhead (*Caretta caretta*) and critically endangered Hawksbill (*Eretmochelys imbricate*) sea turtles, which all nest on Utila's beaches, as well as in the whale shark population (*Rhincodon typus,* locally referred to as "Old Tom"). The IUCN recently classified the whale shark as a vulnerable species. Utila hosts one of the world's only year-round whale shark populations.

With the growth of tourism, Utila's impressive mix of ecosystems includes mangroves, beaches, and coral reef ecosystems have come under intense development pressure. Environmental concerns on the island are numerous, including mangrove and coral reef destruction, overfishing, illegal capture, consumption and sales of endangered and endemic species (including sea turtle and iguana), and inadequate solid waste disposal and waste water treatment systems (Canty, 2007; M. Fernández, personal communication, April 25, 2013). The island also has limited freshwater and is now suffering from declining water quality and the contamination of seawater associated with the pressures of tourism development (Canty, 2007; M. Fernández, personal communication, April 25, 2013). Mangrove wetlands were deforested in the 1990s when the local government approved a program to sell plots in mangrove wetlands to the island's poor and mainland migrants. It began as a shanty development with substandard infrastructure and services, and associated health concerns (e.g. water borne illnesses and skin condition); the development continues to grow as more and more mestizo mainlanders move to the island. Additional mangrove forest sections were removed along the island's shores to accommodate tourism infrastructure and new housing developments for foreign-owned vacation homes.

The arrival of lionfish (*Pterois volitans*) to the Caribbean has significantly altered coral cover and reduced native fish populations. Native to Indo-Pacific, lionfish proliferate at an extremely rapid rate, and consume native fish species. With reproductive maturity under one year and high fecundity (female lionfish can produce floating, unpalatable egg masses of 20–30 thousand eggs every four days all year round), they have spread rapidly and are responsible for the reduction of important native fish and crustacean species (Andradi-Brown et al., 2007).

There are five locally active conservation organizations that work to mitigate ecosystem degradation and promote opportunities to engage in the science of conservation. Two are NGOs: Bay Island Conservation Association (BICA) and *Fundación Islas de la Bahia* (FIB), which runs the operations for the Iguana Research and Breeding Station. Three private conservation businesses also host conservation volunteers on the island: (1) Whale Shark and Oceanic Research Centre (WSORC); (2) Operation Wallacea (Opwall), a UK-based company with sites across the globe, and (3) Kanahau Conservation Research Facility, a smaller operation modelled similarly to Opwall (and started by past Opwall volunteer scientists), but based solely on the island of Utila. These organizations focus on recruiting volunteers from university settings who are interested in gaining field-based experience in ecological or environmental subjects. Opwall, for instance, only promotes their programs within universities, and more than 95% of their volunteers are undergraduates or masters students working on research theses. Coral Reef Alliance, a global NGO focused on community-based conservation of coral reef regions, is also represented, with a full-time project manager based on the island. All of the various organizations work cooperatively with one another and with the Municipality's environmental office, *Unidad Municipal Ambiental* (UMA).

The first and longest-standing organization is BICA, which was founded in 1990 on Roatan by influential island residents concerned about conservation on the Bay Islands. Chapters were then established on the other two Bay Islands, Guanaja and Utila. The origin of BICA-Roatan aligned with Honduras' push to aggressively grow its tourism economy in the early 1990s through the creation of Tourism Zones. Conservation and tourism development became intertwined from then on for the Bay Islands, and conservation organizations played central roles in promoting tourism. In her study of conservation and tourism development on the islands in the 1990s, Stonich (2000, p. 157) found that "the vast majority of BICA programs are aimed at maintaining an environment conducive to the promotion of tourism". Conservation voluntourism is an extension of this approach.

Supporting the entanglement of conservation and development, Honduras established a number of protected areas based on the IUCN classification system, as well as passed legislation to protect key species from overexploitation. The Bay Islands waters (extending 12 nautical miles around the coasts of the islands Guanaja, Roatan, and Utila) were designated as a National Park in 1997 under the Executive Agreement 005-97 of the Honduran government. In 2009, Executive Agreement 142-2009 defined zoned categories for management, including two marine zones where fishing is limited to line fishing only, and one wildlife refuge. It is illegal to hunt Utila's *C. bakeri* and to harvest many marine species, including sea turtles and their eggs, conch, juvenile lobster; spearfishing is banned from all areas of the National Park. It is also illegal to clear mangroves; however, limited funding for monitoring and enforcement combined with a high level of corruption has meant that illegal logging of mangroves is common (Canty, 2007; Harborne, Afzal, & Andrews, 2001). Not all clearance is illegal either; certain development projects have been granted permits for mangrove clearance, including the controversial Oyster Bed Lagoon project which began construction of luxury villas in 2015 (http://oysterbedlagoon.com).

Limited government resources for conservation have also meant that local organizations are understaffed and frequently unable to conduct ecological studies on their own. Financially strapped, they often spend the bulk of their human resources applying to grants and hosting scientists and conservation voluntourists from other parts of the globe. The latter are a central revenue source keeping local organizations afloat. Utila's conservation organizations are all working to enhance their volunteer programs in ways that will lead to financial sustainability. Volunteers are incorporated into all ecological research. Both FIB and Kanahau have active terrestrial programs; the largest for each focuses on the *C. bakeri*. BICA and WSORC are more marine-based in their work, with one of BICA's largest efforts surrounding turtle monitoring, and WSORC focusing on whale shark tracking. BICA and WSORC work together with Opwall on lionfish ecology. All organizations have a variety of other ongoing marine and terrestrial research programs as well. The conservation scientists that come to advance ecological research become critical brokers to enabling affective encounters for general conservation voluntourists with other species.

Methods

The data I draw on comes from several short-term research trips (4–10 weeks each, totaling approximately one year) across at three major time periods (2001–2002, 2011, 2016–2018) and intermittent weekend visits to the island while I was in Honduras working on an earlier project on Afro Indigenous territorial rights (Brondo 2007, 2010, 2011; Brondo and Woods 2007; Brondo and Bown 2011). In addition to ethnographic fieldwork, I worked collaboratively with another anthropologist and staff of conservation organizations on several web-based surveys of conservation voluntourists (2011, 2015, 2016, 2017), exploring topics including the relationship between gender, motivations for volunteering, and environmental values (totaling 168 surveys of conservation volunteers), as well as more evaluative surveys to help the organizations improve their programming (Brondo et al., 2016). Since 2011, I have also been tracking and analyzing the organizational marketing materials for all of the island's conservation organizations, and tourism-related websites and blogs.

Ethnographically, over the years I spent my time "deep hanging out" (Geertz, 1998), both virtually, in the many online mediums that bring Utilians and those who work, study or play on Utila together, and in person, in islander shops, homes, municipal buildings, volunteer barracks, and boats; in and under the sea, deep in the red and black mangroves, on top of hillsides, and along the shores. The 2011 summer fieldwork was focused on volunteers and the work of conservation organizations and the municipality's environmental agency. I conducted in-depth semi-structured interviews with local conservationists (NGO staff and officials from the Municipality) and dozens of volunteers at the Iguana Research and Breeding Station run by FIB. I also worked as a volunteer with the Iguana Station, participating in a range of activities including gathering and preparing food to feed captive iguanas (i.e. fruits, hibiscus flowers, termites nests, and crabs), cleaning cages, leading visitor tours, searching mangrove forests to capture and release swampers (which were weighed, measured, and tagged for further tracking). In 2016 and 2017, I increased the amount of time I spent hanging around in mangrove swamps, in the sea, and in the hardwood forest, working to enhance my ability to comprehend and theorize multispecies encounters, expanding participant observation from the Iguana Station to all four conservation organizations. Between 2016 and 2017, I led a research team focused on exploring the relationship between cultural values and conservation objectives. Our team participated in the range of biological and ecological studies that the organizations were working on, as well conducted a total of 53 individual interviews with conservation staff, local government, and key informants from the community, and 80 door-to-door sustainable livelihood surveys. Several interviews from that joint fieldwork inform this manuscript. All ethnography and interpretation is my own.

Utila's multispecies encounters

Embracing the "affective turn" in the social sciences, decolonial feminist scholars are offering new ways of thinking through the urgencies of the Anthropocene (e.g. Barad, 2007; Haraway, 2016; Tsing, 2015). In urging us to "stay with trouble", Donna Haraway (2016) upends the Anthropocene by introducing a conceptual shift from human exceptionalism and utilitarian individualism to making and becoming-with. Conceptually, this means approaching the now as multispecies kin relations. Haraway (2016) argues that "living with and dying-with each other … can be a fierce reply to the dictates of both Anthropos and Capital" (p. 2). Conservation voluntourism is a site where we can locate multispecies kin relations, where people come together to exchange affect, to become-with.

To be sure, humans have always been entangled in multispecies assemblages. We have always been becoming-with, but how we assemble and how we become-with is shifting through the emergence of new affective economies. The economies of affect that are developing through multispecies conservation voluntourism are distinct from the economies of affect that previously constituted multispecies encounters on and around Utila. In earlier years, the affective component of socioecological assemblages was largely related to a close familiarity with various species that islanders relied on for material benefit (for sustenance or sale) as well as for their cultural importance as part of local dishes served during family gatherings and holidays. Families hunted together to capture iguana or turtle for local dishes, and fishers drew on their understanding of Old Tom's feeding patterns to help them locate catch. Island children kept small turtles in their families crawl (a fenced in spot in the ocean close to one's home where fishers transferred their catch from their dories). Remembering this practice, the following descendant of one of the first settler families from the Cayman Islands shared:

> And the kids would go catch little minnows and frys and different sardines and all kinda little clams on the beach, and rocks and stuff, to feed their baby turtles. So we got the islanders involved in conservation in a very simple way by just working with the kids.

Once the baby turtle matured, it would be released into the sea. Islanders enjoy what the above man describes as the "sweet delicious meat" of green turtles, and the above carework can be understood as a practice that ensured the species survived into adulthood, enabling the eventual harvest of eggs or capture of occasional turtle for consumption. Today, harvesting turtle and its eggs is illegal, and thus, nobody talks openly about eating them; they do however, reminisce about the taste from their childhoods.

While the above multispecies assemblages continue to exist in part, with the growth of conservation and tourism development, there are now new actors transforming how species assemble (including, for instance, migrants, tourists, volunteers, and lionfish). The shifting entanglements under the affect economy associated with conservation and nature tourism fetishize nature, reducing vulnerable species to commodities to serve the emotional needs of tourists and volunteers. Just as is the case of humanitarian volunteerism where the images produced by humanitarians focus on a suffering subject to generate compassion and justify intervention, commodifying a particular category of people (AIDS orphans, hurricane victims, etc.), conservation voluntourism rests on the sanitization and depoliticization of other forms of sentient beings (Coghlan, 2007; Freidus, 2016, 2010; Vodopivec & Jaffe, 2011). Utilians, migrants, ecologists, and biologists, voluntourists and nature tourists now assemble through exchanges of affect mediated by conservation organizations and the dive tourism industry, who create opportunities for affective encounters unique from those that marked the past. There is a desire among humans to have a closer, less alienated relationship to other species; the affect economy that marks conservation voluntourism enables the purchase of such experiences for some, which produces cultural capital that can later be exchanged for additional material benefits (e.g. resume building leading to future educational or employment opportunities). Small nonprofit environmental organizations and local people who work with them benefit through the creation of these experiences, but it also shifts the primary material ways in which some local people had assembled with species in the past (i.e. as food sources). I advance this argument through three examples: whale shark encounters, lionfish hunts, and iguana monitoring.

Save a whale, drink a beer

Posters at the WSORC fundraiser for 2013 International Whale Shark Day encouraged tourists to "feed me your lempiras" and to "save a whale, drink a beer!" (http://wsorc.org/international-whale-shark-day-celebrations/). Contrary to what this slogan suggests, whale sharks are fish (sharks), not whales. WSORC holds the only permit in Honduras to study whale sharks, and its staffs were responsible for developing the legal guidelines for whale shark encounters, approved by the Honduran government in 2008. In addition to hosting conservation voluntourists who help collect photographs of spot patterning for Wild Book, the international NGO that collates global whale sightings such that users can track movements (whale shark spots are like unique, similar to a human fingerprint), WSORC also runs "Ocean Safaris" to take tourists out for responsible whale shark encounters. WSORC operates these excursions through the Bay Islands College of Diving (BICD) and Utila Lodge; all three (WSORC, BICD, and Utila Lodge) are owned by the same business owner, and were founded by the late Jim Engel who moved to the island from the United States. BICD and Utila Lodge constitute one of the largest dive resorts on the island, and they offer several different all-inclusive diving and whale shark sighting packages. Independent boat operators also charter trips for whale shark encounters. These emerging multispecies encounters have both dispossessed Utila's fishers and created new opportunities for them, revealed in the following two quotes from two different fishermen in the documentary film "Big Fish Utila".[5]

> We didn't know too much about the Whale Shark. We knew about the old Tom. We called him here the Old Tom. So the Whale Shark came up since the tourists been coming around.

> The reason why the fishermen are so interested in protecting Old Tom is because the Old Tom is the life of the fisherman. [If] there's no whale shark, there's no bonito. When the whale shark go, he take all the bonito with him.

Long before marine biologists and divers brought the "whale shark" to the island, Utilian fishermen had established a positive affective relationship with Old Tom. Unlike in other parts of the world, Utilian fishermen did not attempt to hunt the whale shark for its fin or meat. Old Tom was a sign of good fishing. When Old Tom is present, that means bonito tuna are present feeding on plankton. Large boils in the water are a good indication that whale sharks are near. In recent times, Old Tom has brought about new possibilities for Utila's boat captains, who continue to be, are formerly were, fisherman or mariners – human/whale encounters. Now, Old Tom is not only a sign of bonito; Old Tom also brings tourists.

Tourist desires and voluntourist experiences are influenced both by what they know and what they wish to demonstrate to others that they know, and what they know is influenced by the spectacle of nature. Take for example the following blog entry by an intern/conservation voluntourist at WSORC; here she reflects on her experience with mainstream tourists, as they were both on excursions to encounter whale sharks, albeit with different hosts (Van Landeghem, 2017).

> When the boat sailed out the excitement only grew. Looking at the horizon for birds or a sign for a tuna, we waited for those magical words "There's a whale shark!" Then it was time to get our gear on, waiting in tension for the 'GO GO GO' of our captain. Getting into the water as calm as possible, seeing that majestic creature eating, swimming underneath you and eventually watching him disappear into the big blue ocean. In silence we go back to the boat, then our excitement comes out…. It's still the most incredible thing ever!
>
> Limited amount of people with the animal, no in-water noise, no touching, keeping distance, no stress for the animal & maximal encounter time …
>
> This is how it should be.
>
> This is how it could be.
>
> But unfortunately this is not reality.
>
> Up to 4 other boats came up close. Doesn't matter how many people are already in the water, every one on those boats jumps in, splashing & screaming around the animal, duck-diving to get that one 100-likes-Facebook-selfie, boats blowing their horns or even cruising in between the snorkelers. And this story repeated itself several times.
>
> Whale shark gone.
>
> Boil gone.
>
> Magic gone.
>
> I'm done.

The title of this entry is "when you're aware, you care". The entry does not end with her being "done". Rather, the author's experiences invoked in her a sense of responsibility to blog and communicate publicly the rationale behind whale shark encounter regulations. Her post documents the lived reality for whale sharks as they indeed endure stress by the dozens of loud tourists in close proximity, who are trying to get photographs (and sometimes even photos to appear as if they were riding a whale). Tour companies on the island feature videos on their Facebook accounts promoting whale shark trips with "#TheGentleGiant". For instance, Bush's Bay Islands Adventure Tours Facebook page features videos where you can hear tourists asking once the boat has sidled up to the whale sharks if they can go in, someone then saying, "Go ahead! Jump on in. Slide on in. He ain't going anywhere". The camera then lands on a man snorkeling directly on top of the whale shark, with several other tourists surrounding nearby creatures.

The blog author likely encountered a tour of this nature. She likens the experience to being a "superstar, eating at a restaurant, [when] suddenly 15 screaming paparazzi come to your table". Of course, you'd take off, and quite possibly never come back to that restaurant, she notes. Hoping to spread awareness, this conservation voluntourist explains that by learning and abiding by the regulatory code, informed tourists can engage in multispecies encounters marked by "care". And, future conservation voluntourists at WSORC can translate their affective labor into scientific data collection, taking photographs to add to the Wild Book database.

While the emergence of whale shark tourism and the conservation science connected to it contributes to global conservation initiatives, it has also shifted relationships to Old Tom for Utila's boat captains. One outcome is that tourists (both regular and conservation volunteers) now make decisions about who to hire for a whale shark encounter based on endorsements of conservation voluntourists who engaged with the private organization that produced the regulatory code (WSORC). It further serves to note that the guidelines for responsible encounters that have been produced by foreign conservation voluntourists and have taken shape in the format of an industry that encourages in water multispecies encounters, a sharp turn from the fisherman's appreciation of Old Tom's presence from inside a boat. Replacing earlier multispecies assemblages of Old Tom, bonito and fisherman, with whale shark tourists, conservation scientists and boat captains, has material and affective effects. The Utilian fisherman had (and continue to have) a relationship to, and knowledge of, Old Tom by living so closely among them, fishing daily in an economy that is affective (they know and care for Old Tom's well-being), and material (Old Tom helped fishers procure food). The emerging affect economy under conservation voluntourism produces different multispecies assemblages and material outcomes. Conservation tourists are purchasing cultural capital through an experience of caring for the whale shark, which they will later exchange for material outcomes. One might also argue that fishers were and are more in sync with the whale shark – they know the rhythm of the whale shark; they respect their presence as connected to a wider assemblage of beings, including bonito, plankton, corals, and other marine life, something lost to those focused on snapping selfies.

Killing lionfish to save the seas

> Honestly, nothing beats killing a lionfish. It's such a thrill! I think you guys will get a such a kick out of it.[6]
> —Volunteer trainer, introduction to culling program

Three organizations partner on lionfish culling and lionfish ecology studies on Utila: WSORC, BICA, and Opwall. Opwall brings paying conservation voluntourists to assist with dissections for stomach content analyses, and population studies, but they do not allow their volunteers to spear lionfish.[7] Conservation volunteers and dive tourists who undergo BICA training to obtain a permit can engage in culling with a Hawaiian sling. Lionfish culling has become part of Utila's protected area management suite of strategies, and only those certified through the training can participate in hunts. The organizations work collaboratively with Utila's dive centers to train interested dive tourists. In practice this means Utila's fishermen could be fined for spearfishing lionfish (because all forms of spearfishing are illegal in the Bay Islands) while volunteers who can pay for the training and permit can participate in culling and lionfish derbies. Thus, the organizations collaborate to hunt and kill lionfish, which are then sold in local restaurants to tourists. This program is one of just a handful that brings some unrestricted funding to the financially struggling conservation organizations.

Marine ecologists studying mesophotic reefs suggest that this management approach may not be effective, largely due our inability to appreciate the life cycle of the lionfish. Mesophotic coral ecosystems are found at depths ranging from about 30 to 150 m and are marked by communities of corals, sponges, and algae. As this marine ecologist shared,

> I came out here to work on mesophotic reefs, and everywhere I looked I saw lots and lots of lionfish in really deep reefs … lionfish exist from really shallow reefs down to 150 to 250 meters, so in Utila if you snorkel in these shallow reefs you very rarely see lionfish because we've built such good culling programs. All the dive centers have people going out culling. As soon as you go below 30 meters they're everywhere; there's huge numbers of them … also their lifecycle is when they spawn they produce these eggs that then float to the surface; young lionfish then move to shallow habitats, such as mangroves, seagrass, and shallow back reefs, and then as they mature they migrate out to the reef crest, and then down the slope as they get more mature … Are we leaving this big lionfish refuge down there that are constantly spawning, and coming back up to the shallows? So, is the management plan really working? Or are we leaving this big population? Because in fish, the more mature they are, the more fecund they are … So not only are these deep lionfish being missed by the culling, but they're also, because of this lifecycle, the most mature ones that are producing the most eggs.

The culling programs may not be producing the conservation goal they hope to in terms of long-term lionfish management, but they are producing value for both conservation volunteers and Utila's hosts. The lionfish culling program creates data for conservation volunteers who are studying lionfish feeding preferences through dissection and gut content analysis. They also contribute to Utila's lionfish derbies, where island restaurants engage in a cook-off competition and party attended by tourists at least twice a year, sometimes more. The lionfish derbies, the culling programs, and the ecological science attached to island conservation organizations all produce value in Utila's affect economy, as a tourist attraction when voluntourists care for the survival of some species through the death of another. The program has also been instrumental in bringing BICA, Opwall, and WSORC closer together to collaborate on joint programming and data sharing. As well, the lionfish science and culling activities rationalize and legitimize the presence of tourist-visitors, who unlike fishermen families from Utila's cays or migrants from the mainland are cast in the language of belonging. They are there to contribute to the economy and to manage an invasive species, whereas the former are more frequently subject to claims of overexploitation of species and lack of belonging (through the language of invasion, much like the lionfish). Moore (2012) found the same in her study of lionfish and fisheries management in The Bahamas:

> This figure of the invasive arrival, into which the lionfish fits all too nicely, is contrasted with that of the overly welcome visitor, the tourist and offshore finance investor who bring resources with them to be captured...

As in The Bahamas, the lionfish in Utila stands to become something of "emergent keystone species", which will then legitimate more conservation science research, invasive species management work and voluntourism opportunities (Moore, 2012, p. 679). Spectacle plays a role in advancing this narrative and this industry. It also shapes the contours of multispecies encounters that volunteers will have with lionfish.

If I ever had a doubt, the attraction was made crystal clear during an all-day volunteer orientation I attended in 2017 when the room of 20-something young women erupted in excited giggles and a chorus of cheerful commentary whenever the topic of lionfish training came up. One young woman from the United States frequently interrupted the volunteer training sessions to talk about her passion for the marine life, exclaimed, "I want to get a long string [to hang my catch] and be able to hold them all out [she poses how she would for a photo] and say 'yeah! I got these!'" This woman already knew how she was going to pose in photos to share her multispecies encounters through social media, celebrating her love for the sea through the care of killing.

As the lionfish culling programs create new opportunities in for the island's economy, they also bring other potentially harmful changes with them. For one, lionfish hunting by dive tourists and conservation volunteers changes the multispecies dive encounter from passive to active, creating opportunities for instrument mishandling resulting in speared corals (and this in indeed why Opwall does not allow their voluntourists to spear hunt). Further, while several parties are indeed capitalizing on the affective labor of volunteer tourists who experience life through death, caring for some species while killing others, these practices dispossess some people (particularly unregistered fishers) of multispecies relationships in favor of others.

Saving the swamper

> I accompany five researchers, 15–25 years my junior, to the undeveloped mangrove-covered shoreline where the team often encounters juvenile *Ctenosaura bakeri*. I spend the first bit of my time trying – unsuccessfully – to locate and capture juveniles that are hiding among the groundcover. About twenty minutes into our excursion, I see that Alice (a pseudonym), the organization's lead herpetologist, has found a nest on the shore and is busily excavating the babies, recording their measurements. She grants me the role of research assistant as she soothes the sixth of the seven juveniles the nest produces. "It's okay, I have you, this won't hurt..." she coos, inserting a long probe into the pockets of the swamper's genitals to determine his sex. "Another male!" I mark the sex in Alice's field notebook, along with the length and

weight measurements we have taken. Next we will cut off its left toenail for genetic testing, and eventually Alice will know if the probing gave us an accurate read. Another volunteer takes photos of our activity along the way. Once this spiny-tailed iguana has been measured, weighed, and sexed, we release him onto a mangrove stump nearby. He doesn't move. "He's in shock. They do this a lot – just stay in the place we release them; I think he is stunned … can you imagine if this was how you experienced your first moments of life outside of the nest? Welcome to the world, little guy. Poor thing … ."

The above field notes excerpt was the second of the two excursions that group made this day, ending with more than a dozen newly documented, and sometimes tagged,[8] swampers to add to the organization's dataset, and to the data pool that will be drawn upon by Alice and two others for their master's and doctoral theses. We also produced dozens of photos, of swampers, snakes, and interesting insects that we encountered along the way, passed around for volunteers to handle.

The above scene was a familiar sight. Sights like this are in abundance on conservation websites, blogs, and social media. I had seen such images before, and through my research I have become part of it, seeing myself on the social media of my conservation organization partners. These images are frequently accompanied with evocative phrases about imminent threats to important and rare species, followed by heartfelt expressions of caring labor, love emojis, and lots of exclamation points. The consumption of these experiences fund the work of the conservation organizations, which in turn produces the data that the organizations need to justify further grants, to produce more science attached to the species, which then funnels back to the IUCN, informing and legitimating protected status for Utila's interesting species. Spectacle is ever present. Through their own Instagram photos, Twitter and Facebook feeds, and blog posts, individuals are joining the contemporary moment where "green consumption is the new activism", demonstrating their engagement with fixing the problem through participating in the (market) solution (Waitt et al., 2014, p. 169). This economy is fed by the selling of experiences in "becoming with" (Haraway, 2008, p. 3), in shaping humans through their relationships of care with other, rarely encountered species. But as spectacle, other localized relationships are obscured, even to those who are managing the operations and producing the data to support continued conservation science. On more than one occasion, in the years I've been following these organizations, I learned from Utilian-born trackers during treks through the mangroves just how "tasty" the swamper is. Here, you have local guides simultaneously involved in research to preserve the species, leading expeditions for research voluntourists, and then – theoretically – going out again to track, kill, and consume one of the very creatures the organization for which their place of employment is working to protect for human exploitation.[9] Why do it then? Even most of the scientists and conservationists I worked with on the island, like the one quoted below, speculate that adding breeding programs to current conservation efforts may be a better solution to managing the swamper than a command-and-control management approach.

> … This is their past. It brings up memories of their childhood. How can I tell someone not to partake in their cultural heritage? I think we should be breeding the swamper so islanders can eat it, not merely for scientific observation and preservation.[10]
>
> —Senior Conservation Staff of Honduran Descent, Utila NGO

Why do it then? One intrepretation is because organizations have figured out that they can be sustained financially if they operate as "affect generators" to create opportunities for the affective exchanges people are after in the twenty-first century economy. The desire for a less alienated relationship with the natural world is coupled with an appeal of being able to exchange these experiences for status. Money purchases the opportunities to engage in multispecies encounters, and these experiences are then exchanged through social media posts and resume lines for something else, usually occupations with higher income generation.

Conclusions

Political ecologists have paid little attention to conservation voluntourists to date, despite the fact that volunteers are becoming increasingly present in conservation work and unavoidable if one wishes to understand the state of conservation efforts and how such efforts impact local people. Conservation voluntourism is an approach to anthropogenic environmental change that is enmeshed in spectacle, grounded in affective labor, and producing new opportunities for becoming with, through multispecies engagements. In this piece, I have argued two interrelated points: (1) that vulnerable species are commodified in multispecies voluntourism encounters, dispossessing and obscuring local relationships; and (2) that the conservation voluntourism industry is both producing and produced by affect, which is then exchanged for material outcomes by global volunteers, who typically come from a western, educated class.

The opportunities to "become with" that are created in this emerging green economy take on exchange value, keeping Utila integrated into the global market. Yet, the emphasis voluntourists place on getting the right photos to document their engagement, to extracting and recording data from species (toenail clippings, photographs of whale shark spots), suggests that the privileged position of the conservation volunteer to experience and "become with" might be lost due to their conditioning by spectacle. The focus on data, documentation, and dissemination to social media networks may get in the way of being able to fully entrain, to get into sync and to become with, at least in the way conceptualized in the emergent ecologies literature.

Notes

1. Others working in this field have referred to this industry as ecological voluntourism (e.g. Waitt, Figueroa, & Nagle, 2014).
2. Many NGOs that support these activities actively push best practice codes of conduct to try to mitigate these issues. Moreover, those who support ecotourism of this form do so in place of direct harvesting of species. Thus, they rationalize that even with the stress to the species caused from tourist encounters, the species would likely have been being killed, so it is an overall biodiversity conservation gain.
3. According to this convention, the United States and Great Britain may not occupy, fortify or colonize any part of Central America.
4. Research suggests a reduction in the overall female population is connected to a food preference among consumers for gravid females (pregnant, carrying eggs). With each consumed pregnant female, there is an overall loss of 8–16 iguana. Researchers on the island are finding 56% of female iguanas to have broken tails, as compared to 35% of males with broken tails, suggesting failed capture attempts by hunters (D. F. Maryon 2018, personal communication).
5. This film was produced independently by two underwater videographers from Germany and England; the latter worked as a scuba instructor on the island for some time. See http://www.whalesharkfilm.com/eng/index_eng.html) for more details.
6. Taken from a title of an article on "GoNomad" travel website - https://www.gonomad.com/3163-utila-honduras-killing-lionfish-save-seas.
7. Opwall volunteers are not allowed to spear lionfish (only Opwall staff scientists can); dive tourists and conservation tourists who are trained through WSORC and BICA programs can participate in spearing. Opwall does not allow its volunteers to spear lionfish for several reasons, including the risk of them accidentally damaging the reef and because the UK university ethical review committees that they work with require killing procedures of a higher standard than the BICA guidelines; they must ensure a quick death for the lionfish, and often spearing does not result in immediate death, potentially leaving lionfish alive on the spear hours later (Andradi-Brown, personal communication).
8. Researchers on the island tag swampers in two ways in case they or another researcher encounters the same iguana again in the future. The least expensive method is by painting an identification number with nail polish. A slightly more expensive and invasive method is by inserting beaded safety pins with uniquely colored beads into their spines for identification purchases.
9. Similar contradictions were found among locally owned businesses that hosted large groups of conservation voluntourists in that they eliminated specific less environmentally sustainable dishes from their menu only when the conservation voluntourists were on the island.
10. This quote also appears in (Author, 2015)

Acknowledgements

I am grateful to the many volunteers and staff from Utila's conservation organizations who shared their time and perspectives with me over the years. A handful of core individuals let me into their lives on the island, whose friendship I now hold dear; I am deliberately leaving their names off to protect confidentiality. Four colleagues helped me tremendously in thinking through various aspects of this article, each having provided comments on earlier drafts. Dominic Andradi-Brown enhanced my understanding of island ecological studies. Andrea Friedus helped me to think through the connections between humanitarian tourism and multispecies conservation voluntourism, both resting on a sanitized and depoliticized suffering subject. Jim Igoe aided my analysis of the relationship between conservation organizations and affect, giving me the term "affect generators" in our discussions. Melissa Johnson pushed me to better articulate the material and social transformations that have resulted from the growth of an affect economy, confronting the fact that humans on Utila have always been becoming-with, but how they assemble and how they become-with has shifted through the emergence of new affective economies. Ryan Kilfoil and Rachel Stark helped with final formatting. Three anonymous reviewers provided critical feedback that greatly enhanced the piece. Suzanne Kent and Daniel Vacanti were, and continue to be, important research partners on Utila's conservation landscape.

Disclosure statement

No potential conflict of interest was reported by the author.

Funding

Travel for this project was made possible with support from Engaged Scholarship Faculty Research Grant and Dunavant Fellowship from the University of Memphis. Aspects of the research were supported by the Center for Collaborative Conservation Fellows Program at Colorado State University.

References

Andradi-Brown, D. A., Grey, R., Hendrix, A., Hitchner, d., Hunt, C. L, Gress, E., … Exton, D. A. (2017). Depth-dependent effects of culling—do mesophotic lionfish populations undermine current management? Royal Society Open Science. 4, 170027. DOI: 10.1098/rsos.170027.

Arsel, M., & Büscher, B. (2012). Nature™ Inc.: Changes and continuities in neoliberal conservation and market-based environmental policy. Development and Change, 43(1), 53–78.

Barad, K. (2007). Meeting the universe halfway: Quantum physics and the entanglement of matter and meaning. Durham: Duke University Press.

Barad, K. (2007). Meeting the universe halfway: Quantum physics and the entanglement of matter and meaning. Durham: Duke University Press.

Berlanga, M., & Faust, B.B. (2007). We thought we wanted a reserve: One community's disillusionment with government conservation management. Conservation and Society, 5(4), 45–477.

Brockington, D. (2009). Celebrity and the environment: Fame, wealth, and power in conservation. London: Zed.

Brockington, D. (2014). The production and construction of celebrity advocacy in international development. Third World Quarterly, 35(1), 88–108.

Brockington, D., & Duffy, R. (Eds.). (2011). Capitalism and conservation. West Sussex, UK: Wiley-Blackwell.

Brockington, D., Duffy R., & Igoe, J. (2009). Nature unbound: Conservation, capitalism, and the future of protected areas. London: Routledge.

Brondo, K. V. (2013). Land grab: Green neoliberalism, gender, and Garifuna resistance in Honduras. Tuscan, AZ: University of Arizona Press.

Brondo, K. V, & Bown, N. (2011). Neoliberal conservation, Garifuna territorial rights and resource management in the Cayos Cochinos marine protected area. *Conservation and Society, 9*(2), 91–105.

Brondo, Keri Vacanti. (2007). Garifuna Women's Land Loss and Activism in Honduras. *Journal of International Women's Studies, 9*(1), 99–116.

Brondo, Keri Vacanti. (2010). When Mestizo Becomes (Like) Indio … or is it Garifuna?: Negotiating Indigeneity and 'Making Place' on Honduras' North Coast. *Journal of Latin American and Caribbean Anthropology, 15*(1), 171–194.

Brondo, Keri Vacanti & Woods, Laura. (2007). Garifuna Land Rights and Ecotourism as Economic Development in Honduras' Cayos Cochinos Marine Protected Area". Ecological and Environmental Anthropology 3(1):2–18.

Büscher, B., & Dressler, W. (2007). Linking neoprotectionism and environmental governance: On the rapidly increasing tensions between actors in the environment-development nexus. *Conservation and Society, 5*(4), 586–511.

Brondo, K. V., Kent, S., & Hill, A. (2016). Teaching Collaborative Environmental Anthropology: A Case Study Embedding Engaged Scholarship in Critical Approaches to Voluntourism. In T. Copeland & F. Dengah (Eds.), *Involve Me and I Learn: Teaching Anthropology and Research Methods in the Classroom and Beyond*. Annals of Anthropological Practice, 40(2), 182–195.

Canty, S.W.J. (2007). *Positive and negative impacts of dive tourism: The case study of Utila*, Honduras (Unpublished master's thesis). Lund University Centre for Sustainability Studies, Sweden.

Carrier, J., & Macleod, D. (2005). Bursting the bubble: The socio-cultural context of ecotourism. *Journal of the Royal Anthropological Institute, 11*(2), 315–334.

Coghlan, A. (2007). Towards an integrated image-based typology of volunteer tourism organizations. *Journal of Sustainable Tourism, 15*(3), 468–485.

Currin, F.H. (2002). *Transformation of paradise: Geographical perspectives on tourism development on a small Caribbean island (Utila, Honduras)* (Unpublished master's thesis). Louisiana State University and Agricultural and Mechanical College, Louisiana.

Davidson, W.V. (1974). *Historical geography of the Bay Islands, Honduras: Anglo-Hispanic conflict in the western Caribbean*. Birmingham: Southern University Press.

Debord, G. (1995 [1967]). The Society of the Spectacle. New York: Zone Books.

Devine, J. (2014). Counterinsurgency ecotourism in Guatemala's Maya biosphere reserve. *Environmental Planning D, 32*, 984–1001.

Fassin, D. (2011). *Humanitarian reason: A moral history of the present*. Berkeley, CA: University of California Press.

Fletcher, R. (2010). Neoliberal environmentality: Towards a poststructuralism political ecology of the conservation debate. *Conservation and Society, 8*(3), 171–181.

Fletcher, R., Dressler W., & Büscher, B. (2014). Introduction. In B. Büscher, W. Dressler, & R. Fletcher (Eds.), *Nature^TM Inc.: New frontiers of environmental conservation in the neoliberal age* (pp. 3–24). Tucson, AZ: University of Arizona Press.

Felman, Ilana, & Miriam Ticktin. (eds.) (2010). *In the Name of Humaity: The Government of Threat and Care*. Durham: Duke University Press.

Freidus, A. (2010). "Saving" Malawi: FAITHFUL responses to orphans and vulnerable children *North American Practicing Anthropology Bulletin, 33*(1), 50–57.

Freidus, A. (2016). Unanticipated outcomes of voluntourism among Malawi's orphans. *Journal of Sustainable Tourism. 25*(9), 1306–1321. doi:10.1080/09669582.2016.1263308.

Geertz, C. (1998). Deep hanging out. *The New York review of Books*. October 22: 69–72.

Grandia, L. (2007). Between Bolivar and bureaucracy: The Mesoamerican Biological Corridor. *Conservation and Society, 5*(4), 478–503.

Haraway, D. (2008). *When species meet*. Minneapolis, MN: University of Minnesota Press.

Haraway, D. (2016). *Staying with the trouble: Making kin in the chuthulucene*. Durhman: Duke University Press.

Harborne, A.R., Afzal, D.C., & Andrews, M.J. (2001). Honduras: Caribbean coast. *Marine Pollution Bulletin, 42*(12), 1221–1235.

Hoffman, D.M. (2014). Conch, cooperatives, and conflict: Conservation and resistance in the Banco Chinchorro Biosphere Reserve. *Conservation and Society, 12*(2), 120–132.

Holmes, G., & Cavanagh, C. (2016). A review of the social impacts of neoliberal conservation: Formations, inequalities, contestations. *Geoforum, 75*, 199–209.

IUCN. (2017). Red list of threatened species. *Ctenosaura bakeri*. Retrieved from http://www.iucnredlist.org/details/44181/0. Accessed March 28, 2017.

Igoe, J. (2016). *The nature of spectacle: On image, money, and conserving capitalism*. Tuscan, AZ: University of Arizona Press.

Igoe, J. (2017). Afterword. In M. Mostafanezhand et al. (Eds.), *Political ecology of tourism: Community, power and the environment* (pp. 309–316). London and New York: Routledge.

Igoe, J., & Brockington, D. (2007). Neoliberal conservation: A brief introduction. *Conservation and Society, 5*(4), 432–499.

Igoe, J., Neves, K., & Brockington, D. (2010). A spectacular eco-tour around the historic bloc: Theorising the convergence of biodiversity conservation and capitalist expansion. *Antipode, 42*(3), 486–512.

Kirksey, E. (2012). *Freedom in entangled worlds: West Papua and the architecture of global power*. Durham: Duke University Press.

Kirksey, E. (2014). *The multispecies salon*. Durham: Duke University Press.

Kirksey, E. (2015). *Emergent ecologies*. Durham: Duke University Press.

Kohn, E. (2007). How dogs dream: Amazonian natures and the politics of transpecies engagement. *American Ethnologist, 34*(1), 3–24.

Kohn, E. (2013). *How forests think: Toward an anthropology beyond the human*. Berkeley, CA: University of California Press.

Malkki, L. (1996) "Speechless emissaries: Refugees, humanitarianism, and dehistoricization." *Cultural Anthropology, 11*(3), 377–04.

Malkki, L. (2010). *In the name of humanity: The government of threat and care*. Durham: Duke University Press.

Marx, K., [1867] (1957). *Capital (Vol. 1)*. London: J. M. Dent & Sons Ltd.

Moore, A. (2012). The aquatic invaders: Marine management figuring fishermen, fisheries, and lionfish in The Bahamas. *Cultural Anthropology, 27*(4), 667–688.

Mostafanezhad, M. (2013). 'Getting in touch with your inner Angelina': Celebrity humanitarianism and the cultural politics of gendered generosity in volunteer tourism. *Third World Quarterly, 34*(3), 486–499.

Mostafanezhad, M., Norum, R., Shelton, E.J., & Thompson-Carr, A. (2017). Introduction. In M. Mostafanezhad et al. (Eds.), *Political ecology of tourism: Community, power and the environment* (pp. 1–21). London and New York: Routledge.

Neves, K. (2010). Cashing in on cetourism: A critical ecological engagement with E-NGO discourses on whaling, cetacean conservation, and whale watching. *Antipode, 42*(3), 719–741.

Phillips, J.P. (2015). *Honduras in dangerous times: Resistance and resilience*. Lantham: Lexington Books.

Rose, D.B., van Dooren, T., Chrulew, M., Cooke, S., Kearnes, M., & O'Gorman, E. (2012). Thinking through the environment, unsettling the humanities. *Environmental Humanities, 1*(1), 1–5.

Rose, R.H. (1904). *Utila: Past and present*. Dansville, NY: F.A. Owen Publ. Co.

Schulz, K. (2017). Decolonizing political ecology: Ontology, technology and 'critical' enchantment. *Journal of Political Ecology, 24*, 125–143.

Stonich, S.C. (2000). *The other side of paradise: Tourism, conservation, and development in the bay islands*. New York, NY: Cognizant Communication Corporation.

Sullivan, S. (2013). Banking Nature?: The specatular financialisation of environmental conservation. *Antipode, 45*(1), 198–217.

Sullivan, S., Igoe, J., & Büscher, B. (2013). Introducing nature on the move–a triptych. *Proposals: Journal of Marxism and Interdisciplinary Inquiry, 6*(1–2), 15–19.

Tickin, M. (2011). *Casualties of CareImmigration and the Politics of Humanitarianism in France*. Berkeley: University of California Press.

Tsing, A.L. (2012). Unruly edges: Mushrooms as companion species. *Environmental Humanities, 1*, 141–154.

Tsing, A.L. (2015). *The Mushroom at the end of the world: On the possibility of life in capitalist ruins*. Princeton, NJ: Princeton University Press.

Van Landeghem, M. (2017). When you're aware, you care. WSORC blog, April 7, 2017. Retrieved from http://wsorc.org/when-youre-aware-you-care/

Vodopivec, B., & Jaffe, R. (2011). Save the world in a week: Volunteer tourism, development and difference. *European Journal of Development Research, 23*, 111–128

Waitt, G., Figueroa, R., & Nagle, T. (2014). "Paying for proximity: Touching the moral economy of ecological tourism." In M. Mostafanezhad & K. Hannam (Eds.), *Moral encounters in tourism* (pp. 167–181). Farnham, UK: Ashgate Publishing.

Wearing, S. (2002). Re-centering the self in volunteer tourism. In G.S. Dann (Ed.), *The tourist as a metaphor of the social world* (pp. 237–262). Wallingford, CT: CABI.

West, P., Igoe, J., & Brockington, D. (2006) Parks and peoples: The social impacts of protected areas. *Annual Review of Anthropology, 35*, 251–277.

Afterword: Involving Earth – Tourism Matters of Concern

Edward H. Huijbens

This book is premised on understanding the Anthropocene as a fundamental social imaginary in contemporary global society, and skilfully weaves a critical narrative thereto with insights from around the globe in ten chapters. Through multiple and sometimes contradictory thinking, politics and practices that drive the anthropocenic imaginary in both tourist practices and environmental subjectivities, the Anthropocene is indeed telling us a story. It is a story, depending on how it is told and by whom, that will inform our actions, guide us in the everyday and provide for a moral compass to varying degrees (Swyngedouw, 2019). Seeing its dependence on the narrator, the question that confronts us all today is how does this narrative guide and inform? That is, how is the complex and ever burgeoning narrative of the Anthropocene providing for that moral compass and wayfinding in the everyday? The focus of this volume is on tourism and leisure activities and thereby how tourism enrols planetary environmental concerns with our leisure activities and aspirations. In this afterword I want to reflect on two particular points of concern when it comes to the Anthropocene narrative in this context. First, and related to above question of how, is a reflection drawing inspiration from Latour's (2018) work about the profound implications of the Anthropocene narrative. The second concern relates to the first but will outline the ramifications of these implications and how we can use the Anthropocene narrative to guide us towards a climatically and environmentally sound tourism future.

As to the first point of concern, what the Anthropocene narrative has done is bring to the fore the environment in which we live and force us to consider it as part and parcel of our being and everyday actions. And this is a fragile and globally connected environment indeed. Upon the Earth's crust and up towards around 5,000 metres above sea level the global human population has managed to sustain itself and provide for habitat. It is the only place in the universe we know we can do so, and conditions in this biosphere have remained relatively stable since the end of the last ice age some 12,000 years ago. The Anthropocene narrative introduces uncertainty, volatility and change into this fragile home of humanity. As a consequence, people are reconsidering their modes of transport, their consumption patterns and their choices of activities with concern for how it might impact our common climate and local environment. Well, at least some are! Bringing the environment, the climate and even the well-being of our whole planet to centre stage when it comes to everyday activities of people going about their lives is a rather radical transformation of the ideas and aspirations, which generations up until now have been brought up with and indoctrinated. These ideas can roughly be framed as Western consumerism, something to which at times the entire planet's population seems to be currently aspiring. Thoroughly wrapped with enlightenment ideals of progress, consumerism as our contemporary cultural logic dictates that more is merrier. One should always aspire to have more stuff, a bigger house, a better car, the latest designer apparel, the smartest household equipment, and … to travel more and to increasingly exotic places. Now, however, we are all being told that this is wrong. As Lewis and Maslin (2018) inform us, modern-day consumer capitalism, with its legacy of colonial conquest, slavery and all-round exploitation of peoples and resources, has come up against planetary limits. Much like when Copernicus informed us that the Sun did not revolve around the Earth, we are now to realize that our lives revolve around the Earth, not the stuff the paycheck of tomorrow can bring to our mortgage equity, household goods or that long-awaited holiday. This revolution of our aspirations shakes the foundations of our social and political order. Whilst indeed there are some who are embracing the Earth and trying to align their practices with its needs and that of other nonhumans cohabiting the planet, there are others who seek to entrench their gains from the exploitative legacy of capitalism. A reactionary entrenchment that is even seeking to destroy and discredit the platforms for dissenting voices provided by existing liberal institutions, such as the UN. Yet others are in a state of denial and try to carry on as if nothing in particular is out of kilter. Still others feel puzzled and simply do not know which way to turn. The consequences of this foregrounding of the Earth through the Anthropocene narrative are thus disorientation on the geopolitical map we thought so familiar. We already see unscrupulous politicians tending to those in denial, puzzled and even angry and demanding the promises of Western consumer capitalism. To counter these a new map needs to be drawn of the Earth. Not one of solid borders between sovereign nation states and their offshore concerns defined through property and usage rights, but one of our common biospheres. This is a map constituted by relations and matters of concern extending across the globe and including more than humans. Bruno Latour and colleagues put this simply,

> The question no longer seems to be for humans to profit from their freedom in a world made of mere things, but for humans to learn how to swap their agency with countless life forms, each of them having their own ideas about what counts as freedom and about which sort of territory they wish to expand. A geopolitics of life forms has taken over all the questions of rights, freedom, property, responsibility and justice that have been reserved until now only for humans.

(Latour et al., 2019, p. 3)

Here arises my second point of concern. At this juncture I want to harken to my own work with my colleague Martin Gren, whereby we have drawn some of the contours of what it means to foreground the Earth in terms of tourism (Gren and Huijbens, 2012, 2014, 2016). These contours represent the gestation of a geopolitical map of lifeforms, of matter movements and of us imbricated with the whole fragile veneer we call our home on the crust of planet Earth. As opposed to the surveying tactics of the colonial and consumptive enterprise, which is about enclosure and overcoming of any limits to capital accumulation, this is a map that is inclusive and open. Mapmaking is indeed a way to frame narratives and has been the mainstay of my discipline: geography since its nineteenth-century origins – for better or worse. An inclusive map is one that is 'topoecological' as the Swedish geographer Torstein Hägerstrand (2004) would call it. It is open by means of allowing for time for all the elements of the map, a very necessary reorientation of our spatial register at our current climatic juncture. Once adding a temporal dimension to all observed forms, 'we discover a web-like form of trajectories, of which some are stationary in space and some are in motion, while some entities may grow and others shrink in the process'. Hägerstrand (2004, p. 323) goes on to explain that '[t]his condition is the basis for cooperation and conflict and for the human yearning for power over spaces filled with resources or at least over parts of their contents'. At the same time this understanding allows us to recognize the value of connectivity. That is, that everything is related to everything else, and thereby space is fundamentally the 'togetherness' of all phenomena, to borrow a term from Doreen Massey (2005, p. 195). The crucial question is however how we go about the valuing of the more than human that which sits beyond the register of the promises of progress we have been accustomed to? In the context of tourism, Gren and I concluded in our own edited volume from 2016 with a call for a 'hospitality geo-ethics', incorporating geo resources at its core, i.e. the Earth. To facilitate this we outlined three tentative tourism destinies. The first is 'non-carbon tourism', i.e. simply traveling without burning fossil fuels or using carbon. Transport infrastructure powered by green energy becomes the traveling option. This can be achieved even long distances if time is allowed for. Greta Thunberg demonstrated in August 2019 that you can attend international meetings of the highest order without needing a private jet to whisk you there. Sailing across the Atlantic to attend the New York climate week, the solar-powered racing yacht Malizia II got her from Plymouth, England to the big Apple in two weeks flat. Non-carbon tourism, however, spells doom for the aviation industry unless stringent mitigation policies are enforced. The prospect of electric or solar-powered flights is simply a myth (Peeters, 2017). As a consequence, travel will inherently take more time and traveling becomes not only about the destination, but also the journey itself. The second tourism destiny we identify relates to the fact that we do not all need to be traveling all the time in order to gain experiences and learn new things. 'Stay home tourism' is then about learning to appreciate that which is close at hand and in your local surroundings, boosting local economies and senses of community. All the things that make the place we call home special and dear to us can be made more of and engaged with in order to nurture, cultivate and enhance the environment in which we live our everyday lives. Spinning off the appreciation of the close at hand and that which is around is the third tourism destiny, which we call 'destination stewardship'. Becoming sensitized and attuned to the more-than-human rhythms and lifeworlds is key to this notion of stewardship, a holistic sense of oneness and being-with the place in the moment.

These three tourism destinies all have their own manifestations in tourism at current. The Anthropocene narrative should be informed by these examples and promulgate them to contribute to a different map of tourism for the anthropogenic imaginary. More importantly, however, is that tourism destinations start to factor in, design and develop for these and other anthropocenic destinies of humanity as a whole. Designing 'earthly destinations' that allow for time and the journey experience, the foregrounding of the earth and the chance to foster attachments and attuning to more-than-human lifeworlds is a challenge to be presented to all those interested in tourism and its development. At the same time, we have to bear in mind the fact that modern-day mobilities are dominated by the very same parts of humanity which have hitherto contributed by far the greatest to our current climate predicament. In this context, the future of tourism is not about bringing the prospects or benefits of jet-propelled globe-trotting to all the inhabitants of the planet, but about cultivating this new vision of mobility, which will nurture local attachments, necessitate slowing down and encourage an appreciation of that which makes every day worth living.

References

Gren, M. and Huijbens, E. (Eds.) (2016). *Tourism and the Anthropocene*. London: Routledge, tourism series.

Gren, M. and Huijbens, E. (2014). Tourism in the Anthropocene. *Scandinavian Journal of Hospitality and Tourism*, 14(1): 6–22.

Gren, M. and Huijbens, E. (2012). Tourism Theory and the Earth. *Annals of Tourism Research*, 39(1): 155–170.

Hägerstrand, T. (2004). The two vistas. *Geografiska Annaler: Series B, Human Geography*, 86(4), 315–323.

Latour, B. (2018). *Down to Earth – Politics in the New Climatic Regime*. Cambridge: Polity Press.

Latour, B., Weibel, P., Guinard, M. and Korintenberg, B. (2019). *Critical Zones – Observatories for Earthly Politics. Exhibit prospectus 9.5. – 4.10.2020*. Karlsruhe: ZKM | Zentrum für Kunst und Medien.

Lewis, S.L. and Maslin, M. (2018). *The Human Planet. How We Created the Anthropocene*. New Haven: Yale University Press.

Massey, D. (2005). *For Space*. London: Sage.

Peeters, P. (2017). *Tourism's Impact on Climate Change and Its Mitigation Challenges. How can Tourism become 'Climatically Sustainable'*. PhD thesis from the Technical University of Delft.

Swyngedouw, E. (2019). The Anthropo(obs)cene. In the Antipode Editorial Collective (Eds.) *Keywords in Radical Geography: Antipode at 50, First Edition*. Hoboken: Wiley Blackwell, pp. 253–258.

Index